Advanced Thermodynamics for Engineers

Advanced Thermodynamics for Engineers

Desmond E Winterbone

FEng, BSc, PhD, DSc, FIMechE, MSAE

Thermodynamics and Fluid Mechanics Division
Department of Mechanical Engineering
UMIST

A member of the Hodder Headline Group

LONDON • SYDNEY • AUCKLAND
Copublished in North, Central and South America by
John Wiley & Sons, Inc., New York • Toronto

First published in Great Britain 1997 by Arnold,
a member of the Hodder Headline Group,
338 Euston Road, London NW1 3BH

Copublished in North, Central and South America by
John Wiley & Sons, Inc., 605 Third Avenue,
New York, NY 10158–0012

British Library Cataloguing in Publication Data
A catalogue record for this book is available from the British Library

Library of Congress Cataloging-in-Publication Data
A catalog record for this book is available from the Library of Congress

ISBN 0 340 67699 X (pb) 0 470 23718 X (Wiley)

Typeset in 10/12 pt Times by Mathematical Composition Setters Ltd, Salisbury, Wilts

Printed and bound in Great Britain by J W Arrowsmith Ltd, Bristol

Contents

Preface

When reviewing, or contemplating writing, a textbook on engineering thermodynamics, it is necessary to ask: what does this book offer that is not already available? The author has taught thermodynamics to mechanical engineering students, at both undergraduate and post-graduate level, for 25 years and has found that the existing texts cover very adequately the basic theories of the subject. However, by the final years of a course, and at post-graduate level, the material which is presented is very much influenced by the lecturer, and here it is less easy to find one book that covers all the syllabus in the required manner. This book attempts to answer that need, for the author at least.

The engineer is essentially concerned with manufacturing devices to enable tasks to be preformed cost effectively and efficiently. Engineering has produced a new generation of automatic 'slaves' which enable those in the developed countries to maintain their lifestyle by the consumption of fuels rather than by manual labour. The developing countries still rely to a large extent on 'manpower', but the pace of development is such that the whole world wishes to have the machines and quality of life which we, in the developed countries, take for granted: this is a major challenge to the engineer, and particularly the thermodynamicist. The reason why the thermodynamicist plays a key role in this scenario is because the methods of converting any form of energy into power is the domain of thermodynamics: all of these processes obey the four laws of thermodynamics, and their efficiency is controlled by the Second Law. The emphasis of the early years of an undergraduate course is on the First Law of thermodynamics, which is simply the conservation of energy; the First Law does not give any information on the *quality* of the energy. It is the hope of the author that this text will introduce the concept of the quality of energy and help future engineers use our resources more efficiently. Ironically, some of the largest demands for energy may come from cooling (e.g. refrigeration and air-conditioning) as the developing countries in the tropical regions become wealthier – this might require a more basic way of considering energy utilisation than that emphasised in current thermodynamic texts. This book attempts to introduce basic concepts which should apply over the whole range of new technologies covered by engineering thermodynamics. It considers new approaches to cycles, which enable their irreversibility to be taken into account; a detailed study of combustion to show how the chemical energy in a fuel is converted into thermal energy and emissions; an analysis of fuel cells to give an understanding of the direct conversion of chemical energy to electrical power; a detailed study of property relationships to enable more sophisticated analyses to be made of both high and low temperature plant; and irreversible thermodynamics, whose principles might hold a key to new ways of efficiently converting energy to power (e.g. solar energy, fuel cells).

The great advances in the understanding and teaching of thermodynamics came rapidly towards the end of the 19th century, and it was not until the 1940s that these were embodied in thermodynamics textbooks for mechanical engineers. Some of the approaches used in teaching thermodynamics still contain the assumptions embodied in the theories of heat engines without explicitly recognising the limitations they impose. It was the desire to remove some of these shortcomings, together with an increasing interest in what limits the efficiency of thermodynamic devices, that led the author down the path that has culminated in this text.

I am still a strong believer in the pedagogical necessity of introducing thermodynamics through the traditional route of the Zeroth, First, Second and Third Laws, rather than attempting to use the Single-Axiom Theorem of Hatsopoulos and Keenan, or The Law of Stable Equilibrium of Haywood. While both these approaches enable thermodynamics to be developed in a logical manner, and limit the reliance on cyclic processes, their understanding benefits from years of experience – the one thing students are lacking. I have structured this book on the conventional method of developing the subject. The other dilemma in developing an advanced level text is whether to introduce a significant amount of *statistical thermodynamics*; since this subject is related to the particulate nature of matter, and most engineers deal with systems far from regions where molecular motion dominates the processes, the majority of the book is based on *equilibrium thermodynamics*; which concentrates on the macroscopic nature of systems. A few examples of statistical thermodynamics are introduced to demonstrate certain forms of behaviour, but a full understanding of the subject is not a requirement of the text.

The book contains 17 chapters and, while this might seem an excessive number, these are of a size where they can be readily incorporated into a degree course with a modular structure. Many such courses will be based on two hours lecturing per week, and this means that most of the chapters can be presented in a single week. Worked examples are included in most of the chapters to illustrate the concepts being propounded, and the chapters are followed by exercises. Some of these have been developed from texts which are now not available (e.g. Benson, Haywood) and others are based on examination questions. Solutions are provided for all the questions. The properties of gases have been derived from polynomial coefficients published by Benson: all the parameters quoted have been evaluated by the author using these coefficients and equations published in the text – this means that all the values are self-consistent, which is not the case in all texts. Some of the combustion questions have been solved using computer programs developed at UMIST, and these are all based on these gas property polynomials. If the reader uses other data, e.g. JANAF tables, the solutions obtained might differ slightly from those quoted.

Engineering thermodynamics is basically *equilibrium thermodynamics*, although for the first two years of the conventional undergraduate course these words are used but not often defined. Much of the thermodynamics done in the early years of a course also relies heavily on *reversibility*, without explicit consideration of the effects of irreversibility. Yet, if the performance of thermodynamic devices is to be improved, it is the irreversibility that must be tackled. This book introduces the effects of irreversibility through considerations of availability (exergy), and the concept of the endoreversible engine. The thermal efficiency is related to that of an ideal cycle by the rational efficiency – to demonstrate how closely the performance of an engine approaches that of a reversible one. It is also shown that the Carnot efficiency is a very artificial yardstick against which to compare real engines: the internal and external reversibilities imposed by the cycle mean that it produces zero power at the maximum achievable efficiency. The approach by Curzon and Ahlborn to define the efficiency of an endoreversible engine producing maximum power output is introduced: this shows the effect of *external irreversibility*. This analysis also introduces the concept of *entropy generation* in a manner readily understandable by the engineer; this concept is the cornerstone of the theories of *irreversible thermodynamics* which are at the end of the text.

Whilst the laws of thermodynamics can be developed in isolation from consideration of the property relationships of the system under consideration, it is these relationships that enable the equations to be *closed*. Most undergraduate texts are based on the evaluation of the fluid properties from the simple perfect gas law, or from tables and charts. While this approach enables typical engineering problems to be solved, it does not give much insight into some of the phenomena which can happen under certain circumstances. For example, is the specific heat at constant volume a function of temperature alone for gases in certain regions of the state diagram? Also, why is the assumption of constant stagnation, or even static, temperature valid for flow of a perfect gas through a throttle, but never for steam? An understanding of these effects can be obtained by examination of the more complex equations of state. This immediately enables methods of gas liquefaction to be introduced.

An important area of engineering thermodynamics is the combustion of hydrocarbon fuels. These fuels have formed the driving force for the improvement of living standards which has been seen over the last century, but they are presumably finite, and are producing levels of pollution that are a constant challenge to engineers. At present, there is the threat of global warming due to the build-up of carbon dioxide in the atmosphere: this requires more efficient engines to be produced, or for the carbon–hydrogen ratio in fuels to be reduced. Both of these are major challenges, and while California can legislate for the Zero Emissions Vehicle (ZEV) this might not be a worldwide solution. It is said that the ZEV is an electric car running in Los Angeles on power produced in Arizona! – obviously a case of exporting pollution rather than reducing it. The real challenge is not what is happening in the West, although the energy consumption of the USA is prodigious, but how can the aspirations of the East be met. The combustion technologies developed today will be necessary to enable the Newly Industrialised Countries (NICs) to approach the level of energy consumption we enjoy. The section on combustion goes further than many general textbooks in an attempt to show the underlying general principles that affect combustion, and it introduces the interaction between thermodynamics and fluid mechanics which is so important to achieving clean and efficient combustion. The final chapter introduces the thermodynamic principles of fuel cells, which enable the direct conversion of the Gibbs energy in the fuel to electrical power. Obviously the fuel cell could be a major contributor to the production of 'clean' energy and is a goal for which it is worth aiming.

Finally, a section is included on irreversible thermodynamics. This is there partly as an intellectual challenge to the reader, but also because it introduces concepts that might gain more importance in assessing the performance of advanced forms of energy conversion. For example, although the fuel cell is basically a device for converting the Gibbs energy of the reactants into electrical energy, is its efficiency compromised by the thermodynamics of the steady state that are taking place in the cell? Also, will photo-voltaic devices be limited by phenomena considered by irreversible thermodynamics?

I have taken the generous advice of Dr Joe Lee, a colleague in the Department of Chemistry, UMIST, and modified some of the wording of the original text to bring it in line with more modern chemical phraseology. I have replaced the titles Gibbs free energy and Helmholtz free energy by Gibbs and Helmholtz energy respectively: this should not cause any problems and is more logical than including the word 'free'. I have bowed, with some reservations, to using the internationally agreed spelling sulfur, which again should not cause problems. Perhaps the most difficult concept for engineers will be the replacement of the terms 'mol' and 'kmol' by the term 'amount of substance'. This has been common practice in chemistry for many years, and separates the general concept of a quantity of matter from the units of that quantity. For example, it is common to talk of a mass of substance without defining whether it is in grams, kilograms, pounds, or whatever system of units is appropriate. The use of the phrase 'amount of substance' has the same generalising effect when dealing with quantities based on molecular equivalences. The term mol will still be retained as the adjective and hence molar enthalpy is the enthalpy per unit amount of substance in the appropriate units (e.g. kJ/mol, kJ/kmol, Btu/lb-mol, etc).

I would like to acknowledge all those who have helped and encouraged the writing of this text. First, I would like to acknowledge the influence of all those who attempted to teach me thermodynamics; and then those who encouraged me to teach the subject, in particular Jim Picken, Frank Wallace and Rowland Benson. In addition, I would like to acknowledge the encouragement to develop the material on combustion which I received from Roger Green during an Erskine Fellowship at the University of Canterbury, New Zealand. Secondly, I would like to thank those who have helped in the production of this book by reading the text or preparing some of the material. Amongst these are Ed Moses, Marcus Davies, Poh Sung Loh, Joe Lee, Richard Pearson and John Horlock; whilst they have read parts of the text and provided their comments, the responsibility for the accuracy of the book lies entirely in my hands. I would also like to acknowledge my secretary, Mrs P Shepherd, who did some of the typing of the original notes. Finally, I must thank my wife, Veronica, for putting up with lack of maintenance in the house and garden, and many evenings spent alone while I concentrated on this work.

D E Winterbone

Structure of book

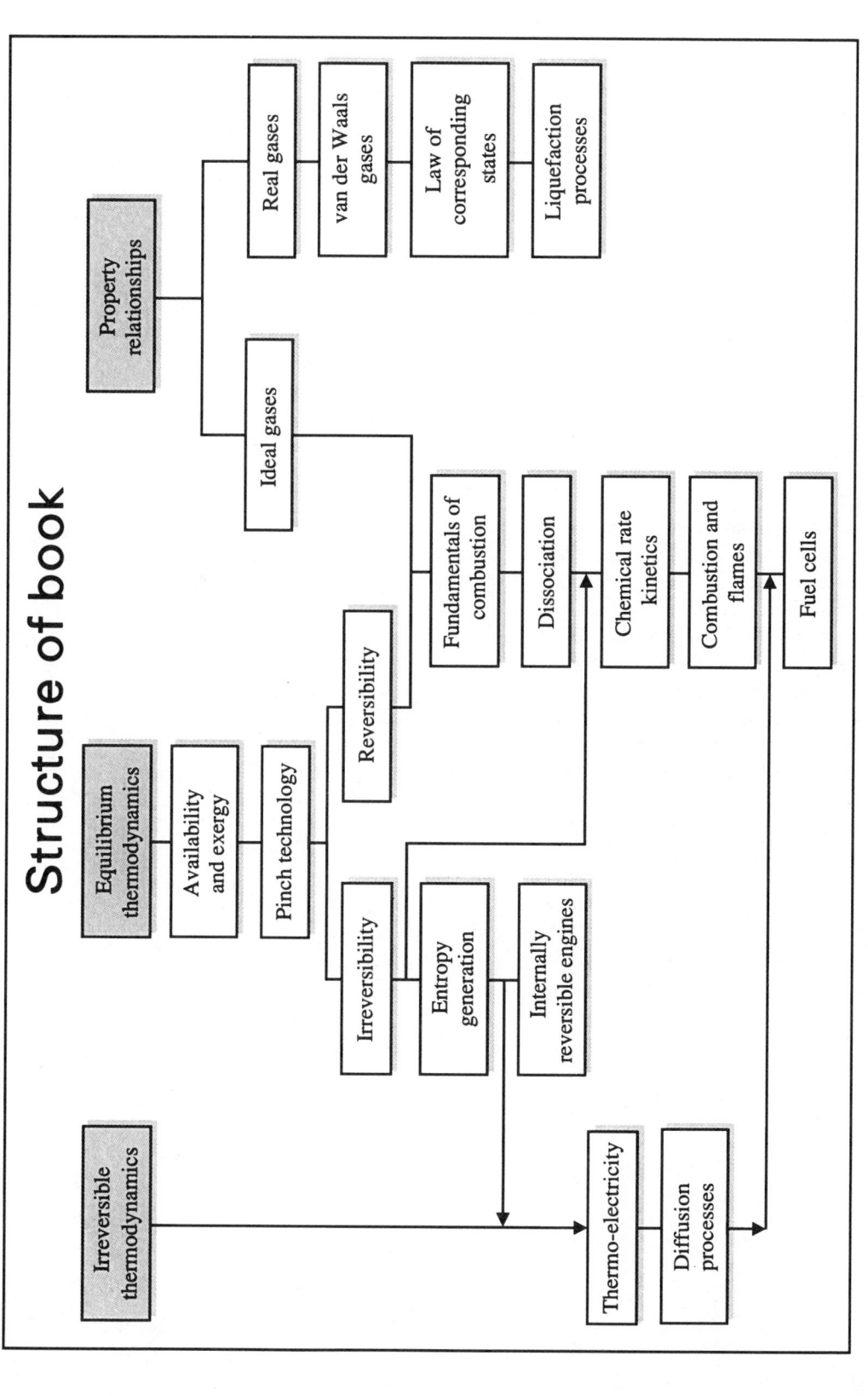

Symbols

a	activity coefficient
a	specific non-flow availability
a	coefficient in van der Waals' equation
a_f	specific flow availability
a_i	enthalpy coefficient for gas properties
A	non-flow availability
A	area
b	specific exergy
b	coefficient in van der Waals' equation
B	exergy
c_p	specific heat capacity at constant pressure (sometimes abbreviated to specific heat at constant pressure)
$\bar{c}_p$	mean value of specific heat capacity at constant pressure over a range of temperatures
$c_{p,m}$	molar specific heat capacity at constant pressure (i.e. specific heat capacity at constant pressure based on *mols*)
c_v	specific heat capacity at constant volume (sometimes abbreviated to specific heat at constant volume)
$c_{v,m}$	molar specific heat capacity at constant volume (i.e. specific heat capacity at constant volume based on *mols*)
C	conductivity for heat flow into engines
d	increment in – usually used for definite integral, e.g. property, etc
D	mass diffusivity
e	specific internal energy
E	internal energy
E	activation energy
E	electromotive force of a cell (emf)
E^0	standard emf of a cell
E_{oc}	open circuit voltage
f	specific Helmholtz energy (specific Helmholtz function)
F	Helmholtz energy (Helmholtz function)
F	force

F	Faraday constant (charge carried by kmol of unit positive valency [96 487 kC/kmol])
g	specific Gibbs energy (specific Gibbs function)
g_0	specific Gibbs energy at datum temperature (or absolute zero)
g	acceleration due to gravity
G	Gibbs energy (Gibbs function)
h	specific enthalpy
h_0	specific enthalpy at datum temperature (or absolute zero)
h	height
H	enthalpy
I	irreversibility
I	electrical current
J	thermodynamic velocity, or flow
J_I	electrical flow rate
J_Q	heat flow rate
J_S	entropy flow rate
k	isothermal compressibility, isothermal bulk modulus
k	thermal conductivity
k	Boltzmann constant [$1.380\,62 \times 10^{-23}$ J/K]
k_s	adiabatic, or isentropic, compressibility
k	rate of reaction
K	Karlovitz number [$(u'/l_T)(\delta_l/u_l)$]
K_p	equilibrium constant
l	length
l_T	Taylor microscale
L	coefficient relating thermodynamic force and velocity
Le	Lewis number [$\lambda/\rho c_p D$]
m	mass
m_w	molecular weight
n	polytropic index
n	amount of substance, chemical amount (sometimes referred to as number of mols)
n	reaction order
$\tilde{N}$	Avogadro constant (6.023×10^{26} kmol^{-1})
p	pressure
p_i	partial pressure of component i
p_0	datum pressure (often 1 bar or 1 atmosphere)
P_r	Prandtl number [$c_p\mu/k$]
q	specific heat (energy) transfer
q_I	electrical charge
Q	heat (energy) transfer
Q^*	heat of transport
Q_p	enthalpy of reaction (energy of reaction at constant pressure)
Q_p'	calorific value (at constant pressure) $= -Q_p$
Q_v	internal energy of reaction (energy of reaction at constant volume)
Q_v'	calorific value (at constant volume) $= -Q_v$
R	specific gas constant
R	rate of formation, rate of reaction

R	radical
R	electrical resistance
$\mathfrak{R}$	universal gas constant
Re	Reynolds number $[\rho v l / \mu]$
s	specific entropy
s_0	specific entropy at datum temperature (or absolute zero)
S	entropy
S^*	entropy of transport
t	temperature on discontinuous scale
t	time
T	temperature on absolute scale (thermodynamic temperature)
u	specific intrinsic internal energy
u_0	specific intrinsic internal energy at datum temperature (or absolute zero)
u'	turbulence intensity
u_l	laminar burning velocity
u_t	turbulent burning velocity
U	intrinsic internal energy
U	overall heat transfer coefficient
v	specific volume
V	volume
V	velocity
V	voltage
$\bar{V}_p$	mean piston speed (for reciprocating engines)
w	specific work
$\hat{w}$	*maximum* specific work
W	work
x	dryness fraction (quality)
x	molar fraction
x	distance
X	thermodynamic force
y	mass fraction
z	valency

Greek characters

α	degree of dissociation
α	$[A]/[A]_e$
α	branching multiplication coefficient
α	molecular thermal diffusivity
α	crank angle (in internal combustion engines)
β	coefficient of thermal expansion
β	$[B]/[B]_e$
δ	increment in – usually used for indefinite integral, e.g. work (W), heat (Q)
δ	$[D]/[D]_e$
Δ	increment in – usually used for indefinite integral, e.g. work (W), heat (Q)
ΔH_a	atomisation energy
ΔH_f	enthalpy of formation
$\Delta H(\ )$	dissociation energy

ε	fraction of reaction
ε	potential difference, voltage
$\varepsilon_{A,B}$	Seebeck coefficient for material pair A, B
ε	eddy diffusivity
κ	ratio of specific heats (c_p/c_v)
λ	electrical conductivity
μ	dynamic viscosity
μ	Joule–Thomson coefficient
μ	chemical potential
$\bar{\mu}$	electrochemical potential
ν	kinematic viscosity
$\pi_{A,B}$	Peltier coefficient for material pair A, B
θ	entropy generation per unit volume
σ	Thomson coefficient
ν	stoichiometric coefficient
γ	$[C]/[C]_e$
η	efficiency
ϕ	equivalence ratio
ρ	density
ξ	specific exergy
ψ	inner electric potential of a phase
τ	temperature ratio
Ξ	exergy

Suffices

0	dead state conditions
actual	value from actual cycle, as opposed to ideal cycle
av	available (as in energy)
b	backward (reaction)
b	burned (products of combustion)
B	boiler
c	at critical point, e.g. pressure, temperature, specific volume
C	cold (as in temperature of reservoir)
C	compressor
[ ]$_e$	equilibrium molar density
f	forward (reaction)
f	value for saturated liquid, e.g. h_f = enthalpy of saturated liquid
fg	difference between properties on saturated vapour and saturated liquid lines, i.e. $h_{fg} = h_g - h_f$
g	value for saturated gas, e.g. h_g = enthalpy of saturated gas
g	gaseous state (as in reactants or products)
h	constant enthalpy
H	hot (as in temperature of reservoir)
i	inversion
i	ith constituent
in	into system
isen	isentropic (as in a process)

l	latent energy required for evaporation (equals h_{fg} or u_{fg})
l	liquid state (as in reactants or products)
l	laminar
max	maximum value, or maximum useful work
mol	specific property based on *mols*
net	net (as in work output from system)
O	overall
oc	open circuit
out	out of system
p	at constant pressure
P	products
P	pump
rejected	not used by cycle (usually energy)
res	energy contained in molecules by resonance
R	rational (as in efficiency)
R	reactants
R	reduced properties (in Law of Corresponding States)
R	reversible
s	shaft (as in work)
s	at constant entropy
surr	surroundings
sys	system
T	turbine
T	at constant temperature
th	thermal
u	useful (as in work)
u	unburned (as of reactants)
unav	unavailable (as in energy)
univ	universe (i.e. system + surroundings)
use	useful (as in work)
v	at constant specific volume

1

State of Equilibrium

Most texts on thermodynamics restrict themselves to dealing exclusively with equilibrium thermodynamics. This book will also focus on equilibrium thermodynamics but the effects of making this assumption will be explicitly borne in mind. The majority of processes met by engineers are in thermodynamic equilibrium, but some important processes have to be considered by non-equilibrium thermodynamics. Most of the combustion processes that generate atmospheric pollution include non-equilibrium effects, and carbon monoxide (CO) and oxides of nitrogen (NO_x) are both the result of the inability of the system to reach thermodynamic equilibrium in the time available.

There are four kinds of equilibrium, and these are most easily understood by reference to simple mechanical systems (see Fig 1.1).

(i) Stable equilibrium

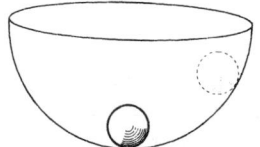

Marble in bowl.
For stable equilibrium $\Delta S)_E^* < 0$ and $\Delta E)_S > 0$.
(ΔS is the sum of Taylor's series terms).
Any deflection causes motion back towards equilibrium position.
*Discussed later.

(ii) Neutral equilibrium

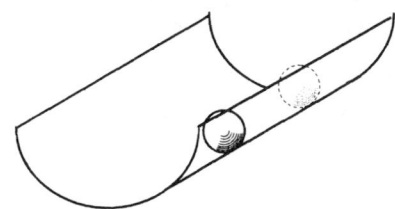

Marble in trough.
$\Delta S)_E = 0$ and $\Delta E)_S = 0$ along trough axis. Marble in equilibrium at any position in x-direction.

(iii) Unstable equilibrium

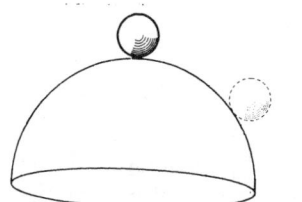

Marble sitting on maximum point of surface.
$\Delta S)_E > 0$ and $\Delta E)_S < 0$.
Any movement causes further motion from 'equilibrium' position.

Fig. 1.1 States of equilibrium

(iv) Metastable equilibrium

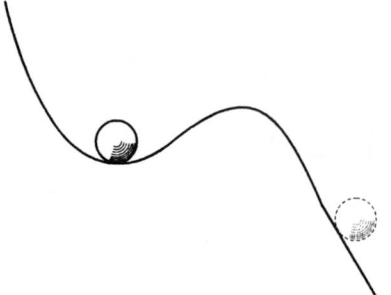

Marble in higher of two troughs. Infinitesimal variations of position cause return to equilibrium – larger variations cause movement to lower level.

Fig. 1.1 *Continued*

*****The difference between ΔS and dS**

Consider Taylor's theorem

$$\Delta S = \frac{dS}{dx} \Delta x + \frac{1}{2} \frac{d^2S}{dx^2} \Delta x^2 + \frac{1}{6} \frac{d^3S}{dx^3} \Delta x^3 + \cdots$$

$$= dS + \frac{1}{2} \frac{d^2S}{dx^2} \Delta x^2 + \cdots$$

Thus dS is the first term of the Taylor's series only. Consider a circular bowl at the position where the tangent is horizontal. Then

$$\left(\frac{dS}{dx} \right)_{(x=0)} = 0$$

However $\Delta S = dS + \dfrac{1}{2} \dfrac{d^2S}{dx^2} \Delta x^2 + \cdots \neq 0$, because $\dfrac{d^2S}{dx^2}$ etc are not zero.

Hence the following statements can be derived for certain classes of problem

stable equilibrium	$(dS)_E = 0$
	$(\Delta S)_E < 0$
neutral equilibrium	$(dS)_E = 0$
	$(\Delta S)_E = 0$
unstable equilibrium	$(dS)_E = 0$
	$(\Delta S)_E > 0$

(see Hatsopoulos and Keenan, 1972).

1.1 Equilibrium of a thermodynamic system

The type of equilibrium in a mechanical system can be judged by considering the variation in energy due to an infinitesimal disturbance. If the energy (potential energy) increases

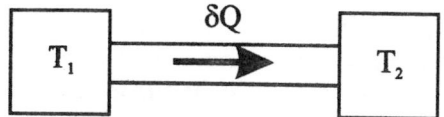

Fig. 1.2 Heat transfer between two blocks

then the system will return to its previous state, if it decreases it will not return to that state.

A similar method for examining the equilibrium of thermodynamic systems is required. This will be developed from the Second Law of Thermodynamics and the definition of entropy. Consider a system comprising two identical blocks of metal at different temperatures (see Fig 1.2), but connected by a conducting medium. From experience the block at the higher temperature will transfer 'heat' to that at the lower temperature. If the two blocks together constitute an isolated system the energy transfers will not affect the total energy in the system. If the high temperature block is at an temperature T_1 and the other at T_2 and if the quantity of energy transferred is δQ then the change in entropy of the high temperature block is

$$dS_1 = -\frac{\delta Q}{T_1} \tag{1.1}$$

and that of the lower temperature block is

$$dS_2 = +\frac{\delta Q}{T_2} \tag{1.2}$$

Both eqns (1.1) and (1.2) contain the assumption that the heat transfers from block 1, and into block 2 are reversible. If the transfers were irreversible then eqn (1.1) would become

$$dS_1 > -\frac{\delta Q}{T_1} \tag{1.1a}$$

and eqn (1.2) would be

$$dS_2 > +\frac{\delta Q}{T_2} \tag{1.2a}$$

Since the system is isolated the energy transfer to the surroundings is zero, and hence the change of entropy of the surroundings is zero. Hence the change in entropy of the system is equal to the change in entropy of the universe and is, using eqns (1.1) and (1.2),

$$dS = dS_1 + dS_2 = -\frac{\delta Q}{T_1} + \frac{\delta Q}{T_2} = \delta Q\left(\frac{1}{T_2} - \frac{1}{T_1}\right) \tag{1.3}$$

Since $T_1 > T_2$, then the change of entropy of both the system and the universe is $dS = (\delta Q / T_2 T_1)(T_1 - T_2) > 0$. The same solution, namely $dS > 0$, is obtained from eqns (1.1a) and (1.2a).

The previous way of considering the equilibrium condition shows how systems will tend to go towards such a state. A slightly different approach, which is more analogous to the one used to investigate the equilibrium of mechanical systems, is to consider these two blocks of metal to be in equilibrium and for heat transfer to occur spontaneously (and reversibly) between one and the other. Assume the temperature change in each block is δT, with one temperature increasing and the other decreasing, and the heat transfer is δQ. Then the change of entropy, dS, is given by

$$dS = \frac{\delta Q}{T + \delta T} - \frac{\delta Q}{T - \delta T} = \frac{\delta Q}{(T + \delta T)(T - \delta T)}(T - \delta T - T - \delta T)$$

$$= \frac{\delta Q}{T^2 + \delta T^2}(-2\delta T) \approx -2\delta Q \frac{dT}{T^2} \tag{1.4}$$

This means that the entropy of the system would have decreased. Hence maximum entropy is obtained when the two blocks are in equilibrium and are at the same temperature. The general criterion of equilibrium according to Keenan (1963) is as follows.

For stability of any system it is necessary and sufficient that, in all possible variations of the state of the system which do not alter its energy, the variation of entropy shall be negative.

This can be stated mathematically as

$$\Delta S)_E < 0 \tag{1.5}$$

It can be seen that the statements of equilibrium based on energy and entropy, namely $\Delta E)_S > 0$ and $\Delta S)_E < 0$, are equivalent by applying the following simple analysis. Consider the marble at the base of the bowl, as shown in Fig 1.1(i): if it is lifted up the bowl, its potential energy will be increased. When it is released it will oscillate in the base of the bowl until it comes to rest as a result of 'friction', and if that 'friction' is used solely to raise the temperature of the marble then its temperature will be higher after the process than at the beginning. A way to ensure the end conditions, i.e. the initial and final conditions, are identical would be to cool the marble by an amount equivalent to the increase in potential energy before releasing it. This cooling is equivalent to lowering the entropy of the marble by an amount ΔS, and since the cooling has been undertaken to bring the energy level back to the original value this proves that $\Delta E)_S > 0$ and $\Delta S)_E < 0$.

Equilibrium can be defined by the following statements:

(i)   if the properties of an isolated system change spontaneously there is an increase in the entropy of the system;

(ii)   when the entropy of an isolated system is at a maximum the system is in equilibrium;

(iii)   if, for all the possible variations in state of the isolated system, there is a negative change in entropy then the system is in stable equilibrium.

These conditions may be written mathematically as:

(i) $dS)_E > 0$ spontaneous change (unstable equilibrium)
(ii) $dS)_E = 0$ equilibrium (neutral equilibrium)
(iii) $\Delta S)_E < 0$ criterion of stability (stable equilibrium)

1.2 Helmholtz energy (Helmholtz function)

There are a number of ways of obtaining an expression for Helmholtz energy, but the one based on the Clausius derivation of entropy gives the most insight.

In the previous section, the criteria for equilibrium were discussed and these were derived in terms of $\Delta S)_E$. The variation of entropy is not always easy to visualise, and it would be more useful if the criteria could be derived in a more tangible form related to other properties of the system under consideration. Consider the arrangements in Figs 1.3(a) and (b). Figure 1.3(a) shows a System A, which is a general system of constant composition in which the work output, δW, can be either shaft or displacement work, or a combination of both. Figure 1.3(b) is a more specific example in which the work output is displacement work, $p\,\delta V$; the system in Fig 1.3(b) is easier to understand.

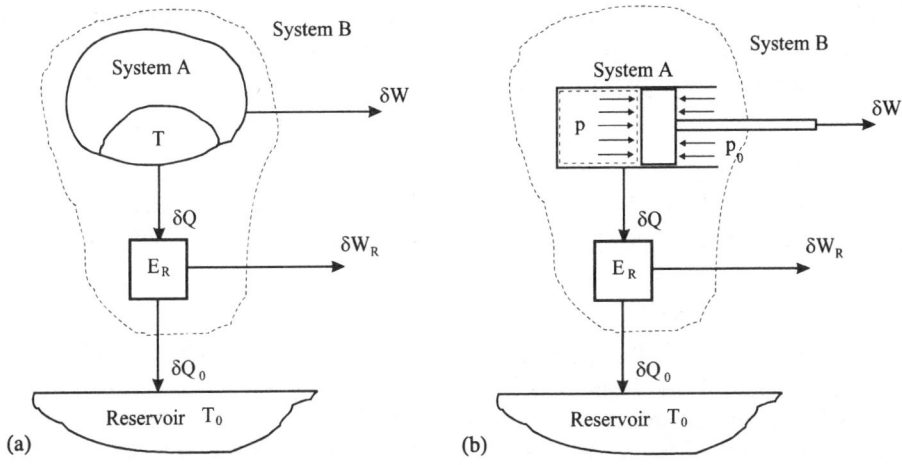

Fig. 1.3 Maximum work achievable from a system

In both arrangements, System A is a closed system (i.e. there are no mass transfers), which delivers an infinitesimal quantity of heat δQ in a ***reversible manner*** to the heat engine E_R. The heat engine then rejects a quantity of heat δQ_0 to a reservoir, e.g. the atmosphere, at temperature T_0.

Let dE, dV and dS denote the changes in internal energy, volume and entropy of the system, which is of constant, invariant composition. For a specified change of state these quantities, which are changes in properties, would be independent of the process or work done. Applying the First Law of Thermodynamics to System A gives

$$\delta W = -dE + \delta Q \tag{1.6}$$

If the heat engine (E_R) and System A are considered to constitute another system, System B, then, applying the First Law of Thermodynamics to System B gives

$$\delta W_{net} = \delta W + \delta W_R = -dE + \delta Q_0 \tag{1.7}$$

where $\delta W + \delta W_R = $ net work done by the heat engine and System A. Since the heat engine is internally reversible, and the entropy flow on either side is equal, then

$$\frac{\delta Q_0}{T_0} = \frac{\delta Q}{T} \tag{1.8}$$

and the change in entropy of System A during this process, because it is reversible, is $dS = \delta Q_R/T$. Hence

$$\left.\begin{aligned} \delta W_{net} &= -dE + T_0\,dS \\ &= -d(E - T_0 S) \end{aligned}\right\} \text{ because } T_0 = \text{constant} \tag{1.9}$$

The expression $E - T_0 S$ is called the *Helmholtz energy* or *Helmholtz function*. In the absence of motion and gravitational effects the energy, E, may be replaced by the intrinsic internal energy, U, giving

$$\delta W_{net} = -d(U - T_0 S) \tag{1.10}$$

The significance of δW_{net} will now be examined. The changes executed were considered to be reversible and δW_{net} was the net work obtained from System B (i.e. System A + heat engine E_R). Thus, δW_{net} must be the maximum quantity of work that can be obtained from the combined system. The expression for δW is called the change in the Helmholtz energy, where the Helmholtz energy is defined as

$$F = U - TS \tag{1.11}$$

Helmholtz energy is a property which has the units of energy, and indicates the maximum work that can be obtained from a system. It can be seen that this is less than the internal energy, U, and it will be shown that the product TS is a measure of the unavailable energy.

1.3   Gibbs energy (Gibbs function)

In the previous section the maximum work that can be obtained from System B, comprising System A and heat engine E_R, was derived. It was also stipulated that System A could change its volume by δV, and while it is doing this it must perform work on the atmosphere equivalent to $p_0\,\delta V$, where p_0 is the pressure of the atmosphere. This work detracts from the work previously calculated and gives the *maximum useful work*, δW_u, as

$$\delta W_u = \delta W_{net} - p_0\,dV \tag{1.12}$$

if the system is in pressure equilibrium with surroundings.

$$\begin{aligned} \delta W_u &= -d(E - T_0 S) - p_0\,dV \\ &= -d(E + p_0 V - T_0 S) \end{aligned}$$

because $p_0 = \text{constant}$. Hence

$$\delta W_u = -d(H - TS) \tag{1.13}$$

The quantity $H - TS$ is called the *Gibbs energy*, *Gibbs potential*, or the *Gibbs function*, G. Hence

$$G = H - TS \tag{1.14}$$

Gibbs energy is a property which has the units of energy, and indicates the maximum useful work that can be obtained from a system. It can be seen that this is less than the enthalpy, H, and it will be shown that the product TS is a measure of the unavailable energy.

1.4   The use and significance of the Helmholtz and Gibbs energies

It should be noted that the definitions of Helmholtz and Gibbs energies, eqns (1.11) and (1.14), have been obtained for systems of invariant composition. The more general form of these basic thermodynamic relationships, in differential form, is

$$dU = T\,dS - p\,dV + \Sigma\,\mu_i\,dn_i$$
$$dH = T\,dS + V\,dp + \Sigma\,m_i\,dn_i$$
$$dF = -S\,dT - p\,dV + \Sigma\,\mu_i\,dn_i$$
$$dG = -S\,dT + V\,dp + \Sigma\,\mu_i\,dn_i \tag{1.15}$$

The additional term, $\Sigma\,\mu_i\,dn_i$, is the product of the chemical potential of component i and the change of the amount of substance (measured in moles) of component i. Obviously, if the amount of substance of the constituents does not change then this term is zero. However, if there is a reaction between the components of a mixture then this term will be non-zero and must be taken into account. Chemical potential is introduced in Chapter 12 when dissociation is discussed; it is used extensively in the later chapters where it can be seen to be the driving force of chemical reactions.

1.4.1   HELMHOLTZ ENERGY

(i)   The change in Helmholtz energy is the maximum work that can be obtained from a closed system undergoing a reversible process whilst remaining in temperature equilibrium with its surroundings.

(ii)   A decrease in Helmholtz energy corresponds to an increase in entropy, hence the minimum value of the function signifies the equilibrium condition.

(iii) A decrease in entropy corresponds to an increase in F; hence the criterion $dF)_T > 0$ is that for stability. This criterion corresponds to work being done on the system.

(iv)   For a constant volume system in which $W = 0$, $dF = 0$.

(v)   For reversible processes, $F_1 = F_2$; for all other processes there is a decrease in Helmholtz energy.

(vi)   *The minimum value of Helmholtz energy corresponds to the equilibrium condition.*

1.4.2   GIBBS ENERGY

(i)   The change in Gibbs energy is the maximum useful work that can be obtained from a system undergoing a reversible process whilst remaining in pressure and temperature equilibrium with its surroundings.

(ii)   The equilibrium condition for the constraints of constant pressure and temperature can be defined as:

(1) $dG)_{p,T} < 0$ spontaneous change

(2) $dG)_{p,T} = 0$ equilibrium

(3) $\Delta G)_{p,T} > 0$ criterion of stability.

(iii) *The minimum value of Gibbs energy corresponds to the equilibrium condition.*

1.4.3 EXAMPLES OF DIFFERENT FORMS OF EQUILIBRIUM MET IN THERMODYNAMICS

Stable equilibrium is the most frequently met state in thermodynamics, and most systems exist in this state. Most of the theories of thermodynamics are based on stable equilibrium, which might be more correctly named thermostatics. The measurement of thermodynamic properties relies on the measuring device being in equilibrium with the system. For example, a thermometer must be in thermal equilibrium with a system if it is to measure its temperature, which explains why it is not possible to assess the temperature of something by touch because there is heat transfer either to or from the fingers – the body 'measures' the heat transfer rate. A system is in a stable state if it will permanently stay in this state without a tendency to change. Examples of this are a mixture of water and water vapour at constant pressure and temperature; the mixture of gases from an internal combustion engine when they exit the exhaust pipe; and many forms of crystalline structures in metals. Basically, stable equilibrium states are defined by state diagrams, e.g. the $p-v-T$ diagram for water, where points of stable equilibrium are defined by points on the surface; any other points in the $p-v-T$ space are either in unstable or metastable equilibrium. The equilibrium of mixtures of elements and compounds is defined by the state of maximum entropy or minimum Gibbs or Helmholtz energy; this is discussed in Chapter 12. The concepts of stable equilibrium can also be used to analyse the operation of fuel cells and these are considered in Chapter 17.

Another form of equilibrium met in thermodynamics is metastable equilibrium. This is where a system exists in a 'stable' state without any tendency to change until it is perturbed by an external influence. A good example of this is met in combustion in spark-ignition engines, where the reactants (air and fuel) are induced into the engine in a pre-mixed form. They are ignited by a small spark and convert rapidly into products, releasing many thousands of times the energy of the spark used to initiate the combustion process. Another example of metastable equilibrium is encountered in the Wilson 'cloud chamber' used to show the tracks of α particles in atomic physics. The Wilson cloud chamber consists of super-saturated water vapour which has been cooled down below the dew-point without condensation – it is in a metastable state. If an α particle is introduced into the chamber it provides sufficient perturbation to bring about condensation along its path. Other examples include explosive boiling, which can occur if there are not sufficient nucleation sites to induce sufficient bubbles at boiling point to induce normal boiling, and some of the crystalline states encountered in metallic structures.

Unstable states cannot be sustained in thermodynamics because the molecular movement will tend to perturb the systems and cause them to move towards a stable state. Hence, unstable states are only transitory states met in systems which are moving towards equilibrium. The gases in a combustion chamber are often in unstable equilibrium because they cannot react quickly enough to maintain the equilibrium state, which is defined by minimum Gibbs or Helmholtz energy. The 'distance' of the unstable state from the state of stable equilibrium defines the rate at which the reaction occurs; this is referred to as *rate kinetics*, and will be discussed in Chapter 14. Another example of unstable 'equilibrium' occurs when a partition is removed between two gases which are initially separated. These gases then mix due to diffusion, and this mixing is driven by the difference in *chemical potential* between the gases; chemical potential is introduced in Chapter 12 and the process of mixing is discussed in Chapter 16. Some thermodynamic situations never achieve stable equilibrium, they exist in a steady state with energy passing between systems in stable equilibrium, and such a situation can be analysed using the techniques of *irreversible thermodynamics* developed in Chapter 16.

1.4.4 SIGNIFICANCE OF THE MINIMUM GIBBS ENERGY AT CONSTANT PRESSURE AND TEMPERATURE

It is difficult for many engineers readily to see the significance of Gibbs and Helmholtz energies. If systems are judged to undergo change while remaining in temperature and pressure equilibrium with their surroundings then most mechanical engineers would feel that no change could have taken place in the system. However, consideration of eqns (1.15) shows that, if the system were a multi-component mixture, it would be possible for changes in Gibbs (or Helmholtz) energies to take place if there were changes in composition. For example, an equilibrium mixture of carbon dioxide, carbon monoxide and oxygen could change its composition by the carbon dioxide breaking down into carbon monoxide and oxygen, in their stoichiometric proportions; this breakdown would change the composition of the mixture. If the process happened at constant temperature and pressure, in equilibrium with the surroundings, then an increase in the Gibbs energy, G, would have occurred; such a process would be depicted by Fig 1.4. This is directly analogous to the marble in the dish, which was discussed in the introductory remarks to this section.

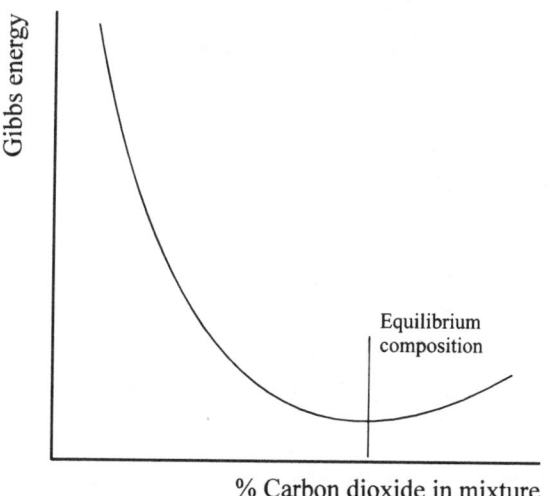

Fig. 1.4 Variation of Gibbs energy with chemical composition, for a system in temperature and pressure equilibrium with the environment

1.5 Concluding remarks

This chapter has considered the state of equilibrium for thermodynamic systems. Most systems are in equilibrium, although non-equilibrium situations will be introduced in Chapters 14, 16 and 17.

It has been shown that the change of entropy can be used to assess whether a system is in a stable state. Two new properties, Gibbs and Helmholtz energies, have been introduced and these can be used to define equilibrium states. These energies also define the maximum amount of work that can be obtained from a system.

PROBLEMS

1 Determine the criteria for equilibrium for a thermally isolated system at (a) constant volume; (b) constant pressure. Assume that the system is:
   (i) constant, and invariant, in composition;
   (ii) variable in composition.

2 Determine the criteria for isothermal equilibrium of a system at (a) constant volume, and (b) constant pressure. Assume that the system is:
   (i) constant, and invariant, in composition;
   (ii) variable in composition.

3 A system at constant pressure consists of 10 kg of air at a temperature of 1000 K. This is connected to a large reservoir which is maintained at a temperature of 300 K by a reversible heat engine. Calculate the maximum amount of work that can be obtained from the system. Take the specific heat at constant pressure of air, c_p, as 0.98 kJ/kg K.

[3320.3 kJ]

4 A thermally isolated system at constant pressure consists of 10 kg of air at a temperature of 1000 K and 10 kg of water at 300 K, connected together by a heat engine. What would be the equilibrium temperature of the system if
(a) the heat engine has a thermal efficiency of zero;
(b) the heat engine is reversible?

{Hint: consider the definition of equilibrium defined by the entropy change of the system.}

Assume

   for water: $c_v = 4.2$ kJ/kg K
                $\kappa = c_p/c_v = 1.0$
   for air: $c_v = 0.7$ kJ/kg K
                $\kappa = c_p/c_v = 1.4$

[432.4 K; 376.7 K]

5 A thermally isolated system at constant pressure consists of 10 kg of air at a temperature of 1000 K and 10 kg of water at 300 K, connected together by a heat engine. What would be the equilibrium temperature of the system if the maximum thermal efficiency of the engine is only 50%?

Assume

   for water: $c_v = 4.2$ kJ /kg K
                $\kappa = c_p/c_v = 1.0$
   for air: $c_v = 0.7$ kJ/kg K
                $\kappa = c_p/c_v = 1.4$

[385.1 K]

6 Show that if a liquid is in equilibrium with its own vapour and an inert gas in a closed vessel, then

$$\frac{dp_v}{dp} = \frac{\rho_v}{\rho_l}$$

where p_v is the partial pressure of the vapour, p is the total pressure, ρ_v is the density of the vapour and ρ_1 is the density of the liquid.

7 An incompressible liquid of specific volume v_1, is in equilibrium with its own vapour and an inert gas in a closed vessel. The vapour obeys the law

$$p(v - b) = \Re T$$

Show that

$$\ln\left(\frac{p_v}{p_0}\right) = \frac{1}{\Re T} \{(p - p_0)v_1 - (p_v - p_0)b\}$$

where p_0 is the vapour pressure when no inert gas is present, and p is the total pressure.

8 (a) Describe the meaning of the term thermodynamic equilibrium. Explain how entropy can be used as a measure of equilibrium and also how other properties can be developed that can be used to assess the equilibrium of a system.

If two phases of a component coexist in equilibrium (e.g. liquid and vapour phase H_2O) show that

$$T\frac{dp}{dT} = \frac{l}{v_{fg}}$$

where

T = temperature,

p = pressure,

l = latent heat,

and v_{fg} = difference between liquid and vapour phases.

Show the significance of this on a phase diagram.

(b) The melting point of tin at a pressure of 1 bar is 505 K, but increases to 508.4 K at 1000 bar. Evaluate

    (i) the change of density between these pressures, and

    (ii) the change in entropy during melting.

The latent heat of fusion of tin is 58.6 kJ/kg.

$$[254\ 100\ \text{kg/m}^3;\ 0.1157\ \text{kJ/kg K}]$$

9 Show that when different phases are in equilibrium the specific Gibbs energy of each phase is equal.

Using the following data, show the pressure at which graphite and diamond are in equilibrium at a temperature of 25°C. The data for these two phases of carbon at 25°C and 1 bar are given in the following table:

	Graphite	Diamond
Specific Gibbs energy, $g/(kJ/kg)$	0	269
Specific volume, $v/(m^3/kg)$	0.446×10^{-3}	0.285×10^{-3}
Isothermal compressibility, $k/(bar^{-1})$	2.96×10^{-6}	0.158×10^{-6}

It may be assumed that the variation of kv with pressure is negligible, and the lower value of the solution may be used.

[17990 bar]

10 Van der Waals' equation for water is given by

$$p = \frac{0.004619T}{v - 0.0016891} - \frac{0.017034}{v^2}$$

where p = pressure (bar), v = specific volume $(m^3/kmol)$, T = temperature (K).

Draw a p–v diagram for the following isotherms: 250°C, 270°C, 300°C, 330°C, 374°C, 390°C.

Compare the computed specific volumes with Steam Table values and explain the differences in terms of the value of $p_c v_c / \Re T_c$.

2

Availability and Exergy

Many of the analyses performed by engineers are based on the First Law of Thermo-dynamics, which is a law of energy conservation. Most mechanical engineers use the Second Law of Thermodynamics simply through its derived property – entropy (S). However, it is possible to introduce other 'Second Law' properties to define the maximum amounts of work achievable from certain systems. Previously, the properties Helmholtz energy (F) and Gibbs energy (G) were derived as a means of assessing the equilibrium of various systems. This section considers how the maximum amount of work available from a system, when interacting with surroundings, can be estimated. This shows, as expected, that all the energy in a system cannot be converted to work: the Second Law stated that it is impossible to construct a heat engine that does not reject energy to the surroundings.

2.1 Displacement work

The work done by a system can be considered to be made up of two parts: that done against a resisting force and that done against the environment. This can be seen in Fig 2.1. The pressure inside the system, p, is resisted by a force, F, and the pressure of the environment. Hence, for System A, which is in equilibrium with the surroundings

$$pA = F + p_0 A \tag{2.1}$$

where A is the area of cross-section of the piston.

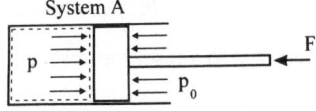

Fig. 2.1 Forces acting on a piston

If the piston moves a distance $\mathrm{d}x$, then the work done by the various components shown in Fig 2.1 is

$$pA\ \mathrm{d}x = F\ \mathrm{d}x + p_0 A\ \mathrm{d}x \tag{2.2}$$

where $pA\ \mathrm{d}x = p\ \mathrm{d}V = \delta W_{\mathrm{sys}}$ = work done by the fluid in the system, $F\ \mathrm{d}x = \delta W_{\mathrm{use}}$ = work done against the resisting force, and $p_0 A\ \mathrm{d}x = p_0\ \mathrm{d}V = \delta W_{\mathrm{surr}}$ = work done against the surroundings.

Hence the work done by the system is not all converted into useful work, but some of it is used to do displacement work against the surroundings, i.e.

$$\delta W_{sys} = \delta W_{use} + \delta W_{surr} \qquad (2.3)$$

which can be rearranged to give

$$\delta W_{use} = \delta W_{sys} - \delta W_{surr} \qquad (2.4)$$

2.2 Availability

It was shown above that not all the displacement work done by a system is available to do useful work. This concept will now be generalised to consider all the possible work outputs from a system that is not in thermodynamic and mechanical equilibrium with its surroundings (i.e. not at the ambient, or dead state, conditions).

Consider the system introduced earlier to define Helmholtz and Gibbs energy: this is basically the method that was used to prove the Clausius inequality.

Figure 2.2(a) shows the general case where the work can be either displacement or shaft work, while Fig 2.2(b) shows a specific case where the work output of System A is displacement work. It is easier to follow the derivation using the specific case, but a more general result is obtained from the arrangement shown in Fig 2.2(a).

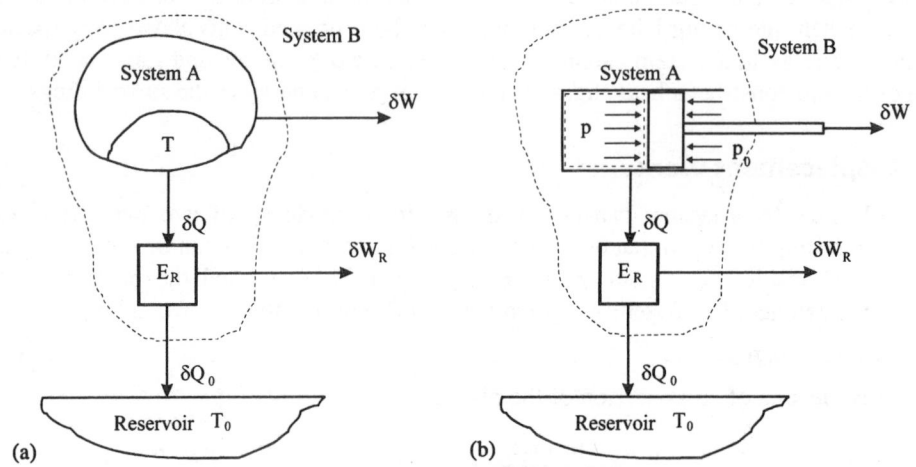

Fig. 2.2 System transferring heat to a reservoir through a reversible heat engine

First consider that System A is a *constant volume* system which transfers heat with the surroundings via a small reversible heat engine. Applying the First Law to the System A

$$dU = \delta Q - \delta W = \delta Q - \delta W_S \qquad (2.5)$$

where δW_S indicates the shaft work done (e.g. System A could contain a turbine).

Let System B be System A plus the heat engine E_R. Then applying the First Law to system B gives

$$dU = \Sigma \, (\delta Q - \delta W) = \delta Q_0 - (\delta W_S + \delta W_R) \qquad (2.6)$$

As System A transfers energy with the surroundings it under goes a change of entropy defined by

$$dS = \frac{\delta Q}{T} = \frac{\delta Q_0}{T_0} \tag{2.7}$$

because the heat engine transferring the heat to the surroundings is reversible, and there is no change of entropy across it. Hence

$$(\delta W_S + \delta W_R) = -(dU - T_0\, dS) \tag{2.8}$$

As stated previously, $(\delta W_S + \delta W_R)$ is the maximum work that can be obtained from a *constant volume, closed system* when interacting with the surroundings. If the volume of the system was allowed to change, as would have to happen in the case depicted in Fig 2.2(b), then the work done against the surroundings would be $p_0\, dV$, where p_0 is the pressure of the surroundings. This work, done against the surroundings, reduces the maximum useful work from the system in which a change of volume takes place to $\delta W + \delta W_R - p_0\, dV$, where δW is the sum of the shaft work and the displacement work.

Hence, the *maximum useful work* which can be achieved from a closed system is

$$\delta W + \delta W_R = -(dU + p_0\, dV - T_0\, dS) \tag{2.9}$$

This work is given the symbol dA. Since the surroundings are at fixed pressure and temperature (i.e. p_0 and T_0 are constant) dA can be integrated to give

$$A = U + p_0 V - T_0 S \tag{2.10}$$

A is called the *non-flow availability function*. Although it is a combination of properties, A is not itself a property because it is defined in relation to the arbitrary datum values of p_0 and T_0. Hence it is not possible to tabulate values of A without defining both these datum levels. The datum levels are what differentiates A from Gibbs energy G. Hence the maximum useful work achievable from a system changing state from 1 to 2 is given by

$$W_{max} = -\Delta A = -(A_2 - A_1) = A_1 - A_2 \tag{2.11}$$

The specific availability, a, i.e. the availability per unit mass is

$$a = u + p_0 v - T_0 s \tag{2.12a}$$

If the value of a were based on unit amount of substance (i.e. kmol) it would be referred to as the molar availability.

The change of specific (or molar) availability is

$$\begin{aligned} \Delta a = a_2 - a_1 &= (u_2 + p_0 v_2 - T_0 s_2) - (u_1 + p_0 v_1 - T_0 s_1) \\ &= (h_2 + v_2(p_0 - p_2)) - (h_1 + v_1(p_0 - p_1)) - T_0(s_2 - s_1) \end{aligned} \tag{2.12b}$$

2.3 Examples

Example 1: reversible work from a piston cylinder arrangement (this example is based on Haywood (1980))

System A, in Fig 2.1, contains air at a pressure and temperature of 2 bar and 550 K respectively. The pressure is maintained by a force, F, acting on the piston. The system is taken from state 1 to state 2 by the reversible processes depicted in Fig 2.3, and state 2 is equal to the dead state conditions with a pressure, p_0, and temperature, T_0, of 1 bar and 300 K respectively. Evaluate the following work terms assuming that the air is a perfect gas and that $c_p = 1.005$ kJ/kg K and the ratio of specific heats $\kappa = 1.4$.

(a)   The air follows the process 1-a-2 in Fig 2.3, and transfers heat reversibly with the environment during an isobaric process from 1-a. Calculate the following specific work outputs for processes 1-a and a-2:

(i)	the work done by the system, δW_{sys};
(ii)	the work done against the surroundings, δW_{surr};
(iii)	the useful work done against the resisting force F, δW_{use};
(iv)	the work done by a reversible heat engine operating between the system and the surroundings, δW_R.

Then evaluate for the total process, 1-2, the following parameters:

(v)	the gross work done by the system, $\Sigma(\delta W_{sys} + \delta W_R)$;
(vi)	the net useful work output against the environment, $\Sigma(\delta W_{use} + \delta W_R)$;
(vii)	the total displacement work against the environment, $\Sigma(\delta W_{surr})$;
(viii)	the work term $p_0(v_2 - v_1)$.

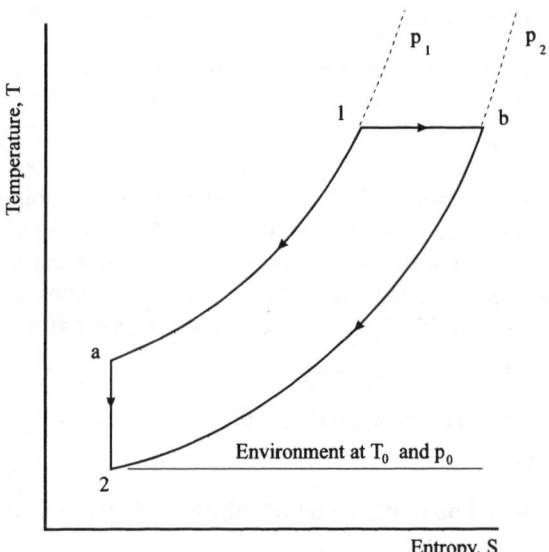

Fig. 2.3 Processes undergone by the system

Solution

It is necessary to evaluate the specific volume at the initial condition, state 1.

$$p_1 v_1 = RT_1, \quad \text{hence } v_1 = RT_1/p_1$$

The gas constant, R, is given by $R = ((\kappa - 1)/\kappa)c_p = 0.4 \times 1.005/1.4 = 0.287$ kJ/kg K, and hence

$$v_1 = \frac{0.287 \times 10^3 \times 550}{2 \times 10^5} = 0.7896 \text{ m}^3/\text{kg}$$

The intermediate temperature, T_a, should now be evaluated. Since the process a-2 is isentropic then T_a must be isentropically related to the final temperature, T_2. Hence

$$T_a = T_2 \left(\frac{p_a}{p_2}\right)^{(\kappa - 1)/\kappa} = 300 \left(\frac{2}{1}\right)^{0.4/1.4} = 365.7 \text{ K}$$

Thus $v_a = \dfrac{p_1 T_a}{p_a T_1} v_1 = \dfrac{365.7}{550} v_1 = 0.6649 v_1$

Consider process 1-a:

$$w_{\text{sys}}|_{1-a} = \int_1^a p \, dv = p_a(v_a - v_1), \text{ for an isobaric process}$$

$$= \frac{2 \times 10^5}{10^3} (0.6649 - 1.0) \times 0.7896 = -52.96 \text{ kJ/kg}$$

$$w_{\text{surr}}|_{1-a} = \int_1^a p_0 \, dv = \frac{1 \times 10^5}{10^3} (0.6649 - 1.0) \times 0.7896$$

$$= -26.46 \text{ kJ/kg}$$

The useful work done by the system against the resisting force, F, is the difference between the two work terms given above, and is

$$w_{\text{use}}|_{1-a} = w_{\text{sys}}|_{1-a} - w_{\text{surr}}|_{1-a} = -26.46 \text{ kJ/kg}$$

The work terms derived above relate to the mechanical work that can be obtained from the system as it goes from state 1 to state a. However, in addition to this mechanical work the system could also do thermodynamic work by transferring energy to the surroundings through a reversible heat engine. This work can be evaluated in the following way. In going from state 1 to state a the system has had to transfer energy to the surroundings because the temperature of the fluid has decreased from 550 K to 365.7 K. This heat transfer could have been simply to a reservoir at a temperature below T_a, in which case no useful work output would have been achieved. It could also have been to the environment through a reversible heat engine, as shown in Fig 2.2(b). In the latter case, useful work would have been obtained, and this would have been equal to

$$\delta w_R = -\eta_R \delta Q,$$

where η_R is the thermal efficiency of a reversible heat engine operating between T and T_0, and δQ is the heat transfer to the system. Thus

$$\delta w_R = -\frac{T - T_0}{T} \delta Q$$

Hence, the work output obtainable from this reversible heat engine as the system changes state from 1 to a is

$$w_R = -\int_1^a \frac{T - T_0}{T} \delta Q = -\int_1^a \frac{T - T_0}{T} c_p \, dT$$

$$= -c_p \left\{ (T_a - T_1) - T_0 \ln \frac{T_a}{T_1} \right\}$$

$$= -1.005 \times \left\{ (365.7 - 550) - 300 \ln \frac{365.7}{500} \right\} = 62.18 \text{ kJ/kg}$$

Now consider process a-2. First, evaluate the specific volume at 2, v_2:

$$v_2 = \frac{p_1 T_2}{p_2 T_1} v_1 = \frac{2 \times 300}{1 \times 550} v_1 = 1.0909 v_1$$

The expansion from a to 2 is isentropic and hence obeys the law $pv^\kappa = \text{constant}$. Thus the system work is

$$W_{\text{sys}}|_{a-2} = \frac{(1 \times 1.0909 - 2 \times 0.6649)}{1 - 1.4} \times \frac{0.7896 \times 10^5}{10^3} = 47.16 \text{ kJ/kg}$$

$$W_{\text{surr}}|_{a-2} = \int_a^2 p_0 \, dv = \frac{1 \times 10^5}{10^3} (1.0909 - 0.6649) \times 0.7896 = 33.63 \text{ kJ/kg}$$

$$W_{\text{use}}|_{a-2} = W_{\text{sys}}|_{a-2} - W_{\text{surr}}|_{a-2} = 13.52 \text{ kJ/kg}$$

The energy available to drive a reversible heat engine is zero for this process because it is adiabatic; hence

$$W_R|_{a-2} = 0$$

The work terms for the total process from 1 to 2 can be calculated by adding the terms for the two sub-processes. This enables the solutions to questions (v) to (viii) to be obtained.

The gross work done by the system is

$$W_{\text{gross}} = \Sigma(W_{\text{sys}} + W_R) = -52.96 + 47.16 + 62.18 + 0 = 56.38 \text{ kJ/kg}$$

The net useful work done by the system is

$$W_{\text{u.net}} = \Sigma(W_{\text{use}} + W_R) = -26.46 + 13.52 + 62.18 = 49.24 \text{ kJ/kg}$$

The total displacement work done against the surroundings is

$$W_{\text{surr}} = -26.46 + 33.63 = 7.17 \text{ kJ/kg}$$

The displacement work of the surroundings evaluated from

$$p_0(v_2 - v_1) = 1 \times 10^5 \times (1.0909 - 1) \times 0.7896/10^3 = 7.17 \text{ kJ/kg}$$

Example 2: change of availability in reversible piston cylinder arrangement

Calculate the change of availability for the process described in Example 1. Compare the value obtained with the work terms evaluated in Example 1.

Solution

The appropriate definition of availability for System A is the non-flow availability function, defined in eqn (2.12a), and the maximum useful work is given by eqns (2.11) and (2.12b). Hence, w_{use} is given by

$$W_{\text{use}} = -\Delta a = -(a_2 - a_1) = a_1 - a_2$$
$$= u_1 - u_2 + p_0(v_1 - v_2) - T_0(s_1 - s_2)$$

The change of entropy, $s_1 - s_2$, can be evaluated from

$$ds = c_p \frac{dT}{T} - R \frac{dp}{p}$$

which for a perfect gas becomes

$$s_1 - s_2 = c_p \ln \frac{T_1}{T_2} - R \ln \frac{p_1}{p_2}$$

$$= 1.005 \times \ln \frac{550}{300} - 0.287 \times \ln \frac{2}{1} = 0.41024 \text{ kJ/kgK}$$

Substituting gives

$$W_{use} = c_v(T_1 - T_2) + p_0(v_1 - v_2) - T_0(s_1 - s_2)$$

$$= (1.005 - 0.287)(550 - 300) + \frac{1 \times 10^5}{10^3} (1 - 1.0909) \times 0.7896$$

$$- 300 \times 0.41024$$

$$= 179.5 - 7.17 - 123.07 = 49.25 \text{ kJ/kg}$$

This answer is identical to the net useful work done by the system evaluated in part (vi) above; this is to be expected because the change of availability was described and defined as the maximum useful work that could be obtained from the system. Hence, the change of availability can be evaluated directly to give the maximum useful work output from a number of processes without having to evaluate the components of work output separately.

Example 3: availability of water vapour

Evaluate the specific availability of water vapour at 30 bar, 450°C if the surroundings are at $p_0 = 1$ bar; $t_0 = 35$°C. Evaluate the maximum useful work that can be obtained from this vapour if it is expanded to (i) 20 bar, 250°C; (ii) the dead state.

Solution

From Rogers and Mayhew tables:

at $p = 30$ bar; $t = 450$°C
$u = 3020$ kJ/kg; $h = 3343$ kJ/kg; $v = 0.1078$ m^3/kg; $s = 7.082$ kJ/kg K

Hence the specific availability at these conditions is

$$a = u + p_0 v - T_0 s$$

$$= 3020 + 1 \times 10^5 \times 0.1078/10^3 - 308 \times 7.082$$

$$= 849.5 \text{ kJ/kg}$$

(i)   Change of availability if expanded to 20 bar, 250°C.
    From Rogers and Mayhew tables:

    at $p = 20$ bar; $t = 250$°C
    $u = 2681$ kJ/kg; $h = 2904$ kJ/kg; $v = 0.1115$ m^3/kg; $s = 6.547$ kJ/kg K

The specific availability at these conditions is

$$a = u + p_0 v - T_0 s$$
$$= 2681 + 1 \times 10^5 \times 0.1115/10^3 - 308 \times 6.547$$
$$= 675.7 \text{ kJ/kg}$$

Thus the maximum work that can be obtained by expanding the gas to these conditions is

$$w_{max} = a_1 - a_2 = 849.5 - 675.7 = 173.8 \text{ kJ/kg}$$

This could also be evaluated using eqn (2.12b), giving

$$w_{max} = -\Delta a = h_1 + v_1(p_0 - p_1) - (h_2 + v_2(p_0 - p_2)) - T_0(s_1 - s_2)$$
$$= 3343 - 2904 + 0.1078 \times (1 - 30) \times 10^2 - 0.1115 \times (1 - 20) \times 10^2$$
$$- 308 \times (7.082 - 6.547)$$
$$= 439 - 312.6 + 211.9 - 164.78 = 173.5 \text{ kJ/kg}$$

The values obtained by the different approaches are the same to the accuracy of the figures in the tables.

(ii)   Change of availability if expanded to datum level at $p_0 = 1$ bar; $t_0 = 35°C$

$$u = 146.6 \text{ kJ/kg}; \ h = 146.6 \text{ kJ/kg}; \ v = 0.1006 \times 10^{-2} \text{ m}^3/\text{kg};$$
$$s = 0.5045 \text{ kJ/kg K}$$

The specific availability at the dead state is

$$a = u + p_0 v - T_0 s$$
$$= 146.6 + 1 \times 10^5 \times 0.001006/10^3 - 308 \times 0.5045$$
$$= -8.7 \text{ kJ/kg}$$

and the maximum work that can be obtained by expanding to the dead state is

$$w_{max} = a_1 - a_2 = 849.5 - (-8.7) = 858.2 \text{ kJ/kg}$$

Again, this could have been evaluated using eqn (2.12b) to give

$$w_{max} = -\Delta a = h_1 + v_1(p_0 - p_1) - (h_2 + v_2(p_0 - p_2)) - T_0(s_1 - s_2)$$
$$= 3343 - 146.6 + 0.1078 \times (1 - 30) \times 10^2 - 0.001006 \times (1 - 1) \times 10^2$$
$$- 308 \times (7.082 - 0.5045)$$
$$= 3196.4 - 312.6 + 0 - 2025.9 = 857.9 \text{ kJ/kg}$$

This shows that although the energy available in the system was 3020 kJ/kg (the internal energy) it is not possible to convert all this energy to work. It should be noted that this 'energy' is itself based on a datum of the triple point of water, but the dead state is above this value. However, even if the datum levels were reduced to the triple point the maximum useful work would still only be

$$w_{max} = -\Delta a = (3343 - 0) + 0.1078 \times (0.00612 - 30) \times 10^2 - 273(7.082 - 0)$$
$$= 1086.3 \text{ kJ/kg}$$

It is also possible to evaluate the availability of a steady-flow system and this is defined in a similar manner to that used above, but in this case the reversible heat engine extracts energy from the flowing stream. If the kinetic and potential energies of the flowing stream are negligible compared with the thermal energy then the steady flow availability is

$$a_f = h - T_0 s \qquad (2.13\text{a})$$

The change in steady flow availability is

$$\Delta a_f = a_{f2} - a_{f1} = h_2 - T_0 s_2 - (h_1 - T_0 s_1) \qquad (2.13\text{b})$$

Thus, the maximum work that could be obtained from a steady-flow system is the change in flow availability, which for the previous example is

$$\begin{aligned} w_{max} = -\Delta a_f &= -(a_{f2} - a_{f1}) = h_1 - T_0 s_1 - (h_2 - T_0 s_2) \\ &= (3343 - 146.6) - 308 \times (7.082 - 0.5045) \\ &= 1170.4 \text{ kJ/kg} \end{aligned}$$

2.4   Available and non-available energy

If a certain portion of energy is available then obviously another part is unavailable – the unavailable part is that which must be thrown away. Consider Fig 2.4; this diagram indicates an internally reversible process from a to b. This can be considered to be made up of an infinite number of strips 1-m-n-4-1 where the temperature of energy transfer is essentially constant, i.e. $T_1 = T_4 = T$. The energy transfer obeys

$$\frac{\delta Q}{T} = \frac{\delta Q_0}{T_0}$$

where δQ = heat transferred to system and δQ_0 = heat rejected from system, as in an engine (E_R) undergoing an infinitesimal Carnot cycle.

In reality δQ_0 is the minimum amount of heat that can be rejected because processes 1 to 2 and 3 to 4 are both isentropic, i.e. adiabatic and *reversible*.

Hence the amount of energy that must be rejected is

$$E_{unav} = \int dQ_0 = T_0 \int \left. \frac{dQ}{T} \right|_R = T_0 \Delta S \qquad (2.14)$$

Note that the quantity of energy, δQ, can be written as a definite integral because the process is an isentropic (reversible) one. Then E_{unav} is the energy that is unavailable and is given by *cdefc*. The available energy on this diagram is given by *abcda* and is given by

$$E_{av} = Q - E_{unav} = Q - T_0 \Delta S \qquad (2.15)$$

where Q is defined by the area *abfea*.

2.5   Irreversibility

The concept of reversible engines has been introduced and these have operated on reversible cycles, e.g. isentropic and isothermal reversible processes. However, all real processes are irreversible and it is possible to obtain a measure of this irreversibility using the previous analysis. This will be illustrated by two examples: a turbine, which produces a work output; and a compressor, which absorbs a work input.

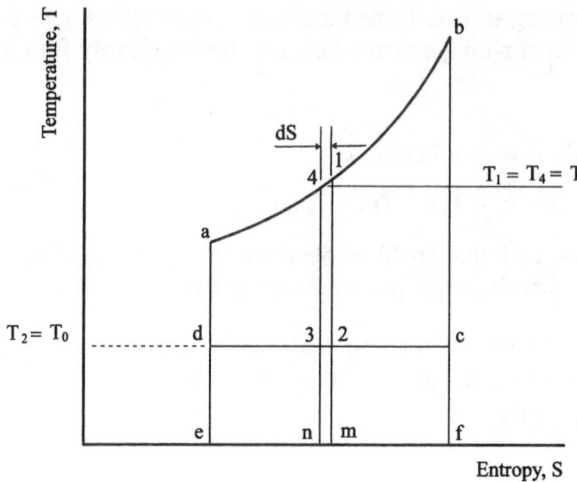

Fig. 2.4 Available and unavailable energy shown on a $T-s$ diagram

Example 4: a turbine

An aircraft gas turbine with an isentropic efficiency of 85% receives hot gas from the combustion chamber at 10 bar and 1000°C. It expands this to the atmospheric pressure of 1 bar. If the temperature of the atmosphere is 20°C, determine (a) the change of availability of the working fluid, and the work done by the turbine *if the expansion were isentropic*. Then, for the actual turbine, determine (b) the change of availability and the work done, (c) the change of availability of the surroundings, and (d) the net loss of availability of the universe (i.e. the irreversibility).

Assume that the specific heat at constant pressure, $c_p = 1.100$ kJ/kg K, and that the ratio of specific heats, $\kappa = 1.35$.

Solution

The processes involved are shown in Fig 2.5.

(a) Isentropic expansion

From the steady flow energy equation the specific work done in the isentropic expansion is

$$w_T|_{\text{isen}} = c_p(T_1 - T_{2'})$$

For the isentropic expansion, $T_{2'} = T_1(p_2/p_1)^{(\kappa-1)/\kappa} = 1273(1/10)^{0.35/1.35} = 700.8$ K, giving

$$w_T|_{\text{isen}} = 1.100 \times (1273 - 700.8) = 629.5 \text{ kJ/kg}$$

The change in availability of the working fluid is given by eqn (2.13b), $\Delta a = a_{f2'} - a_{f1}$, where

$$a_{f2'} = h_{2'} - T_0 s_{2'}$$
$$a_{f1} = h_1 - T_0 s_1$$

For an isentropic change, $s_{2'} = s_1$, and hence $\Delta a = h_{2'} - h_1 = -w_T|_{\text{isen}} = -629.5$ kJ/kg.

The change of availability of the surroundings is 629.5 kJ/kg, and hence the change of availability of the universe is zero for this *isentropic* process. This means that the energy can be transferred between the system and the environment without any degradation, and the processes are reversible – this would be expected in the case of an isentropic process.

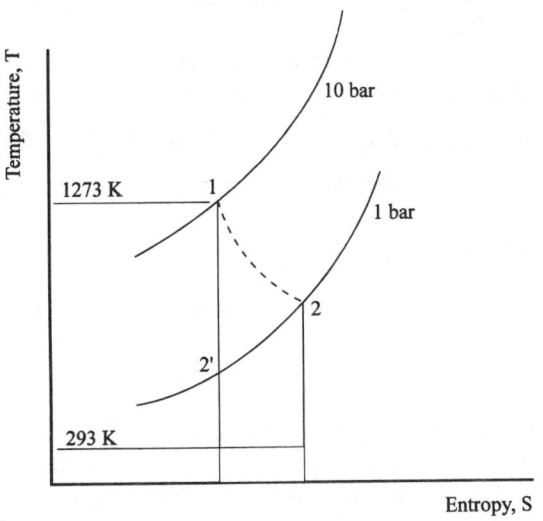

Fig. 2.5 Turbine expansion on a T–s diagram

(b) Non-isentropic expansion
   If the isentropic efficiency of the turbine is 85%, i.e. $\eta_T = 0.85$, then the turbine work is

$$w_T = \eta_T w_T |_{\text{isen}} = 0.85 \times 629.5 = 535.0 \text{ kJ/kg}$$

The temperature at the end of the expansion is

$$T_2 = T_1 - w_T / c_p = 1273 - 535.0/1.100 = 786 \text{ K.}$$

The change of availability is given by eqn (2.13b) as

$$\Delta a = a_{f2} - a_{f1}$$
$$= h_2 - T_0 s_2 - (h_1 - T_0 s_1)$$
$$= h_2 - h_1 - T_0 (s_2 - s_1) = -w_T - T_0 (s_2 - s_1)$$

The change of entropy for a perfect gas is given by

$$\Delta s_{12} = s_2 - s_1 = c_p \ln \frac{T_2}{T_1} - R \ln \frac{p_2}{p_1}$$

where, from the perfect gas law,

$$R = \frac{(\kappa - 1)c_p}{\kappa} = \frac{(1.35 - 1) \times 1.100}{1.35} = 0.285 \text{ kJ/kg K}$$

Thus the change of entropy during the expansion process is

$$s_2 - s_1 = 1.1 \ln \frac{786}{1273} - 0.285 \ln \frac{1}{10} = -0.5304 + 0.6562 = 0.1258 \text{ kJ/kg K}$$

Thus $\Delta a = -535 - T_0(s_2 - s_1) = -535 - 293 \times 0.1258 = -571.9 \text{ kJ/kg}$

(c) The change of availability of the surroundings is equal to the work done, hence $\Delta a_{\text{surroundings}} = 535 \text{ kJ/kg}$

(d) The irreversibility is the change of availability of the universe, which is the sum of the changes of availability of the system and its surroundings, i.e.

$$\Delta a_{\text{universe}} = \Delta a_{\text{system}} + \Delta a_{\text{surroundings}} = -571.9 + 535.0 = -36.9 \text{ kJ/kg}$$

This can also be calculated directly from an expression for irreversibility

$$I = T_0 \, \Delta S$$
$$= 293 \times 0.1258 = 36.86 \text{ kJ/kg}$$

Note that the irreversibility is positive because it is defined as the loss of availability.

Example 5: an air compressor

A steady flow compressor for a gas turbine receives air at 1 bar and 15°C, which it compresses to 7 bar with an efficiency of 83%. Based on surroundings at 5°C, determine (a) the change of availability and the work for isentropic compression. For the actual process evaluate (b) the change of availability and work done, (c) the change of availability of the surroundings and (d) the irreversibility.

Treat the gas as an ideal one, with the specific heat at constant pressure, $c_p = 1.004 \text{ kJ/kg K}$, and the ratio of specific heats, $\kappa = 1.4$.

Solution

The processes are shown in Fig 2.6.

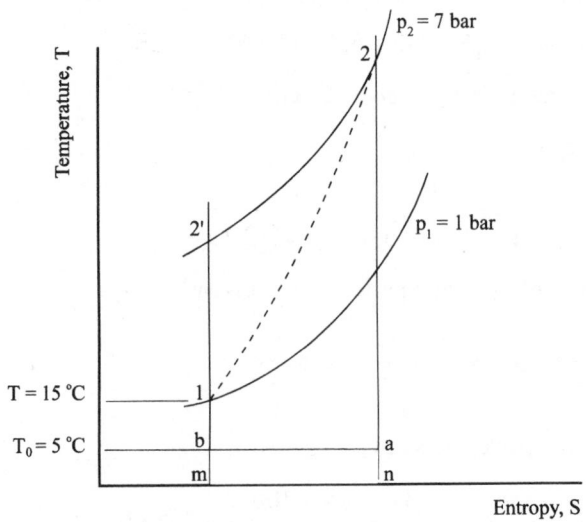

Fig. 2.6 Compression process on a T–s diagram

(a) Isentropic compression:

$$T_{2'} = T_1 \left(\frac{p_2}{p_1}\right)^{(\kappa-1)/\kappa} = 288 \times 7^{0.286} = 502.4 \text{ K}$$

The work done can be calculated from the steady flow energy equation, giving

$$w_C|_{\text{isen}} = -\Delta h = -215.3 \text{ kJ/kg}$$

The change of availability is given by eqn (2.13b) as

$$\Delta a_f = a_{f2'} - a_{f1} = h_{2'} - h_1 - T_0(s_{2'} - s_1) = \Delta h - T_0 \Delta S$$

If the process is isentropic, $\Delta S = 0$, and then

$$\Delta a_f = c_p(T_{2'} - T_1) = 1.004 \times (502.4 - 288) = 215.3 \text{ kJ/kg}$$

(b) The actual work done, $w_C = w_C|_{\text{isen}}/\eta_C = -215.3/0.83 = -259.4$ kJ/kg. Hence $h_2 = h_1 - w_C = 1.004 \times 288 - (-259.4) = 548.6$ kJ/kg, and the temperature at 2, T_2, is given by $T_2 = h_2/c_p = 546.4$ K.

$$\Delta s_{12} = c_p \ln\frac{T_2}{T_1} - R \ln\frac{p_2}{p_1} = 0.6411 - 0.5565 = 0.08639 \text{ kJ/kg K}$$

Hence $\Delta a_f = a_{f2} - a_{f1} = \Delta h - T_0 \Delta s$
$$= 259.4 - 278 \times 0.08639 = 235.4 \text{ kJ/kg}$$

There has been a *greater increase* in availability than in the case of reversible compression. This reflects the higher temperature achieved during irreversible compression.

(c) The change of availability of the surroundings is equal to the enthalpy change of the system, i.e. $\Delta a_{\text{sur}} = -259.4$ kJ/kg.

(d) The process an adiabatic one, i.e. there is no heat loss or addition. Hence the irreversibility of the process is given by

$$I = T_0 \Delta s = 278 \times 0.08639 = 24.02 \text{ kJ/kg}$$

This is the negative of the change of availability of the universe (system + surroundings) which is given by

$$\Delta a_{\text{univ}} = \Delta a_{\text{system}} + \Delta a_{\text{surroundings}} = 235.4 - 259.4 = -24.0 \text{ kJ/kg}$$

Hence, while the available energy in the fluid has been increased by the work done on it, the change is less than the work done. This means that, even if the energy in the gas after compression is passed through a *reversible* heat engine, it will not be possible to produce as much work as was required to compress the gas. Hence, the *quality of the energy in the universe has been reduced*, even though the quantity of energy has remained constant.

2.6 Graphical representation of available energy, and irreversibility

Consider the energy transfer from a high temperature reservoir at T_H through a heat engine (not necessarily reversible), as shown in Fig 2.7.

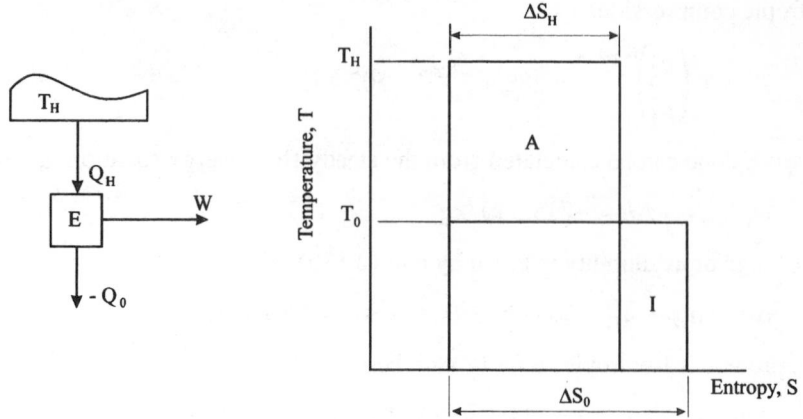

Fig. 2.7 Representation of available energy, and irreversibility

The available energy flow from the hot reservoir is

$$E_H = Q_H - T_0 \Delta S_H \tag{2.16}$$

The work done by the engine is

$$W = Q_H - Q_0$$

The total change of entropy of the universe is

$$\Sigma \Delta S = \Delta S_H + \Delta S_0 = \frac{Q_H}{T_H} - \frac{Q_0}{T_0} \tag{2.17}$$

Hence the energy which is unavailable due to irreversibility is defined by

$$\begin{aligned} E_{irrev} = E_H - W &= Q_H - T_0 \Delta S_H - W \\ &= Q_H - T_0 \Delta S_H (Q_H - Q_0) = Q_0 - T_0 \Delta S_H \\ &= T_0 (\Delta S_0 - \Delta S_H) \end{aligned} \tag{2.18}$$

In the case of a reversible engine $\Sigma \Delta S = 0$ because entropy flow is conserved, i.e.

$$\frac{Q_H}{T_H} = \frac{Q_0}{T_0} \tag{2.19}$$

Hence the unavailable energy for a reversible engine is $T_0 \Delta S_H$ while the irreversibility is zero. However, for all other engines it is non-zero. The available energy is depicted in Fig 2.7 by the area marked 'A', while the energy 'lost' due to irreversibility is denoted 'I' and is defined

$$E_{irrev} = T_0 (\Delta S_0 - \Delta S_H) \tag{2.20}$$

2.7 Availability balance for a closed system

The approaches derived previously work very well when it is possible to define the changes occurring inside the system. However, it is not always possible to do this and it is useful to derive a method for evaluating the change of availability from 'external' parameters. This can be done in the following way for a closed system.

If a closed system goes from state 1 to state 2 by executing a process then the changes in that system are:

from the First Law:
$$U_2 - U_1 = \int_1^2 (\delta Q - \delta W) = \int_1^2 \delta Q - W \qquad (2.21)$$

from the Second Law:
$$S_2 - S_1 = \int_1^2 \frac{\delta Q}{T} + \sigma \qquad (2.22)$$

where σ is the internal irreversibility of the system, and T is the temperature at which the heat transfer interactions with the system occur (see Fig 2.2(a)). Equations (2.21) and (2.22) can be written in terms of availability (see eqn (2.10)), for a system which can change its volume during the process, as

$$A_2 - A_1 = U_2 - U_1 - T_0(S_2 - S_1) + p_0(V_2 - V_1)$$

$$= \int_1^2 \delta Q - W + p_0(V_2 - V_1) - T_0 \int_1^2 \frac{\delta Q}{T} - T_0\sigma \qquad (2.23)$$

Equation (2.23) can be rearranged to give

$$A_2 - A_1 = \int_1^2 \left(1 - \frac{T_0}{T}\right)\delta Q - W + p_0(V_2 - V_1) - T_0\sigma \qquad (2.24)$$

The change in specific availability is given by

$$a_2 - a_1 = u_2 - u_1 - T_0(s_2 - s_1) + p_0(v_2 - v_1)$$

$$= \int_1^2 \left(1 - \frac{T_0}{T}\right)\delta q - w + p_0(v_2 - v_1) - T_0\sigma_m \qquad (2.25)$$

where q, w and σ_m are the values of Q, W and σ per unit mass.

The significance of eqn (2.24) can be examined by means of a couple of simple examples.

Example 6: change in availability for a closed system

A steel casting weighing 20 kg is removed from a furnace at a temperature of 800°C and heat treated by quenching in a bath containing 500 kg water at 20°C. Calculate the change in availability of the universe due to this operation. The specific heat of the water is 4.18 kJ/kg K, and that of steel is 0.42 kJ/kg K. Assume that the bath of water is rigid and perfectly insulated from the surroundings after the casting has been dropped in, and take the datum temperature and pressure as 20°C and 1 bar respectively.

Solution

The process can be considered to be a closed system if it is analysed after the casting has been introduced to the bath of water. Hence eqn (2.24) can be applied:

$$A_2 - A_1 = U_2 - U_1 - T_0(S_2 - S_1) + p_0(V_2 - V_1) = \int_1^2 \left(1 - \frac{T_0}{T}\right) \delta Q - W + p_0(V_2 - V_1) - T_0 \sigma$$

In this case, if the combined system is considered, $\delta Q = 0$, $W = 0$ and $p_0(V_2 - V_1) = 0$ because the system is adiabatic and constant volume. Thus

$$A_2 - A_1 = -T_0 \sigma$$

The irreversibility can be calculated in the following manner:

1. final temperature of system

$$T_f = \frac{m_c c_c T_c + m_w c_w T_w}{m_c c_c + m_w c_w} = \frac{20 \times 0.42 \times 1073 + 500 \times 4.18 \times 293}{0.42 \times 20 + 500 \times 4.18}$$

$$= 296.12 \text{ K}$$

2. change of entropy of casting

$$\delta S_c = \int_{T_s}^{T_f} \frac{\delta Q_R}{T} = \int_{T_s}^{T_f} \frac{m_c c_c \, dT}{T} = m_c c_c \ln \frac{T_f}{T_c} = 20 \times 0.42 \times \ln \frac{296.12}{1073}$$

$$= -10.815 \text{ kJ/K}$$

3. change of entropy of water

$$\delta S_w = \int_{T_w}^{T_f} \frac{\delta Q_R}{T} = \int_{T_w}^{T_f} \frac{m_w c_w \, dT}{T} = m_w c_w \ln \frac{T_f}{T_w} = 500 \times 4.18 \times \ln \frac{296.12}{293}$$

$$= 22.154 \text{ kJ/K}$$

4. change of entropy of system (and universe)

$$\delta S = \delta S_c + \delta S_w = -10.815 + 22.154 = 11.339 \text{ kJ/K}$$

5. change of availability

$$A_2 - A_1 = -T_0 \sigma = -293 \times 11.34 = -3323 \text{ kJ}$$

This solution indicates that the universe is less able to do work after the quenching of the casting than it was before the casting was quenched. The loss of availability was analysed above by considering the irreversibility associated with the transfer of energy between the casting and the water. It could have been analysed in a different manner by taking explicit note of the work which could be achieved from each part of the composite system before and after the heat transfer had taken place. This will now be done.

1. Work available by transferring energy from the casting to the environment through a reversible heat engine before the process

$$W_c = \int_{T_s}^{T_0} \left(1 - \frac{T_0}{T}\right) dQ = \int_{T_s}^{T_0} \left(1 - \frac{T_0}{T}\right) m_c c_c \, dT = m_c c_c [T - T_0 \ln T]_{T_c}^{T_0}$$

$$= 20 \times 0.42 \times \left[(293 - 1073) - 293 \times \ln \left(\frac{293}{1073}\right)\right] = 3357.3 \text{ kJ}$$

2. Work available by transferring energy from the water bath to the environment through a reversible heat engine after the process

$$W_w = \int_{T_f}^{T_0}\left(1 - \frac{T_0}{T}\right)dQ = \int_{T_f}^{T_0}\left(1 - \frac{T_0}{T}\right)m_w c_w\, dT = m_w c_w[T - T_0 \ln T]_{T_f}^{T_0}$$

$$= 500 \times 4.18 \times \left[(293 - 296.12) - 293 \times \ln\left(\frac{293}{296.12}\right)\right] = -34.47\ \text{K}$$

3. Change of availability of universe is

$$A_2 - A_1 = -W = -(3357.3 - 34.47) = -3323\ \text{kJ}$$

This is the same solution as obtained by considering the irreversibility.
Equation (2.24) can be considered to be made up of a number of terms, as shown below:

$$A_2 - A_1 = \underbrace{\int_1^2\left(1 - \frac{T_0}{T}\right)\delta Q}_{\substack{\text{availability transfer}\\ \text{accompanying heat transfer}}} - \underbrace{W + p_0(V_2 - V_1)}_{\substack{\text{availability transfer}\\ \text{accompanying work}}} - \underbrace{T_0\sigma}_{\substack{\text{availability destruction}\\ \text{due to irreversibilities}}}$$

$$= \underbrace{\int_1^2\left(1 - \frac{T_0}{T}\right)\delta Q}_{\substack{\text{availability transfer}\\ \text{accompanying heat transfer}}} - \underbrace{W + p_0(V_2 - V_1)}_{\substack{\text{availability transfer}\\ \text{accompanying work}}} - \underbrace{I}_{\substack{\text{availability destruction}\\ \text{due to irreversibilities}}} \qquad (2.26)$$

It is also possible to write eqn (2.26) in the form of a rate equation, in which case the rate of change of availability is

$$\frac{dA}{dt} = \underbrace{\left(1 - \frac{T_0}{T}\right)\dot Q}_{\substack{\text{availability transfer}\\ \text{accompanying heat transfer}}} - \underbrace{\dot W + p_0\frac{dV}{dt}}_{\substack{\text{availability transfer}\\ \text{accompanying work}}} - \underbrace{T_0\dot\sigma}_{\substack{\text{availability destruction}\\ \text{due to irreversibilities}}}$$

$$= \underbrace{\left(1 - \frac{T_0}{T}\right)\dot Q}_{\substack{\text{availability transfer}\\ \text{accompanying heat transfer}}} - \underbrace{\dot W + p_0\frac{dV}{dt}}_{\substack{\text{availability transfer}\\ \text{accompanying work}}} - \underbrace{\dot I}_{\substack{\text{availability destruction}\\ \text{due to irreversibilities}}} \qquad (2.27)$$

Equations (2.26) and (2.27) can be written in terms of specific availability to give

$$a_2 - a_1 = \int_1^2\left(1 - \frac{T_0}{T}\right)\delta q - w + p_0(v_2 - v_1) - T_0\sigma_m$$

$$= \int_1^2\left(1 - \frac{T_0}{T}\right)\delta q - w + p_0(v_2 - v_1) - i \qquad (2.28)$$

and $\qquad \dfrac{da}{dt} = \left(1 - \dfrac{T_0}{T}\right)\dot q - \dot w + p_0\dfrac{dv}{dt} - T_0\dot\sigma_m = \left(1 - \dfrac{T_0}{T}\right)\dot q - \dot w + p_0\dfrac{dv}{dt} - \dot i \qquad (2.29)$

Example 7: change of availability for a closed system in which the volume changes

An internal combustion engine operates on the Otto cycle (with combustion at constant volume) and has the parameters defined in Table 2.1. The data in this example is based on Heywood (1988).

Table 2.1 Operating parameters for Otto cycle

Compression ratio, r_c	12
Calorific value of fuel, $Q'_v/(\text{kJ}/\text{kg})$	44 000
Ratio, $x_a = \Delta A_R/Q'_v$	1.0286
Pressure at start of compression, $p_1/(\text{bar})$	1.0
Temperature at start of compression, $T_1/(^\circ C)$	60
Air/fuel ratio, ε	15.39
Specific heat of air at constant volume, $c_v/(\text{kJ}/\text{kg K})$	0.946
Ratio of specific heats, κ	1.30
Temperature of surroundings, $T_0/(\text{K})$	300
Pressure of surroundings, $p_0/(\text{bar})$	1.0

Calculate the variation in availability of the gases in the cylinder throughout the cycle from the start of compression to the end of expansion. Assume the compression and expansion processes are adiabatic.

Solution

This example introduces two new concepts:

- the effect of change of volume on the availability of the system;
- the effect of 'combustion' on the availability of the system.

Equation (2.24) contains a term which takes into account the change of volume ($p_0(V_2 - V_1)$), but it does not contain a term for the change in availability which occurs due to 'combustion'.

Equation (2.24) gives the change of availability of a system of constant composition as an extensive property. This can be modified to allow for combustion by the addition of a term for the availability of reaction. This is defined as

$$\Delta A_R = \Delta F_R = U_P(T_s) - U_R(T_s) - T_0(S_R(T_s) - S_P(T_s)) \tag{2.30}$$

where T_s is the temperature at which the energies of reaction are evaluated. Hence

$$\Delta A_R = \Delta F_R = \Delta U_R - T_0(S_R(T_s) - S_P(T_s))$$
$$= (Q_v)_s - T_0(S_R(T_s) - S_P(T_s)) \tag{2.31}$$

This term can be added into eqn (2.24) to give

$$A_2 - A_1 = \Delta A_R + (U_2 - U_1) - T_0(S_2 - S_1) + p_0(V_2 - V_1) \tag{2.32}$$

Equation (2.32) can be generalised to give the value at state i relative to that at state 0, resulting in

$$A_i - A_0 = \Delta A_R + (U_i - U_0) - T_0(S_i - S_0) + p_0(V_i - V_0)$$
$$= m_f \Delta A_R + m[(u_i - u_0) - T_0(s_i - s_0) + p_0(v_i - v_0)] \tag{2.33}$$

where m_f = mass of fuel

and m = total mass of mixture = $m_f + m_{air}$

The specific availability, based on the total mass of mixture, is

$$a_i - a_0 = \frac{A_i - A_0}{m} = \frac{m_f}{m} \Delta A_R + [(u_i - u_0) - T_0(s_i - s_0) + p_0(v_i - v_0)]$$

$$= \frac{\Delta A_R}{\varepsilon + 1} + [(u_i - u_0) - T_0(s_i - s_0) + p_0(v_i - v_0)] \tag{2.34}$$

where $\varepsilon =$ overall air–fuel ratio. Examples of the availability of reaction are given in section 2.9.2. The availability of reaction can be related to the internal energy of reaction by

$$\Delta A_R = (Q_v)_s \left(1 - \frac{T_0(S_R(T_s) - S_P(T_s))}{(Q_v)_s} \right) = x_a(Q_v)_s \tag{2.35}$$

which gives

$$a_i - a_0 = \frac{x_a(Q_v)_s}{\varepsilon + 1} + [(u_i - u_0) - T_0(s_i - s_0) + p_0(v_i - v_0)] \tag{2.36}$$

If the energy of reaction per unit mass of mixture is written $q^* = (Q_v)_s/(\varepsilon + 1)$ then

$$\frac{a_i - a_0}{q^*} = \frac{1}{q^*} [(u_i - u_0) + x_a q^* - T_0(s_i - s_0) + p_0(v_i - v_0)] \tag{2.37}$$

Considering the individual terms in eqn (2.37)

$$u_i - u_0 = c_v(T_i - T_0) = c_v T_0 \left(\frac{T_i}{T_0} - 1 \right) \tag{2.38}$$

$$p_0(v_i - v_0) = p_0 v_0 \left(\frac{v_i}{v_0} - 1 \right) = RT_0 \left(\frac{v_i}{v_0} - 1 \right) = (\kappa - 1)c_v T_0 \left(\frac{v_i}{v_0} - 1 \right) \tag{2.39}$$

$$T_0(s_i - s_0) = T_0 \left[c_p \ln \frac{T_i}{T_0} - R \ln \frac{p_i}{p_0} \right] = c_v T_0 \left\{ \kappa \ln \frac{T_i}{T_0} - (\kappa - 1)\ln \frac{p_i}{p_0} \right\} \tag{2.40}$$

Hence,

$$\frac{a_i - a_0}{q^*} = \frac{c_v T_0}{q^*} \left[\left(\frac{T_i}{T_0} - 1 \right) + \frac{x_a q^*}{c_v T_0} - \left\{ \kappa \ln \frac{T_i}{T_0} - (\kappa - 1)\ln \frac{p_i}{p_0} \right\} \right.$$

$$\left. + (\kappa - 1)\left(\frac{v_i}{v_0} - 1 \right) \right] \tag{2.41}$$

Equation (2.41) can be applied around the cycle to evaluate the availability relative to that at the datum state. The values at the state points of the Otto cycle are shown in Tables 2.2 and 2.3, and the variation of availability as a function of crankangle and volume during the cycle (calculated by applying eqn (2.41) in a step-by-step manner) is shown in Figs 2.8 and 2.9 respectively.

Table 2.2 State points around Otto cycle

State	v/v_0	p(bar)	T(K)	a/q^*	$\delta a/q^*$
0	1	1	300	1.0286	
1	1.11	1	333	1.029375	0.000775
2	0.0925	25.28923	701.7762	1.127041	0.097666
3	0.0925	127.5691	3540.043	0.955991	−0.17105
4	1.11	5.044404	1679.787	0.332836	−0.62316

Table 2.3 Terms in eqn (2.41) evaluated around cycle

State	$\left(\dfrac{T_i}{T_0} - 1\right)$	$\dfrac{x_a q^*}{c_v T_0}$	$(\kappa - 1)\left(\dfrac{v_i}{v_0} - 1\right)$	$\left\{\kappa \ln \dfrac{T_i}{T_0} - (\kappa - 1)\ln \dfrac{p_i}{p_0}\right\}$	$\dfrac{a_i}{q^*}$	$\dfrac{\delta a_{i,i-1}}{q^*}$
0	0	9.731469	0	0	1.0286	
1	0.11	9.731469	0.033	0.135668	1.029375	0.000775
2	1.339254	9.731469	−0.27225	0.135668	1.127041	0.097666
3	10.80014	9.731469	−0.27225	1.753948	0.955991	−0.17105
4	4.599289	9.731469	0.033	1.753948	0.332836	−0.62316

The value of availability of the charge at state 0 is based on the ratio of the availability of combustion of octane to the heat of reaction of octane, and this is calculated in section 2.9.2. The calculation of most of the other points on the cycle is straightforward, but it is worthwhile considering what happens during the combustion process. In the Otto cycle combustion takes place instantaneously at top dead centre (tdc), and the volume remains constant. This means that when eqn (2.41) is used to consider the effect of adiabatic, constant volume combustion occurring between points 2 and 3, this gives

$$\frac{a_3 - a_0}{q^*} - \frac{a_2 - a_0}{q^*}$$

$$= \frac{c_v T_0}{q^*}\left[\underbrace{\left[\left(\frac{T_3}{T_0} - 1\right) - \left\{\left(\frac{T_2}{T_0} - 1\right) + \frac{x_a q^*}{c_v T_0}\right\}\right]}_{\substack{\text{change of availability due to combustion as}\\ \text{fuel changes from reactants to products,} = 0}} - \underbrace{\left\{\kappa \ln \frac{T_3}{T_2} - (\kappa - 1)\ln \frac{p_3}{p_2}\right\}}_{\substack{\text{change of availability of gases due}\\ \text{to change of entropy of gases}}}\right] \quad (2.42)$$

Hence, the availability at point 3 is

$$\frac{a_3 - a_0}{q^*} = \frac{a_2 - a_0}{q^*} + \frac{c_v T_0}{q^*}\left[-\left\{\kappa \ln \frac{T_3}{T_2} - (\kappa - 1)\ln \frac{p_3}{p_2}\right\}\right] \quad (2.43)$$

Equation (2.43) is the change in entropy of the working fluid which is brought about by combustion, and since the entropy of the gases increases due to combustion this term reduces the availability of the gas.

Figures 2.8 and 2.9 show how the non-dimensional availability of the charge varies around the cycle. This example is based on a cycle with a premixed charge, i.e. the fuel and

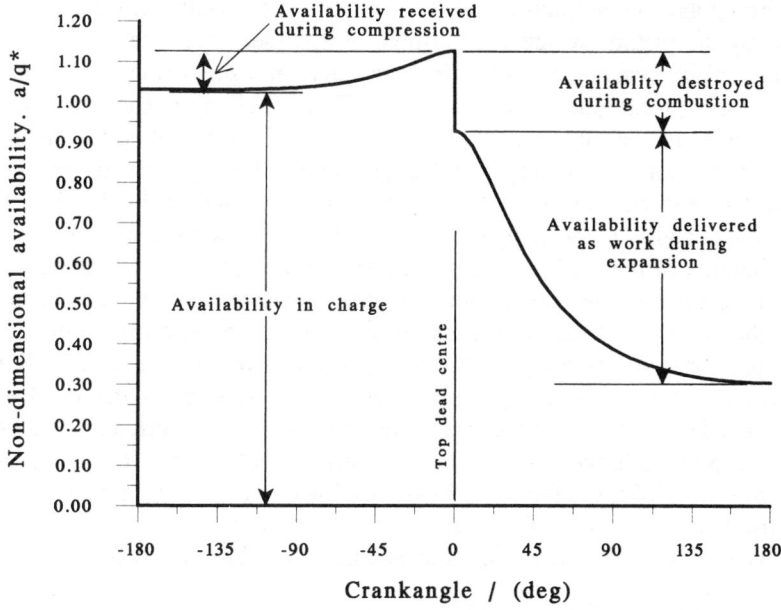

Fig. 2.8 Variation of the non-dimensional availability with crank angle for an Otto cycle

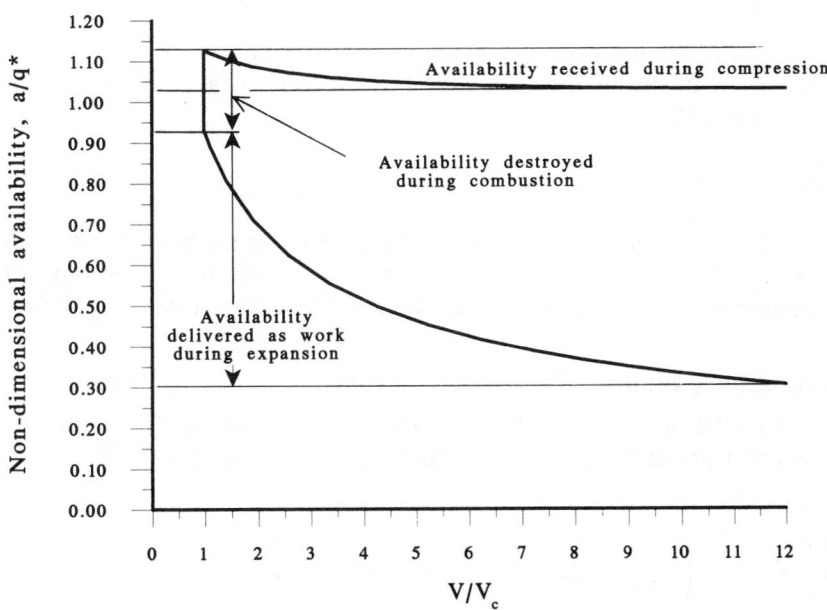

Fig. 2.9 Variation of non-dimensional availability with volume for an Otto cycle

air are induced into the engine through the inlet valve, and the availability of the charge is greater at the beginning of the cycle than the end because the fuel contains availability which is released during the chemical reaction. This is similar to the energy contained in the reactants in a simple combustion process and released as the internal energy (or enthalpy) of reaction of the fuel. Considering Fig 2.8, it can be seen that the availability of the charge increases during the compression stroke ($-180°$ crankangle to $0°$ crankangle) because of the work done by the piston (which is isentropic in this ideal cycle case). The 'combustion' process takes place instantaneously at $0°$ crankangle and the availability contained in the fuel in the form of the chemical bonds (see Chapter 11) is released and converted, at constant volume, to thermal energy. The effect of this is to introduce an irreversibility defined by the change of entropy of the gas (eqn (2.42)), and this causes a loss of availability. The expansion process (which is again isentropic for this ideal cycle) causes a reduction of availability in the charge as it is converted into expansion work.

The availability of the working fluid at the end of the cycle, 4, is a measure of the work that could be obtained from the charge after the cycle in the cylinder has been completed. This availability can be obtained if the working fluid is taken down to the dead state by reversible processes. Turbochargers are used to convert part of this availability into work but cannot convert it all because they can only take the state of the fluid down to the pressure of the dead state, p_0. If the full availability is to be converted it is necessary to use a turbine (e.g. a turbocharger) and then a 'bottoming' cycle (e.g. a Rankine cycle based on a suitable working fluid) to take the exhaust gas down to the dead state temperature, T_0.

The non-dimensional availability, a/q^*, is a measure of the work that can be obtained from the cycle. Hence, the total 'non-dimensional work' that can be achieved is

$$\underbrace{\frac{\delta a_1}{q^*} - \frac{\delta a_4}{q^*}}_{\substack{\text{net work from} \\ \text{availability}}} + \underbrace{\frac{\delta a_{23}}{q^*}}_{\substack{\text{availability} \\ \text{generated}}} = 1.029375 - 0.332836 - 0.17105 = 0.5255$$

This is the ratio of the net work output to the energy supplied (q^*), and is equal to the thermal efficiency of the engine cycle. The thermal efficiency of this Otto cycle, with a compression ratio of 12, is

$$\eta_{\text{Otto}} = 1 - \frac{1}{r^{(\kappa-1)}} = 0.5255$$

Thus, the availability approach has given the same result as the basic equation. Where the availability approach shows its strength is in analysing the effects of finite combustion rates and heat transfer in real cycles; this has been discussed by Patterson and Van Wylen (1964).

2.8 Availability balance for an open system

The availability balance of a closed system was derived in section 2.7. A similar approach can be used to evaluate the change of availability in an open system. The *unsteady flow energy equation* is

$$\sum_j \dot{Q} - \dot{W} = \left(\frac{dE}{dt}\right)_{cv} + \sum_e \dot{m}_e(h_e + V_e^2/2 + gz_e) - \sum_i \dot{m}_i(h_i + V_i^2/2 + gz_i) \qquad (2.44)$$

where $\sum_j \dot{Q}$ is the sum of heat transfers to the control volume over the total number of heat transfer processes, j, and the summations of the inflows and outflows take into account all the inflows and outflows. If the changes in kinetic and potential energies are small then this equation can be reduced to

$$\sum_j \dot{Q} - \dot{W} = \left(\frac{dE}{dt}\right)_{cv} + \sum_e \dot{m}_e h_e - \sum_i \dot{m}_i h_i \qquad (2.45)$$

It is also possible to relate the rate of change of entropy to the heat transfers from the system, and the irreversibilities in the system as

$$\dot{S} = \frac{dS_{cv}}{dt} = \sum_j \frac{\dot{Q}}{T} + \dot{\sigma}_{cv} - \sum_e \dot{m}_e s_e + \sum_i \dot{m}_i s_i \tag{2.46}$$

If the kinetic and potential energies of the control volume do not change then the internal energy in the control volume, $(E)_{cv}$, can be replaced by the intrinsic internal energy of the control volume, $(U)_{cv}$. The availability of the control volume is defined by eqn (2.10), and hence

$$\frac{dA_{cv}}{dt} = \frac{dU_{cv}}{dt} + p_0 \frac{dV_{cv}}{dt} - T_0 \frac{dS_{cv}}{dt} \tag{2.47}$$

Substituting eqns (2.46) and (2.47) into eqn (2.45) gives

$$\frac{dA_{cv}}{dt} = \sum_j \left(1 - \frac{T_0}{T}\right)\dot{Q} - \left(\dot{W} - p_0 \frac{dV}{dt}\right) + \sum_i \dot{m}_i(h_i - T_0 s_i)$$

$$- \sum_e \dot{m}_e(h_e - T_0 s_e) - T_0 \dot{\sigma}_{cv} \tag{2.48}$$

The terms $h - T_0 s$ are the flow availability, a_f, and hence eqn (2.48) becomes

$$\frac{dA_{cv}}{dt} = \sum_j \left(1 - \frac{T_0}{T}\right)\dot{Q} - \left(\dot{W} - p_0 \frac{dV}{dt}\right) + \sum_i \dot{m}_i a_{f_i} - \sum_e \dot{m}_e a_{f_e} - T_0 \dot{\sigma}_{cv}$$

$$= \sum_j \left(1 - \frac{T_0}{T}\right)\dot{Q} - \left(\dot{W} - p_0 \frac{dV}{dt}\right) + \sum_i \dot{m}_i a_{f_i} - \sum_e \dot{m}_e a_{f_e} - \dot{I}_{cv} \tag{2.49}$$

where $\dot{I}_{cv}$ is the irreversibility in the control volume.

Equation (2.49) is the *unsteady flow availability equation*, which is the availability equivalent of the unsteady flow energy equation. Many of the processes considered in engineering are steady state ones, which means that the conditions in the control volume do not change with time. This means that $dA_{cv}/dt = 0$ and $dV/dt = 0$, and then eqn (2.49) can be simplified to

$$0 = \sum_j \left(1 - \frac{T_0}{T}\right)\dot{Q} - \dot{W} + \sum_i \dot{m}_i a_{f_i} - \sum_e \dot{m}_e a_{f_e} - T_0 \dot{\sigma}_{cv}$$

$$= \sum_j \left(1 - \frac{T_0}{T}\right)\dot{Q} - \dot{W} + \sum_i \dot{m}_i a_{f_i} - \sum_e \dot{m}_e a_{f_e} - \dot{I}_{cv} \tag{2.50}$$

Example 8: steady flow availability

Superheated steam at 30 bar and 250°C flows through a throttle with a downstream pressure of 5 bar. Calculate the change in flow availability across the throttle, neglecting the kinetic and potential terms, if the dead state condition is $T_0 = 25°C$ and $p_0 = 1$ bar.

Solution

A throttle does not produce any work, and it can be assumed that the process is adiabatic, i.e. $\dot{Q} = 0$, $\dot{W} = 0$. Hence, eqn (2.50) becomes, taking into account the fact that there is only one inlet and outlet,

$$0 = \dot{m}_i a_{f_i} - \dot{m}_e a_{f_e} - \dot{I}_{cv}$$

The conditions at inlet, i, are $h_i = 2858$ kJ/kg, $s_i = 6.289$ kJ/kg K; and the conditions at exhaust, e, are $h_e = 2858$ kJ /kg, $s_e = 7.0650$ kJ/kg K.

Hence the irreversibility per unit mass flow is

$$i_{cv} = \frac{\dot{I}_{cv}}{\dot{m}} = (h_i - T_0 s_i) - (h_e - T_0 s_e)$$

$$= (2858 - 298 \times 6.289) - (2858 - 298 \times 7.0650) = 231.28 \text{ kJ/kg}$$

The significance of this result is that although energy is conserved in the flow through the throttle the ability of the fluid to do work is reduced by the irreversibility. In this case, because the enthalpy does not change across the throttle, the irreversibility could have been evaluated by

$$T_0(s_2 - s_1) = 298 \times (7.065 - 6.289) = 231.28 \text{ kJ/kg}$$

2.9   Exergy

Exergy is basically the available energy based on datum conditions at a dead state; it was introduced in Example 3. An obvious datum to be used in most calculations is the ambient condition, say p_0, T_0. This datum condition can be referred to as the *dead state* and the system reaches this when it is in thermal and mechanical equilibrium with it. *A state of thermal and mechanical equilibrium is reached when both the temperature and pressure of the system are equal to those of the dead state.* The term exergy was proposed by Rant in 1956, and similar functions had previously been defined by Gibbs and Keenan.

Exergy will be given the symbol B (sometimes it is given the symbol Ξ) and specific exergy will be denoted b (or ξ). The exergy of a system at state 1 is defined by

$$B_1 = A_1 - A_0 \tag{2.51}$$

where A_1 is the available energy at state 1, and A_0 is the available energy at the dead state.

If a system undergoes a process between states 1 and 2 the maximum useful work, or available energy, that may be obtained from it is given by

$$W = A_1 - A_2 = (A_1 - A_0) - (A_2 - A_0) = B_1 - B_2 \tag{2.52}$$

Hence the maximum work, or heat transfer, that can be obtained as a system changes between two states is defined as the difference in exergy of those two states.

Exergy is very similar to available energy and is a quasi- or pseudo-property. This is because it is not defined solely by the state of the system but also by the datum, or dead, state that is used.

A number of examples of the use of exergy will be given.

2.9.1   HEAT TRANSFER

It is possible for energy to be transferred from one body to another without any loss of available energy. This occurs when the heat or energy transfer is ideal or reversible. No real heat transfer process will exhibit such perfection and exergy can be used to show the best way of optimising the effectiveness of such an energy transfer. All actual heat transfer processes are irreversible and the irreversibility results in a loss of exergy.

It should be noted in this section that even though the heat transfer processes for each of the systems are internally reversible, they might also be externally irreversible.

Ideal, reversible heat transfer

Ideal reversible heat transfer can be approached in a counterflow heat exchanger. In this type of device the temperature difference between the two streams is kept to a minimum, because the hot 'source' fluid on entering the heat exchanger is in closest contact to the 'sink' fluid which is leaving the device, and vice versa. The processes involved are depicted by two almost coincident lines from 1 to 2 in Fig 2.10.

In this ideal process it will be assumed that, at all times, the fluid receiving the heat is at temperature T, while the temperature of the source of heat is at all times at temperature $T + \delta T$, i.e. T_{1c} is δT less than T_{2h} etc. From the First Law of Thermodynamics it is obvious that, if the boundaries of the control volume are insulated from the surroundings, the energy transferred from the hot stream must be equal to the energy received by the cold stream. This means that the areas under the curves in Fig 2.10 must be equal: in this case they are identical. The exergy change of the hot stream is then given by

$$\Delta B_h = B_{2h} - B_{1h} = (A_{2h} - A_0) - (A_{1h} - A_0)$$
$$= A_{2h} - A_{1h} = (U + p_0 V - T_0 S)_{2h} - (U + p_0 V - T_0 S)_{1h}$$
$$= (U_{2h} - U_{1h}) - T_0 (S_{2h} - S_{1h}) + p_0 (V_{2h} - V_{1h}) \qquad (2.53)$$

If it is assumed that $V_{1h} = V_{2h}$ then

$$\Delta B_h = (U_{2h} - U_{1h}) - T_0 (S_{2h} - S_{1h}) \qquad (2.54)$$

In eqn (2.54), $U_{2h} - U_{1h} =$ heat transferred from the hot stream, and $S_{2h} - S_{1h} =$ the change of entropy of the hot stream. In Fig 2.10, $U_{2h} - U_{1h} =$ area 1h-2h-6-5-1h and $T_0(S_2 - S_{1h}) =$ area 3-4-6-5-3. In the case of the hot stream the change of energy and the change of entropy will both be negative, and because $(U_{2h} - U_{1h}) > T_0(S_{2h} - S_{1h})$ the change of exergy will also be negative.

A similar analysis can be done for the cold stream, and this gives

$$\Delta B_c = B_{2c} - B_{1c} = (A_{2c} - A_0) - (A_{1c} - A_0)$$
$$= (U_{2c} - U_{1c}) - T_0 (S_{2c} - S_{1c}) \qquad (2.55)$$

In this case $U_{2c} - U_{1c} =$ heat transferred to the cold stream (area 1c-2c-5-6-1c), and this will be positive. The term $T_0(S_{2c} - S_{1c}) =$ unavailable energy of the cold stream (area 4-3-5-6-4), and this will also be positive because of the increase of entropy of the cold stream. Figure 2.10 shows that the heating and cooling processes are effectively identical on the $T-s$ diagram and only differ in their direction. Hence, the changes of energy of the hot and cold streams are equal (and opposite), and also the the unavailable energies are equal (and opposite) for both streams. This means that for an ideal counterflow heat exchanger, in which the heat transfer takes place across an infinitesimal temperature difference (i.e.

reversible heat transfer), the loss of exergy of the hot stream is equal to the gain of exergy of the cold stream. The change of exergy of the universe is zero, and the process is reversible.

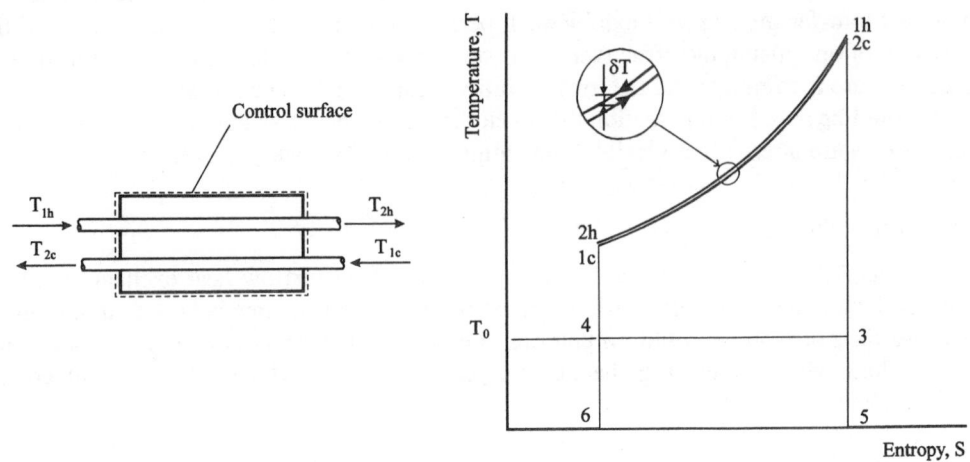

Fig. 2.10 Reversible heat transfer in a counterflow heat exchanger: (a) schematic diagram of heat exchanger; (b) processes shown on a $T-s$ diagram

If the heat exchanger in Fig 2.10 were not reversible, i.e. there is a significant temperature difference between the hot and cold stream, then the $T-s$ diagram would be like that shown in Fig 2.11. Figure 2.11(a) shows the processes for both the hot and cold streams on the same $T-s$ diagram, while Figs 2.11(b) and 2.11(c) show individual $T-s$ diagrams for the hot and cold stream respectively. The first point to recognise is that the energies transferred between two streams (areas 1h-2h-5h-6h-1h and 1c-2c-5c-6c-1c) are equal (and opposite). The change of exergy for the hot stream is denoted by the area 1h-2h-3h-4h-1h, while the unavailable energy is given by 3h-5h-6h-4h-3h. Similar quantities for the cold streams are 1c-2c-3c-4c-1c and 3c-5c-6c-4c-3c. It is obvious from Fig 2.11(a) that the unavailable energy of the cold stream is greater than that for the hot stream because $T_0(S_{2c} - S_{1c}) > T_0(S_{1h} - S_{2h})$. This means that the exergy gained by the cold stream is less than that lost by the hot stream. For the hot stream

$$\Delta B_h = (U_{2h} - U_{1h}) - T_0(S_{2h} - S_{1h}) \tag{2.56}$$

while for the cold stream

$$\Delta B_c = (U_{2c} - U_{1c}) - T_0(S_{2c} - S_{1c}) \tag{2.57}$$

The change of exergy of the universe is given by

$$\Delta B_{univ} = \Delta B_h + \Delta B_c = (U_{2h} - U_{1h}) - T_0(S_{2h} - S_{1h}) + (U_{2c} - U_{1c}) - T_0(S_{2c} - S_{1c})$$
$$= -T_0(S_{2h} - S_{1h} + S_{2c} - S_{1c}) \tag{2.58}$$

The value of $\Delta B_{univ} < 0$ because the entropy change of the cold stream is greater than that of the hot stream. Hence, in this irreversible heat transfer device there has been a loss of exergy in the universe, and energy has been degraded. This means that the maximum useful work available from the cold stream is less than that available from the hot stream, even though the energy contents of the two streams are the same.

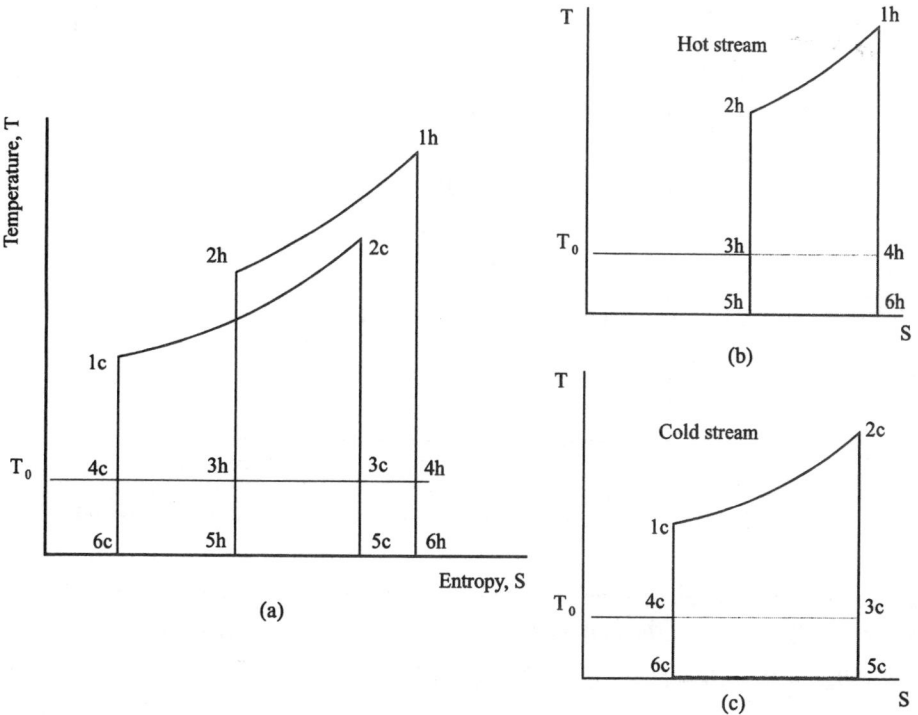

Fig. 2.11 *T–s* diagrams for a counterflow heat exchanger with finite temperature difference between the streams

This result has a further significance, which is that, to obtain the maximum transference of exergy, it is important to reduce the value of $T_0 (S_1 - S_2)$. If the total amount of energy being transferred is kept constant then the loss of exergy can be minimised by increasing the temperature at which the heat is transferred. This is effectively shown in Fig 2.11, where the high temperature stream can be equivalent to a high temperature source of heat, while the low temperature stream is equivalent to a lower temperature source of heat. The quality of the heat (energy) in the high temperature source is better than that in the low temperature one.

Irreversible heat transfer

The processes depicted in Fig 2.12 are those of an infinite heat source transferring energy to a finite sink. The temperature of the source remains constant at T_1 but that of the sink changes from T_5 to T_6. The energy received by the sink will be equal to that lost by the source if the two systems are isolated from the surroundings. The process undergone by the source is one of *decreasing entropy* while that for the sink is one of increasing entropy. Hence, in Fig 2.12, the areas 1-2-8-9-1 and 5-6-10-8-5 are equal. This means that areas

$$(a + b + c) = (b + c + d + e)$$

By definition the exergy change of the source is given by

$$\Delta B_{source} = B_2 - B_1 = (U_2 - U_1) - T_0(S_2 - S_1) \tag{2.59}$$

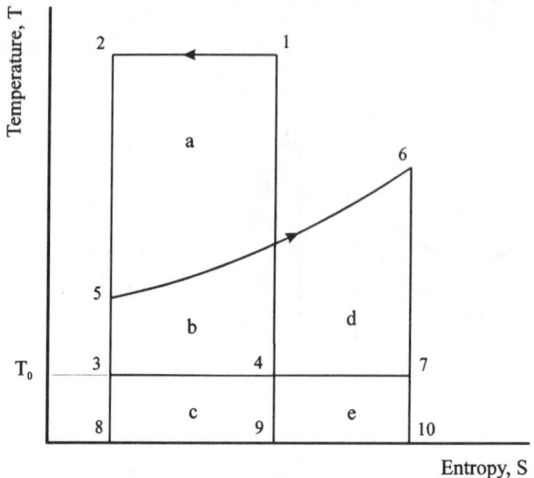

Fig. 2.12 Irreversible heat transfer from an infinite reservoir to a finite sink

This equation assumes that the heat transfer takes place at constant volume. In a similar way the exergy change of the sink is given by

$$\Delta B_{sink} = B_6 - B_5 = (U_6 - U_5) - T_0(S_6 - S_5) \tag{2.60}$$

Since the source and sink are isolated from the surroundings (the remainder of the universe), then the entropy change of the universe is

$$\Delta B_{univ} = \Delta B_{source} + \Delta B_{sink}$$
$$= (U_2 - U_1) - T_0(S_2 - S_1) + [(U_6 - U_5) - T_0(S_6 - S_5)]$$
$$= T_0(S_1 - S_6) \tag{2.61}$$

The term $T_0(S_1 - S_6)$ is depicted by the area marked 'e' in Fig 2.12. Since S_6 is greater than S_1 then the exergy of the universe (that is, its ability to do work) has decreased by this amount. Thus, whilst the energy of the universe has remained constant, the quality of that energy has declined. This is true of all processes which take place irreversibly; that is, all real processes.

2.9.2 EXERGY APPLIED TO COMBUSTION PROCESS

Combustion processes are a good example of irreversible change. In a combustion process the fuel, usually a hydrocarbon, is oxidised using an oxidant, usually air. The structure of the hydrocarbon is broken down as the bonds between the carbon and hydrogen atoms are broken and new bonds are formed to create carbon dioxide, carbon monoxide and water vapour. These processes are basically irreversible because they cannot be made to go in the opposite direction by the addition of a very small amount of energy. This seems to suggest that exergy of the universe is decreased by the combustion of hydrocarbon fuels. The following section describes how combustion can be considered using an exergy approach.

Consider a constant pressure combustion process. When the system is in equilibrium with its surroundings the exergy of component i is:

$$b_i = (h_i - h_0) - T_0(s_i - s_0) \tag{2.62}$$

Exergy of reaction of water

Applying eqn (2.62) to the simple reaction

$$2H_2(g) + O_2(g) \rightarrow 2H_2O(g) \tag{2.63}$$

$$\Delta B = B_2 - B_1 = B_P - B_R = \Sigma(b_i)_P - \Sigma(b_i)_R \tag{2.64}$$

where suffix R indicates reactants and suffix P indicates products.
Thus

$$\Delta B = (h_f^0)_{H_2O} - T_0(s_{H_2O} - s_{H_2} - 0.5s_{O_2}) \tag{2.65}$$

Substituting values, for $T_0 = 25°C$, which is also the standard temperature for evaluating the enthalpy of reaction, gives

$$\Delta B = -241820 - 298 \times (188.71 - 130.57 - 0.5 \times 205.04)$$
$$= -228594.76 \text{ kJ/kmol K} \tag{2.66}$$

This means that the ability of the fuel to do work is 5.5% less than the original enthalpy of formation of the 'fuel', and hence 94.5% of the energy defined by the enthalpy of formation is the maximum energy that can be obtained from it.

Exergy of reaction of methane (CH_4)

The equation for combustion of methane is

$$CH_4 + 2O_2 \rightarrow CO_2 + 2H_2O \tag{2.67}$$

Hence

$$\Delta B = \Sigma(b_i)_P - \Sigma(b_i)_R$$
$$= (\Delta H_R)_{CH_4} - T_0(s_{CO_2} + 2s_{H_2O} - s_{CH_4} - 2s_{O_2})$$
$$= -804.6 \times 10^3 - 298 \times (214.07 + 2 \times 188.16 - 182.73 - 2 \times 204.65)$$
$$= -804.1 \times 10^3 \text{ kJ/kmol}$$

In this case the exergy of reaction is almost equal to the enthalpy of reaction; this occurs because the entropies of the reactants and products are almost equal.

Exergy of reaction of octane (C_8H_{18})

The equation for combustion of octane is

$$C_8H_{18} + 12.5O_2 \rightarrow 8CO_2 + 9H_2O \tag{2.68}$$

Assume that the enthalpy of reaction of octane is -5074.6×10^3 kJ/kmol, and that the entropy of octane at 298 K is 360 kJ/kmol K. Then the exergy of reaction is

$$\Delta B = \Sigma(b_i)_P - \Sigma(b_i)_R$$
$$= (\Delta H_R)_{C_8H_{18}} - T_0(8s_{CO_2} + 9s_{H_2O} - s_{C_8H_{18}} - 12.5s_{O_2})$$
$$= -5074.6 \times 10^3 - 298 \times (8 \times 214.07 + 9 \times 188.16 - 360 - 12.5 \times 204.65)$$
$$= -5219.9 \times 10^3 \text{ kJ/kmol}$$

In this case the exergy of reaction is greater than the enthalpy of reaction by 2.86%.

2.10   The variation of flow exergy for a perfect gas

This derivation is based on Haywood (1980).

The definition of exergy is, from eqn (2.51),

$$B_1 = A_1 - A_0 \tag{2.51}$$

while that for the exergy of a flowing gas is, by comparison with eqn (2.13b) for the availability of a flowing gas,

$$B_{f1} = A_{f1} - A_{f0} = (H_1 - T_0 S_1) - (H_0 - T_0 S_0) \tag{2.69}$$

Equation (2.69) can be expanded to give the specific flow exergy as

$$b_{f1} = a_{f1} - a_{f0} = (h_1 - T_0 s_1) - (h_0 - T_0 s_0) \tag{2.70}$$

Now, for a perfect gas,

$$h = c_p T \tag{2.71}$$

and the change of entropy

$$s_1 - s_0 = c_p \ln \frac{T}{T_0} - R \ln \frac{p}{p_0} \tag{2.72}$$

Hence, the flow exergy for a perfect gas can be written

$$b_{f1} = (h_1 - h_0) - T_0 \left(c_p \ln \frac{T}{T_0} - R \ln \frac{p}{p_0} \right) = c_p(T - T_0) - T_0 \left(c_p \ln \frac{T}{T_0} - R \ln \frac{p}{p_0} \right) \tag{2.73}$$

The flow exergy can be non-dimensionalised by dividing by the enthalpy at the dead state temperature, T_0, to give

$$\frac{b_{f1}}{c_p T_0} = \left(\frac{T}{T_0} - 1 \right) - \ln \frac{T}{T_0} - \frac{\kappa - 1}{\kappa} \ln \frac{p}{p_0} \tag{2.74}$$

Equation (2.74) has been evaluated for a range of temperature ratios and pressure ratios, and the variation of exergy is shown in Fig 2.13.

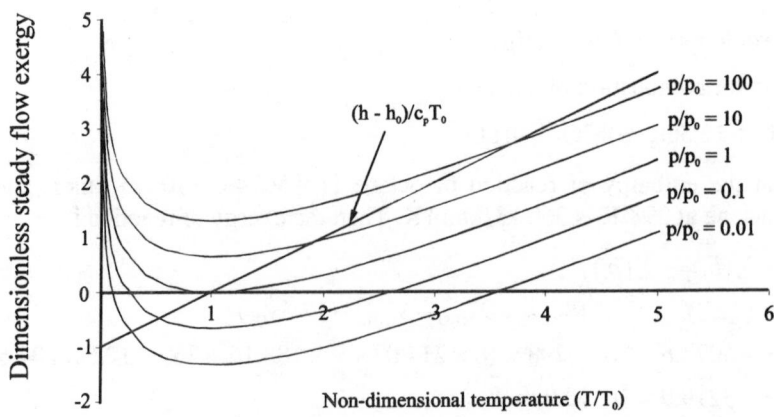

Fig. 2.13 Variation of exergy with temperature and pressure for a perfect gas

Also shown in Fig 2.13, as a straight line, is the variation of enthalpy for the parameters shown. It can be seen that the exergy is sometimes bigger than the enthalpy, and vice versa. This is because the enthalpy is purely a measure of the thermodynamic energy of the gas relative to the datum temperature. When the dimensionless temperature $T/T_0 < 1.0$ the enthalpy is negative, and when $T/T_0 > 1.0$ the enthalpy is positive. This simply means that the thermodynamic energy can be greater than or less than the datum value. While the enthalpy varies montonically with the non-dimensional temperature, the exergy does not. The reason for this is that the exergy term, say at $p/p_0 = 1$, is given by

$$\frac{b_{f1}}{c_p T_0} = \left(\frac{T}{T_0} - 1\right) - \ln\frac{T}{T_0} \tag{2.75}$$

For

$$\frac{T}{T_0} > 1, \qquad \left(\frac{T}{T_0} - 1\right) - \ln\frac{T}{T_0} > 1$$

and also for

$$\frac{T}{T_0} < 1, \qquad \left(\frac{T}{T_0} - 1\right) - \ln\frac{T}{T_0} > 1$$

The physical significance of this is that the system can produce work output as long as there is a temperature difference between the system and the dead state: the sign of the temperature difference does not matter.

2.11 Concluding remarks

This chapter has introduced the concept of the quality of energy through the quasi-properties availability and exergy. It has been shown that energy available at a high temperature has better quality than that at low temperature. The effect of irreversibilities on the quality of energy has been considered, and while the energy of the universe might be considered to remain constant, the quality of that energy will tend to decrease.

It was also shown that the irreversibility of processes can be calculated, and it is this area that should be tackled by engineers to improve the efficiency of energy utilisation in the world.

PROBLEMS

1 A piston-cylinder assembly contains 3 kg of air at 15 bar and 620 K. The environment is at a pressure of 1 bar and 300 K. The air is expanded in a fully reversible adiabatic process to a pressure of 5.5 bar. Calculate the useful work which can be obtained from this process. Also calculate the maximum useful work which can be obtained from the gas in (a) the initial state, and (b) the final state.

Assume that the specific heats for the gas are $c_p = 1.005$ kJ/kg K, and $c_v = 0.718$ kJ/kg K.

[295.5 kJ; 509.6 kJ; 214.1 kJ]

2 Air passes slowly through a rigid control volume C, as shown in Fig P2.2(a), in a hypothetical, *fully reversible*, steady-flow process between specified stable end states

1 and 2 in the presence of an environment at temperature T_0 and pressure p_0. States 1 and 2, and also the values of T_0 and p_0, are defined below:

$$T_1 = 550 \text{ K} \quad \text{and} \quad p_1 = 2 \text{ bar}$$
$$T_2 = T_0 = 300 \text{ K} \quad \text{and} \quad p_2 = p_0 = 1 \text{ bar}$$

The air may be treated as a perfect gas, with $c_p = 1.005 \text{ kJ/kg K}$ and $c_v = 0.718 \text{ kJ/ kg K}$.

(a)   The air follows path 1-a-2 see Fig P2.2(b); calculate the following work *output* quantities in each of the sub-processes 1-a and a-2:

   (i)   The *direct shaft work*, W_x, coming from the control volume as the air passes through the given sub-process.
   (ii)   The *external work*, W_e, produced by any auxiliary cyclic devices required to ensure external reversibility in that exchange with the environment during the given sub-process.

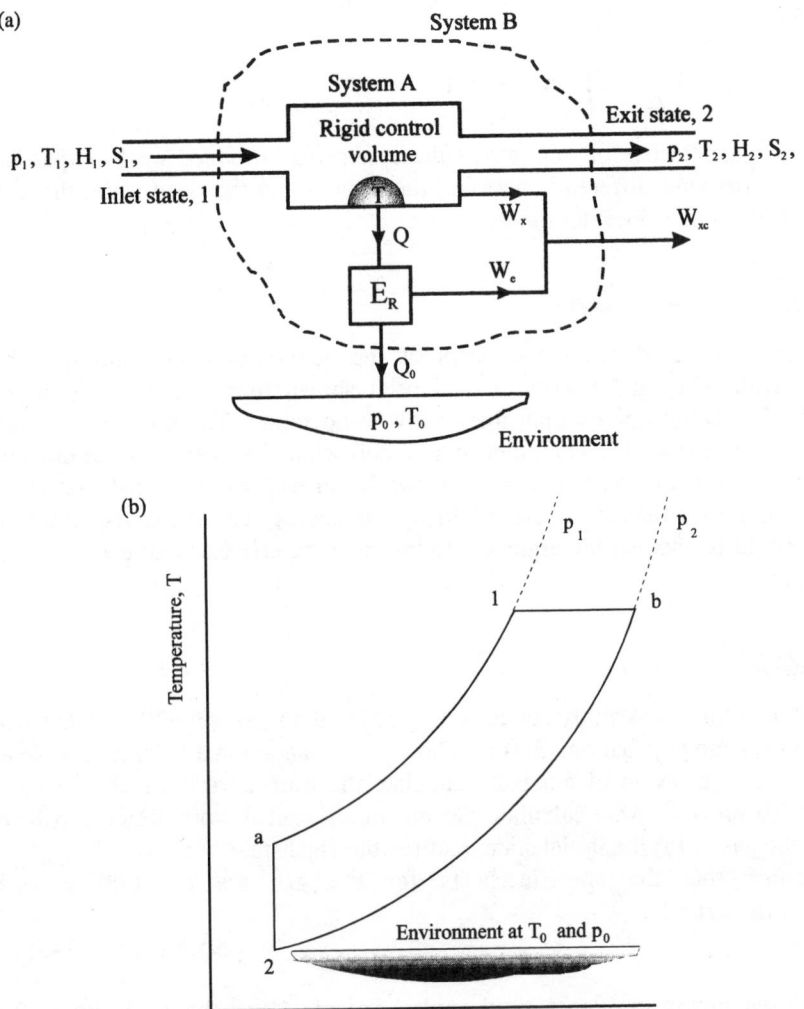

Fig. P2.2

    (iii)   The gross work W_g.
    (iv)   The total useful shaft work W_{xC}.

$$[66.03; 62.2; 128.23; 128.23 \text{ (kJ/kg)}]$$

(b)   Calculate the above work output quantities for each of the sub-processes 1-b and b-2 when the air follows the alternative path 1-b-2.

$$[109.41; 18.77; 128.18; 128.18 \text{ (kJ/kg)}]$$

3 Confirm that item (iv) in Q.2. is equal to $(b_1 - b_2)$, where $b = h - T_0 s$, the specific *steady-flow availability function*. (Note that, since $T_2 = T_0$ and $p_2 = p_0$, state 2 is the *dead state*, so that item (iv) is in this case equal to the *steady-flow exergy* of unit mass of air in state 1 for an environment at $T_0 = 300$ K and $p_0 = 1$ bar.)

$$[128.2 \text{ kJ/kg}]$$

4 A mass of 0.008 kg of helium is contained in a piston-cylinder unit at 4 bar and 235°C. The piston is pushed in by a force, F, until the cylinder volume is halved, and a cooling coil is used to maintain the pressure constant. If the dead state conditions are 1.013 bar and 22°C, determine:

    (i)   the work done on the gas by the force, F;
    (ii)   the change in availability;
    (iii) the heat transfer from the gas;
    (iv) the irreversibility.

Show the heat transfer process defined in part (iii) and the unavailable energy on a T–s diagram, and the work terms on a p–V diagram.

   The following data may be used for helium, which can be assumed to behave as a perfect gas:

    ratio of specific heats, $\kappa = c_p / c_v = 1.667$
    molecular weight, $m_w = 4$
    universal gas constant, $\Re = 8.3143$ kJ/kmol K

$$[-3.154 \text{ kJ}; 1.0967 \text{ kJ}; -10.556 \text{ kJ}; 2.0573 \text{ kJ}]$$

5 A system at constant pressure consists of 10 kg of air at a temperature of 1000 K. Calculate the maximum amount of work that can be obtained from the system if the dead state temperature is 300 K, and the dead state pressure is equal to the pressure in the system. Take the specific heat at constant pressure of air, c_p, as 0.98 kJ/kg K. (See Q.3, Chapter 1.)

$$[3320.3 \text{ kJ}]$$

6 A thermally isolated system at constant pressure consists of 10 kg of air at a temperature of 1000 K and 10 kg of water at 300 K, connected together by a reversible heat engine. What is the maximum work that can be obtained from the system as the temperatures equalise? (See Q.4, Chapter 1.)
   Assume

     for water: $c_v = 4.2$ kJ/kg K
                $\kappa = c_p / c_v = 1.0$
     for air: $c_v = 0.7$ kJ/kg K
                $\kappa = c_p / c_v = 1.4$

$$[2884.2 \text{ kJ}]$$

7 An amount of pure substance equal to 1 kmol undergoes an *irreversible* cycle. Neglecting the effects of electricity, magnetism and gravity state whether each of the following relationships is true or false, giving reasons for your assertion:

   (i) $\oint \delta Q = \oint (du_m + p\,dv_m)$

   (ii) $\oint \dfrac{du_m + p\,dv_m}{T} > 0$

   (iii) $\oint \delta W < \oint p\,dv_m$

   where the suffix m indicates that the quantities are in molar terms.

   [False; false; true]

8 A gas turbine operates between an inlet pressure of 15 bar and an exhaust pressure of 1.2 bar. The inlet temperature to the turbine is 1500 K and the turbine has an isentropic efficiency of 90%. The surroundings are at a pressure of 1 bar and a temperature of 300 K. Calculate, for the turbine alone:

   (i)   the specific power output;
   (ii)   the exhaust gas temperature;
   (iii) the exergy change in the gas passing through the turbine;
   (iv)   the irreversibility or lost work.

   Assume the working substance is an ideal gas with a specific heat at constant pressure of $c_p = 1.005$ kJ/kg K and the specific gas constant $R = 0.287$ kJ/kg K.

   [697.5 kJ/kg; 806 K; 727.7 kJ/kg; 343.6 kJ/kg]

9 It is proposed to improve the energy utilisation of a steelworks by transferring the heat from the gases leaving the blast furnace at 600°C to those entering the furnace at 50°C (before the heat exchanger is fitted). The minimum temperature of the flue gases is limited to 150°C to avoid condensation of sulfurous acid in the pipework at exit pressure of 1 bar.

   Draw a simple schematic diagram of the heat exchanger you would design, showing the hot and cold gas streams. Explain, with the aid of $T-s$ diagrams, why a counterflow heat exchanger is the more efficient. If the minimum temperature difference between the hot and cold streams is 10°C, calculate the minimum loss of exergy for both types of heat exchanger, based on dead state conditions of 1 bar and 20°C.

   [−72.33 kJ/kg; −45.09 kJ/kg]

10 Find the maximum, and maximum useful, specific work (kJ/kg) that could be derived from combustion products that are (a) stationary, and (b) flowing, in an environment under the following conditions.

	p(bar)	$T(K)$	$v(m^3/kg)$	u(kJ/kg)	s(kJ/kg K)
Products (1)	7	1000	0.41	760.0	7.4
Environment (0)	1	298.15	0.83	289.0	6.7

[262.3; 220.3; 466.3; 424.3]

3

Pinch Technology

In recent years a new technology for minimising the energy requirements of process plants has been developed: this has been named *Pinch Technology* or *Process Integration* by its major proponent, Linnhoff (Linnhoff and Senior, 1983; Linnhoff and Turner 1981). Process plants, such as oil refineries or major chemical manufacturing plants, require that heating and cooling of the feed stock take place as the processes occur. Obviously it would be beneficial to use the energy from a stream which requires cooling to heat another which requires heating; in this way the energy that has to be supplied from a high temperature source (or utility) is reduced, and the energy that has to be rejected to a low temperature sink (or utility) is also minimised. Both of these external transfers incur a cost in running the plant. Pinch technology is an approach which provides a mechanism for automating the design process, and *minimising the external heat transfers*.

Pinch situations also occur in power generation plant; for example, in a combined cycle gas turbine (CCGT) plant (see Fig 3.1) energy has to be transferred from the gas turbine exhaust to the working fluid in the steam turbine. A $T-s$ diagram of a CCGT plant is

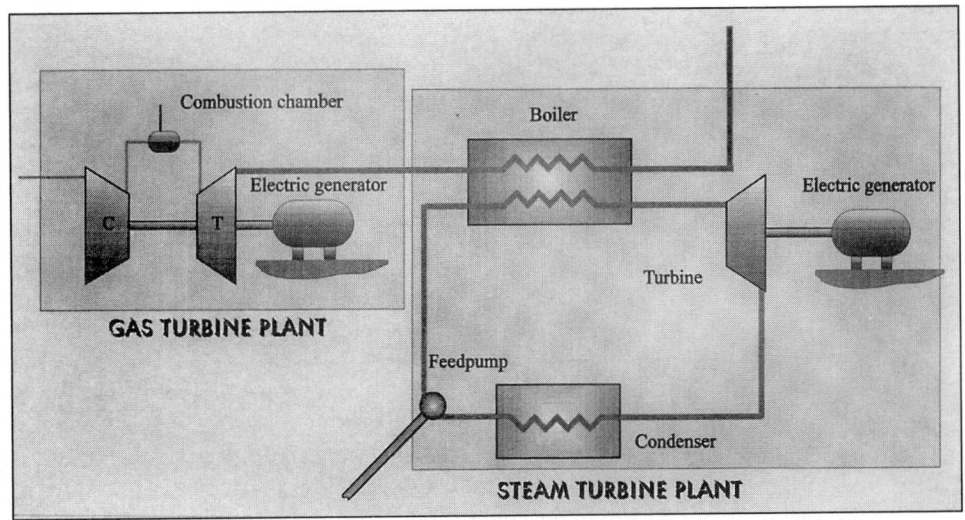

Fig. 3.1 Schematic of CCGT

shown in Fig 3.2, where the heat transfer region is shown: the pinch is the closest approach in temperature between the two lines. It is defined as the minimum temperature difference between the two streams for effective heat transfer, and is due to the difference in the properties of the working fluids during the heat transfer process (namely, the exhaust gas from the gas turbine cools down as a single phase but the water changes phase when it is heated) – this limits the amount of energy that can be taken from the hot fluid. The heat transfer processes are shown on a temperature–enthalpy transfer diagram in Fig 3.3, where the pinch is obvious.

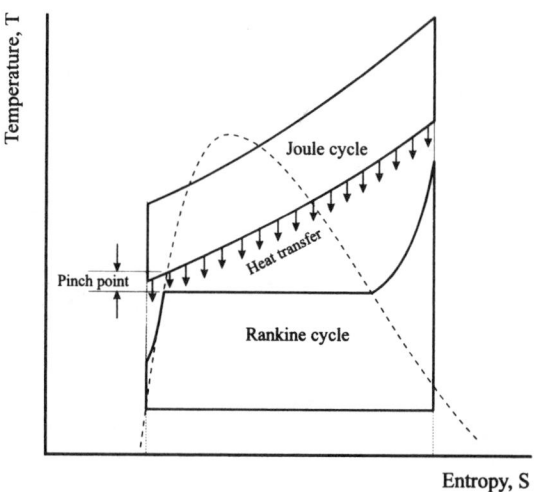

Fig. 3.2 *T–s* diagram of CCGT

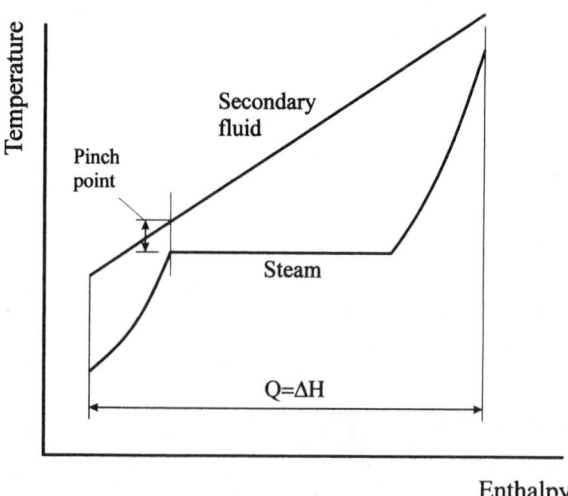

Fig. 3.3 *H–T* diagram of CCGT

Perhaps the easiest way of gaining an understanding of pinch techniques is to consider some simple examples.

3.1 A heat transfer network without a pinch problem

This example has a total of seven streams, three hot and four cold, and it is required to use the heating and cooling potential of the streams to minimise the heat transfer from high temperature utilities, and the heat transfer to low temperature utilities. The parameters for the streams involved in the processes are given in Table 3.1. The supply temperature, T_S, is the initial temperature of the stream, and the target temperature, T_T, is the target final temperature that must be achieved by heat transfer. The heat flow capacity, mC, is the product of the mass flow and the specific heat of the particular stream, and the heat load is the amount of energy that is transferred to or from the streams.

Table 3.1 Specification of hot and cold streams

Stream no	Stream type	Supply temperature $T_S/(°C)$	Target temperature $T_T/(°C)$	Heat flow capacity $mC/(MJ/hK)$	Heat load $Q/(MJ/h)$
1	Cold	95	205	2.88	316.8
2	Cold	40	220	2.88	518.4
3	Hot	310	205	4.28	−449.4
4	Cold	150	205	7.43	408.7
5	Hot	245	95	2.84	−426.0
6	Cold	65	140	4.72	354.0
7	Hot	280	65	2.38	−511.7

In this case there are three streams of fluid which require cooling (the hot streams) and four streams of fluid which require heating (the cold streams). The simplest way of achieving this is to cool the hot streams by transferring heat directly to a cold water supply, and to heat the cold streams by means of a steam supply; this approach is shown in Fig 3.4. This means that the hot utility (the steam supply) has to supply 1597.9 MJ of energy, while the cold utility (a cold water supply) has to remove 1387.1 MJ of energy. Both of these utilities are a cost on the process plant. The steam has to be produced by burning a fuel, and use of the cold water will be charged by the water authority. In reality a *minimum net heat supply* of 1597.9−1387.1 = 210.8 MJ/h could achieve the same result, if it were possible to transfer all the energy available in the hot streams to the cold streams. This problem will now be analysed.

If heat is going to be transferred between the hot and cold streams there must be a temperature difference between the streams: assume in this case that the minimum temperature difference (δT_{min}) is 10°C.

The method of tackling this problem proposed by Linnhoff and Turner (1981) is as follows.

Step 1: Temperature intervals

Evaluate the temperature intervals defined by the 'interval boundary temperatures'. These can be defined in the following way: the unadjusted temperatures of the cold streams can be used, and the hot stream temperatures can be adjusted by subtracting δT_{min} from the actual values. In this way the effect of the minimum temperature difference has been included in the calculation. This results in Table 3.2.

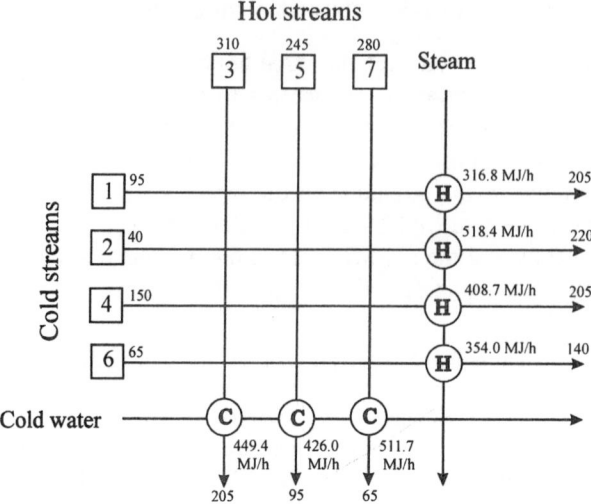

Fig. 3.4 Direct heat transfer between the fluid streams and the hot and cold utilities

Table 3.2 Ordering of hot and cold streams

Stream no	Stream type	Supply temperature $T_S/(°C)$	Target temperature $T_T/(°C)$	Adjusted temperatures		Order
1	Cold	95		95		T_9
			205		205	T_5
2	Cold	40		40		T_{13}
			220		220	T_4
3	Hot	310		300		T_1
			205		195	T_6
4	Cold	150		150		T_7
			205		205	Duplicate
5	Hot	245		235		T_3
			95		85	T_{10}
6	Cold	65		65		T_{11}
			140		140	T_8
7	Hot	280		270		T_2
			65		55	T_{12}

The parameters defining the streams can also be shown on a diagram of temperature against heat load (enthalpy transfer; see Fig 3.5). This diagram has been evaluated using the data in Table 3.2, and is based on the unadjusted temperatures. The hot stream line is based on the *composite* temperature–heat load data for the hot streams, and is evaluated using eqn (3.2); the cold stream line is evaluated by applying the same equation to the cold streams. It can readily be seen that the two lines are closest at the temperature axis, when they are still 25°C apart: this means that there is no 'pinch' in this example because the temperature difference at the pinch point is greater than the minimum value allowable.

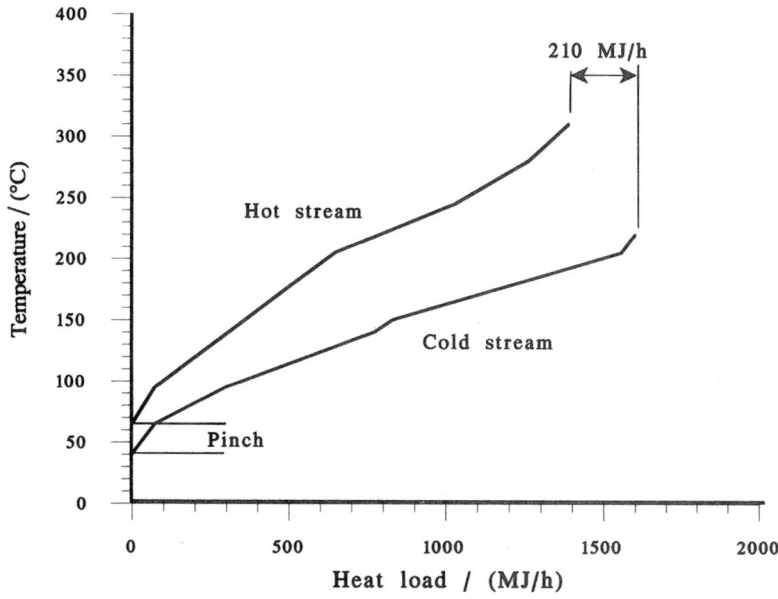

Fig. 3.5 Temperature–heat load diagram

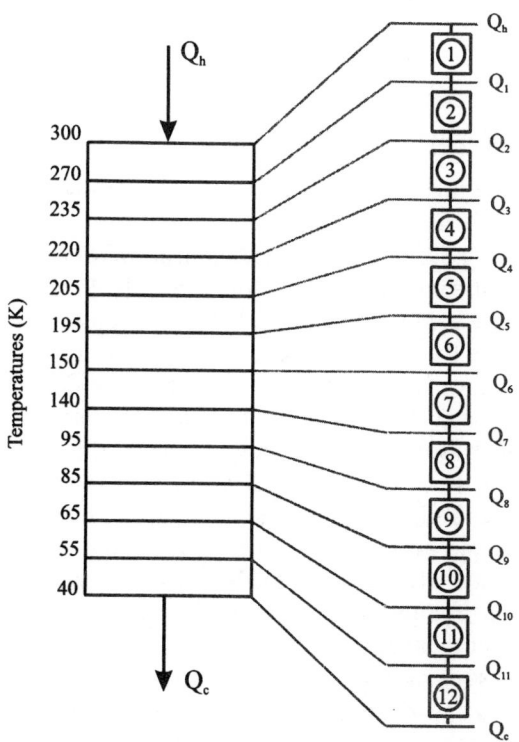

Fig. 3.6 Temperature intervals for heat transfer network

Hence, the problem reduces to transferring energy from the hot streams to the cold streams, and finally adding 210.8 MJ/h from a hot utility. The mechanism for allocating the energy transfers will now be introduced.

Having defined the temperature intervals it is possible to consider the problem as shown in Fig 3.6. The energies flowing into and out of the combined systems, Q_h and Q_c, are those which have to be supplied by and lost to the external reservoirs respectively. It is also apparent that the difference between these values is the difference between the enthalpies of the hot and cold streams, i.e.

$$Q_c - Q_h = \delta H \tag{3.1}$$

Consideration will show that δH is constant, because the difference between the enthalpies of the hot and cold streams is constant, and this means that any additional energy added from the high temperature supplies must be compensated by an equal amount of energy being rejected to the low temperature sinks: hence energy will have just flowed wastefully through the overall system. The heat transfer network can be shown schematically as in Fig 3.7. The heat flows through each of the temperature intervals can be evaluated as shown in the next step.

Step 2: Interval heat balances

Table 3.2 includes the effect of the minimum temperature difference between the streams, δT_{min}, and hence the intervals have been established so that full heat transfer is possible between the hot and cold streams. It is now necessary to apply the First Law to examine the enthalpy balance between the streams, when

$$\delta H_i = \left(\sum_{\substack{Hot \\ i, i+1}} (mC)_h - \sum_{\substack{Cold \\ i, i+1}} (mC)_c \right)(T_i - T_{i+1})$$

where i = initial temperature of the interval   (3.2)
$\quad\quad\quad i + 1$ = final temperature of the interval

Applying this equation to this example results in the heat flows shown by the $\delta H = mC\,\delta T$ values in Fig 3.8.

It can be seen from Fig 3.8 that the individual heat transfers are positive (i.e. from the hot streams to the cold streams) in the first four intervals. In the fifth interval the amount of energy required by the cold streams exceeds that available from the hot streams in that temperature interval, but the energy can be supplied from that available in the higher temperature intervals. However, by the eighth interval the demands of the cold streams exceed the total energy available from the hot streams, and it is at this point that the energy should be added from the hot utility, because this will limit the temperature required in the hot utility. In reality, the 210.8 MJ could be provided from the hot utility at any temperature above 140°C, but the higher the temperature of the energy the more will be the irreversibility of the heat transfer process. It is now useful to look at the way in which the heat can be transferred between the hot and cold streams.

The streams available for the heat transfer processes are shown in Fig 3.9. First, it should be recognised that there is no heat transfer to the cold utility, and thus all the heat transfers from the hot streams must be to cold streams This constrains the problem to ensure that there is always a stream cold enough to receive heat from the hot sources. This means that the temperature of Stream 7 must be cooled to its target temperature of 65°C by transferring heat to a colder stream: the only one available is Stream 2. Hence, the total heat transfer from Stream 7 is passed to Stream 2: an energy balance shows that the

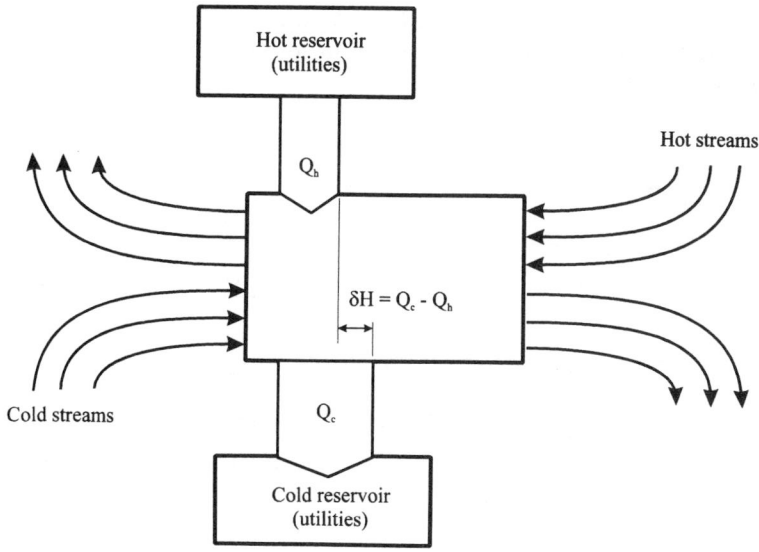

Fig. 3.7 Overall energy for heat transfer network

Temp	Q	Stream	Streams	mCδT	ΣmCδT	
300	Qh					
		①	3	128.4		
270	Q₁				128.4	
		②	3,7	233.1		
235	Q₂				361.5	
		③	3,7,5	142.5		
220	Q₃				504.0	
		④	3,7,5,2	99.3		
205	Q₄				603.3	
		⑤	3,7,5, 2,1,4	-36.9		
195	Q₅				566.4	
		⑥	7,5,2, 1,4	-358.7		
150	Q₆				207.8	
		⑦	7,5,2, 1	-5.4		
140	Q₇				202.4	
		⑧	7,5,2, 1,6	-236.8		Heating 210.8
95	Q₈				-34.4	176.4
		⑨	7,5,2, 6	-23.8		
85	Q₉				-58.2	152.6
		⑩	7,6,2	-104.4		
65	Q₁₀				-162.5	48.2
		⑪	7,2	-5.0		
55	Q₁₁				-167.5	43.2
		⑫	2	-43.2		
40	Qc				-210.8	0

Fig. 3.8 Heat flows

temperature of Stream 2 is raised to 218°C, and there is a residual heat capacity of 6.7 MJ/h before the stream reaches its target temperature In a similar manner it is necessary for Stream 5 to be matched to Stream 6 because this is the only stream cold enough to bring its temperature down to 95°C. In this way it is possible to remove Streams 6 and 7 from further consideration because they have achieved their target temperatures; Streams 2 and 5 must be left in the network because they still have residual energy before they achieve their targets. Figure 3.9 can be modified to Fig 3.10.

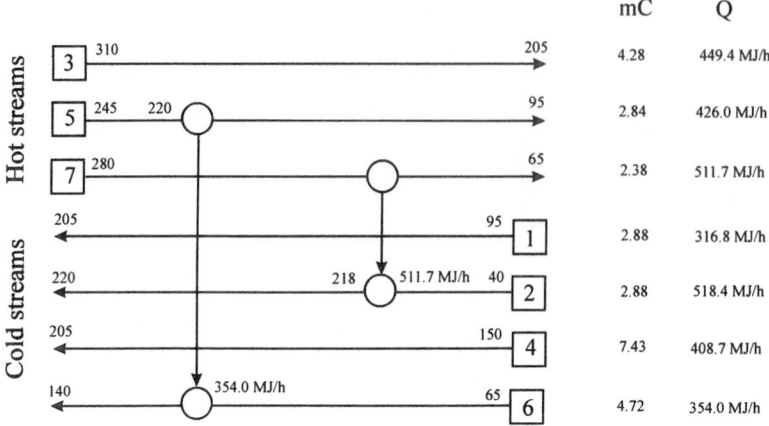

Fig. 3.9 Initial heat transfer: heat transfer from Streams 5 to 6, and 7 to 2

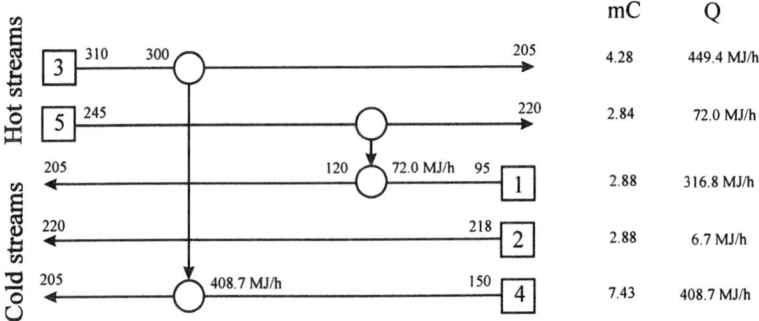

Fig. 3.10 Removing the residuals from Streams 2 and 5

Streams 2 and 5 are represented in this diagram by their residual energies, and by the temperatures that were achieved in the previous processes. It is possible to cool Streams 3 and 5 by transferring energy with either of cold Streams 1 or 4. The decision in this case is arbitrary, and for this case Stream 3 will be matched with Stream 4, and Stream 5 will be matched with Stream 1. This results in the heat transfers shown in Fig 3.10, and by this stage Streams 5 and 4 can be removed from further consideration. This results in another modified diagram, Fig 3.11.

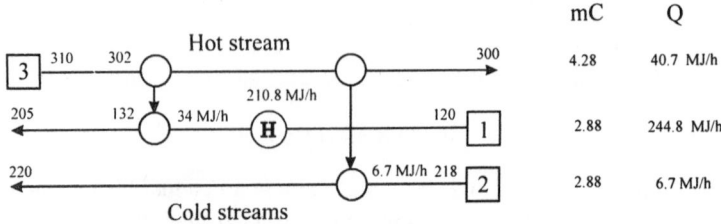

Fig. 3.11 Completing the heat transfer network

By this stage it is necessary to consider adding the heat from the hot utility. In this case, the temperature at which this energy is added is relatively arbitrary, and the heat should be transferred at as low a temperature as possible. This is indicated by the 210.8 MJ/h heat transfer in Fig 3.11.

The previous analysis has considered the problem in discrete parts, but it is now possible to combine all these sub-sections into a composite diagram, and this is shown in Fig 3.12.

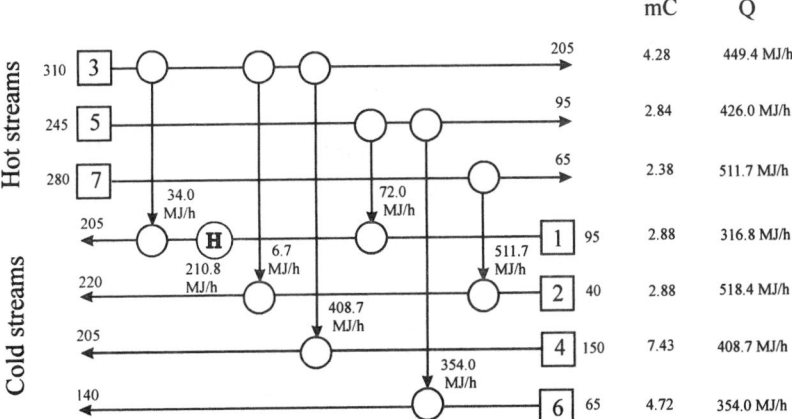

Fig. 3.12 Composite diagram for heat exchanger network

The diagram given in Fig 3.12 suggests that seven heat exchangers are required, but the diagram is not the most succinct representation of the network problem, which is better illustrated as shown in Fig 3.13, which grows directly out of the original arrangement shown in Fig 3.4.

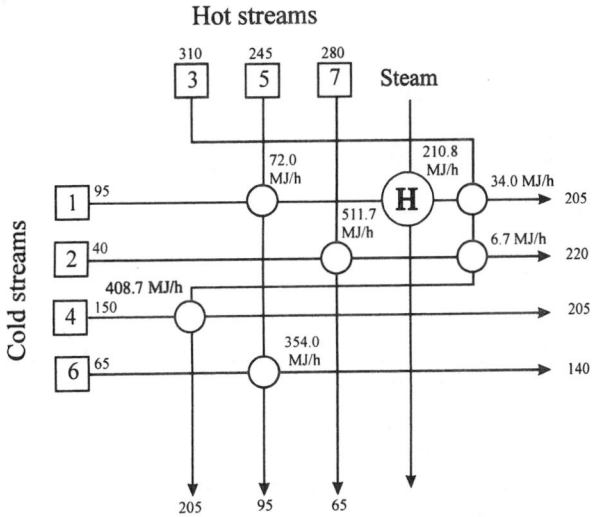

Fig. 3.13 Diagram showing the minimum number of heat exchangers to achieve heat transfer

This case was a relatively straightforward example of a heat exchanger network in which it was always relatively easy to match the streams, because there was always a sufficient temperature difference to drive the heat transfer processes. The next example shows what happens when there is not sufficient temperature difference to drive the heat transfer processes.

3.2   A heat transfer network with a pinch point

Consider there are four streams of fluid with the characteristics given in Table 3.3. Assume that the minimum temperature difference between any streams to obtain acceptable heat transfer is 5°C, i.e. $\delta T_{min} = 5$°C. This value of δT_{min} is referred to as the *pinch point* because it is the closest that the temperatures of the streams are allowed to come. The temperatures of the streams can be ordered in a manner similar to the first case and the results of this are shown in Table 3.4.

Table 3.3   Characteristics of streams

Process stream		Supply temperature $T_S/$(°C)	Target temperature $T_T/$(°C)	Heat capacity flowrate, mC /(MJ/hK)	Heat load $mC(T_S - T_T)$ /(MJ/h)
Number	Type				
1	Cold	50	110	2.0	120
2	Hot	130	70	3.0	180
3	Cold	80	115	4.0	140
4	Hot	120	55	1.5	97.5

Table 3.4 Definition of temperature intervals

Stream no	Stream type	$T_S/$(°C)	$T_T/$(°C)	Adjusted temperatures		Order
1	Cold	50		50		T_6
			110		110	T_3
2	Hot	130		125		T_1
			70		65	T_5
3	Cold	80		80		T_4
			115		115	T_2
4	Hot	120		115		Duplicate
			55		50	Duplicate

These temperature intervals can be depicted graphically as shown in Fig 3.14. It can be seen that there are five intervals in this case, as opposed to the 12 in the first case.

It is now possible to draw the temperature–heat load diagram for this problem, and this is shown in Fig 3.15. It can be seen that the basic cold stream is too close to the hot stream and there will not be a sufficient temperature difference to drive the heat transfer processes. The modified cold stream line has been drawn after undertaking the following analysis, and does produce sufficient temperature difference at the pinch.

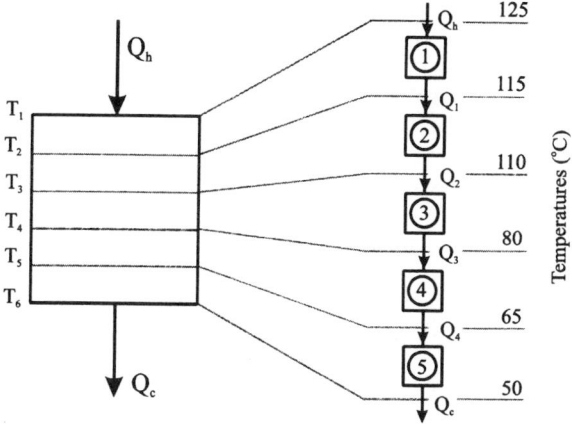

Fig. 3.14 Temperature intervals for heat transfer network

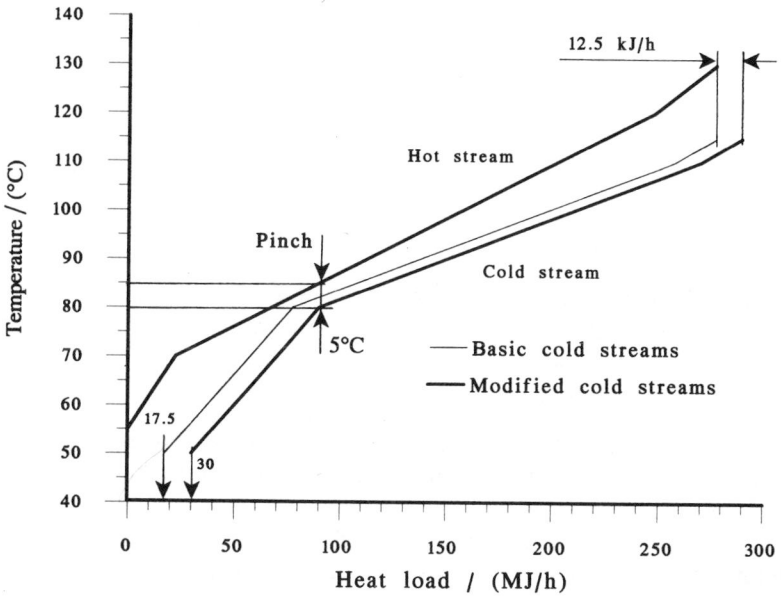

Fig. 3.15 Temperature–heat load diagram, indicating pinch point

Step 3: Heat cascading

This step is an additional one to those introduced previously, and consideration will show why it comes about. Figure 3.16 indicates the heat flows in the various intervals and, in the left-hand column of figures, shows that if the heat flow from the hot utility (or heat source) is zero then there will be a negative heat flow of -12.5 units in temperature interval 3 between 110°C and 80°C. Consideration will show that *this is impossible because it means that the heat will have been transferred against the temperature gradient.* Such a situation can be avoided if sufficient energy is added to the system to make the largest negative heat flow zero in this interval. The result of this is shown in the right-hand column of figures in Fig 3.16, which has been achieved by adding 12.5 units of energy to the system. It can be seen that in both cases the difference between Q_h and Q_c is 17.5 units of energy. If an energy balance is applied to the streams defined in Table 3.3 then

$$\Sigma \, \delta H_i = 3 \times (130 - 70) + 1.5 \times (120 - 55) - 2 \times (110 - 50) - 4 \times (115 - 80)$$
$$= 17.5 \tag{3.3}$$

Hence, as stated previously, the energies obey the steady-flow energy equation for the system shown in Fig 3.16.

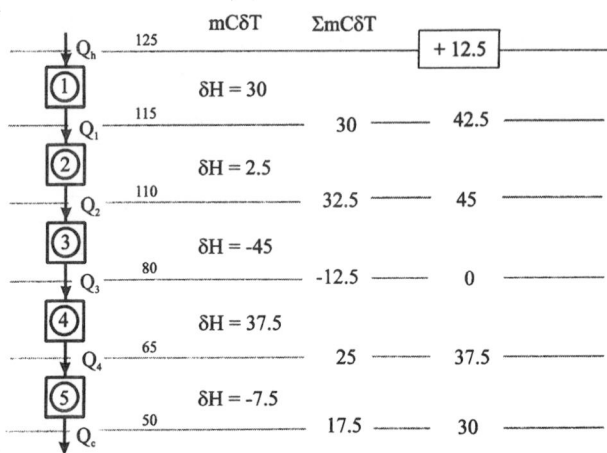

Fig. 3.16 Temperature intervals and heat loading

There is now a point in the temperature range where the heat flow is zero: this point is called *the pinch*. In this example it is at 80°C, which means that the pinch occurs at a cold stream temperature of 80°C and a hot stream temperature of 85°C. There are three important constraints regarding the pinch:

1. Do not transfer heat across the pinch. Any heat flow across the pinch results in the same amount of heat being added to every heat flow throughout the system, and hence increases Q_h and Q_c.
2. Do not use the cold sink above the pinch. If the system has been designed for minimised heat flow it does not reject any heat above the pinch (see Fig 3.16, where the heat rejection has been made zero at the pinch).
3. Do not use the hot source below the pinch. If the system has been designed for minimised heat flow it does not absorb any heat below the pinch.

It is hence possible to reduce the problem into two parts: above the pinch and below the pinch, as shown in Fig 3.17, which is a modification of Fig 3.7.

It is now necessary to break the problem at the pinch, and this results in Fig 3.18, which is the equivalent of Fig 3.9 for the first example.

Now the problem can be analysed, bearing in mind the restrictions imposed by the pinch point. This means that cooling to the utility stream is not allowable above the pinch, and hence the only transfer can be with the hot utility above the pinch. It is now necessary, as far as possible, to match the hot and cold streams above the pinch.

1. Consider Stream 2 is matched with Stream 1, and Stream 4 is matched with Stream 3. Then Stream 2 can transfer 70 MJ/h to Stream 1, and enable Stream 1 to achieve its

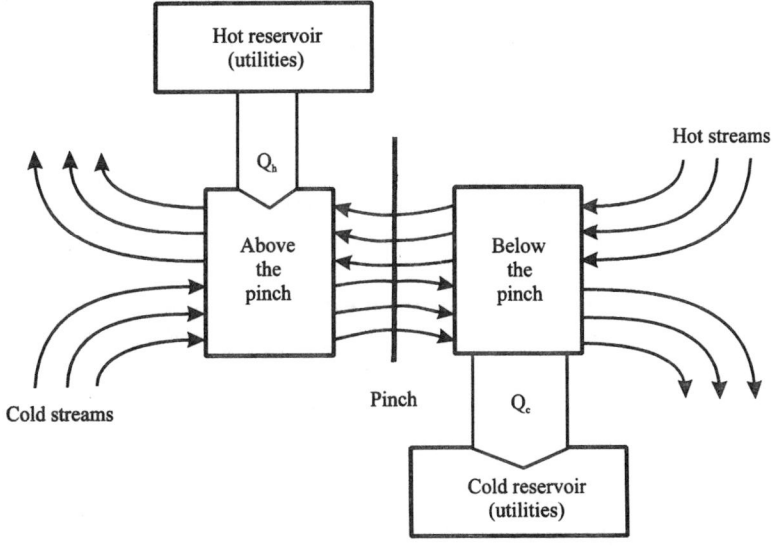

Fig. 3.17 Breaking the problem at the pinch point

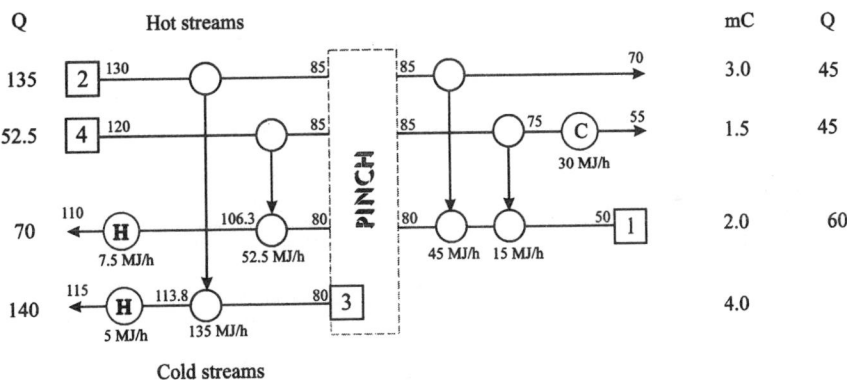

Fig. 3.18 Hot and cold streams with pinch

target temperature, while reducing its own temperature to 106.7°C. Also Stream 4 can transfer its energy to Stream 3 and this will raise the temperature of Stream 3 to 93.1°C. This shows that there is not sufficient energy available in these streams to achieve the target temperatures. The reason for this being an unsuitable approach is because the heat capacity flowrate for Stream 1 above the pinch is greater than that of the cold stream above the pinch. Since both streams have the same temperature at the pinch point, then the higher temperature of the cold stream would have to be higher than that of the hot stream to achieve an energy balance: this would result in an impossible heat transfer situation. Hence, for a satisfactory result to be possible

$$mC_{cold} \geqslant mC_{hot} \qquad (3.4)$$

above the pinch point.

2. If the alternative match is used, namely, Stream 2 matched to Stream 3, and Stream 4 matched to Stream 1, then the answer shown in Fig 3.19 is obtained.

The heat transfers are shown on the diagram. It can be seen that it is not possible to match the hot and cold streams either above or below the pinch. This means that utility heat transfers are required from the hot utility above the pinch, and heat transfers to the cold utility are required below the pinch. This proposal does obey inequality (3.4), and is hence acceptable.

3. It can be seen that, above the pinch point, energy has to be added to the system from the hot utility. This obeys the rules proposed above, and the total energy added is 12.5 MJ/h, which is in agreement with the value calculated in Fig 3.16.

4. Considering the heat transfers below the pinch, it can be seen that Stream 1 can be heated by energy interchange with Streams 2 and 4: neither Stream has sufficient capacity alone to bring Stream 1 to its target temperature. However, it is feasible to bring about the heating because

$$mC_{hot} \geq mC_{cold} \tag{3.5}$$

which is the equivalent of inequality (3.4) for the transfers below the pinch. In this case it was chosen to transfer all the energy in Stream 4 because this results in a lower temperature for heat transfer to the cold utility. The 30 MJ/h transferred to the cold utility is in line with that calculated in Fig 3.16.

These diagrams can now be joined together to give the composite diagram in Fig 3.19.

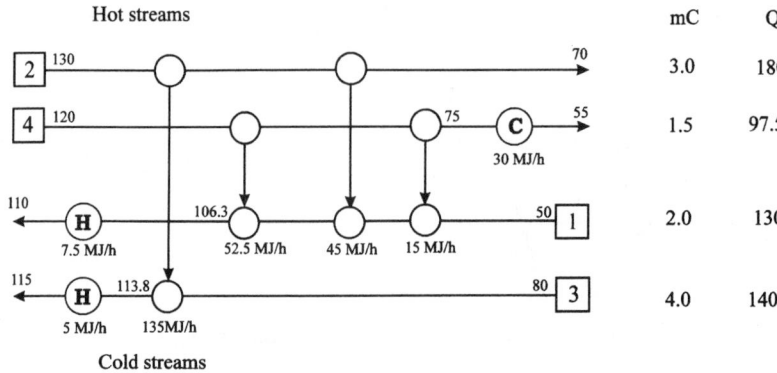

Fig. 3.19 Composite diagram for heat transfer network in Example 2

The heat load against temperature diagram for this problem, before heat transfer from the utilities has been supplied, is shown in Fig 3.15, and was discussed previously. It is now possible to consider the modified diagram, when it can be seen that the energy transfers have produced a sufficient temperature difference to satisfy the constraints of the problem.

3.3 Concluding remarks

A method has been introduced for improving the efficiency of energy transfers in complex plant. It has been shown that in some plants there is a *pinch point* which restricts the freedom to transfer energy between process streams. To ensure that the plant attains its maximum efficiency of energy utilisation, energy should be added to the system only above the pinch, and extracted from it only below the pinch.

PROBLEMS

1 A process plant has two streams of hot fluid and two streams of cold fluid, as defined in Table P3.1. It is required to minimise the energy which must be transferred to hot and cold utilities by transferring energy between the streams. If the minimum temperature difference for effective heat transfer is 20°C, design a network which achieves the requirement, and minimises the transfers to the utilities. Is there a pinch point in this problem, and at what temperature does it occur? Calculate the minimum heat transfers to and from the cold and hot utilities.

Table P3.1 Data related to Q.1

Stream no	Stream type	Supply temperature $T_S/(°C)$	Target temperature $T_T/(°C)$	Heat flow capacity $mC/(MJ/hK)$
1	Hot	205	65	2.0
2	Hot	175	75	4.0
3	Cold	45	180	3.0
4	Cold	105	155	4.5

[105°C (cold stream); $Q_{H_{min}} = 90$ MJ/h; $Q_{C_{min}} = 140$ MJ/h]

2 Some stream data have been collected from a process plant, and these are listed in Table P3.2. Assuming the minimum temperature difference between streams, $\Delta T_{min} = 10°C$,
(a) calculate the data missing from Table P3.2;
(b) analyse this data to determine the minimum heat supplied from the hot utility, the minimum heat transferred to the cold utility, and the pinch temperatures;
(c) draw a schematic diagram of the heat transfer network.

Table P3.2 Data related to Q.2

Stream no	Stream type	Supply temperature $T_S/(°C)$	Target temperature $T_t/(°C)$	Enthalpy change $\Delta H/(kW)$	Heat flow capacity $mC/(kW/K)$
1	Cold	60	180	?	3
2	Hot	180	40	?	2
3	Cold	30	130	220	?2·2
4	Hot	150	40	400	?4·0
5	Cold	60	80	40	?2·0

[360 kW; 280 kW; 2.2 kW/K; 4.0 kW/K; 2.0 kW/K; $Q_{H_{min}} = 60$ kW; $Q_{C_{min}} = 160$ kW; $T_{C_{pinch}} = 140°C$]

3 Figure P3.3 shows a network design using steam, cooling water and some heat recovery.
(a) Does this design achieve the minimum energy target for $\Delta T_{min} = 20°C$?
(b) If the current network does not achieve the targets, show a network design that does.

[(a) $T_{C_{pinch}} = 150°C$; $T_{H_{pinch}} = 170°C$; $Q_C = 480$ kW; $Q_H = 380$ kW; (b) $Q_{C_{min}} = 360$ kW; $Q_{H_{min}} = 260$ kW]

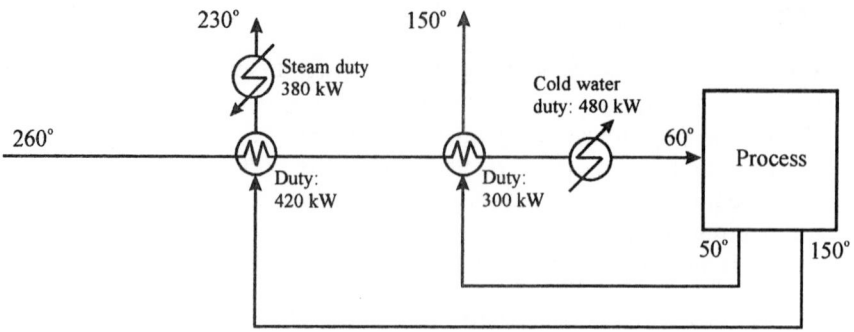

Fig. P3.3 Network for Q.3

4 Figure P3.4 shows two hot streams and two cold streams for heat integration (subject to $\Delta T_{min} = 20°C$).

    (i)   What are the energy targets?

    (ii) Show a network design achieving these targets.

$$[Q_{H_{min}} = 0; \; Q_{C_{min}} = 0]$$

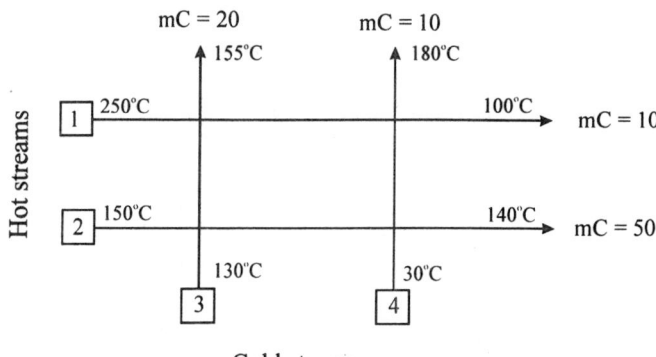

Fig. P3.4 Network for Q.4

5 Figure P3.5 shows an existing design of a process plant, containing two exothermic processes. These require streams of reactants as shown in the diagram, and produce products at the temperatures shown. The plant achieves the necessary conditions by providing 480 kW of heat from a steam source, and rejects a total of 560 kW of energy to cold water utilities; only 460 kW is transferred between the streams.

(a) Show that there is a pinch point, and evaluate the temperature.

(b) Show that the existing plant is inefficient in its use of the energy available.

(c) Calculate the energy targets for $\Delta T_{min} = 20°C$ and show a design that achieves these targets.

$$[(a) \; T_{C_{pinch}} = 110°C; \; T_{H_{pinch}} = 130°C; \; (b) \; Q_C = 560 \text{ kW}; \; Q_H = 480 \text{ kW};$$
$$(c) \; Q_{C_{min}} = 210 \text{ kW}; \; Q_{H_{min}} = 130 \text{ kW}]$$

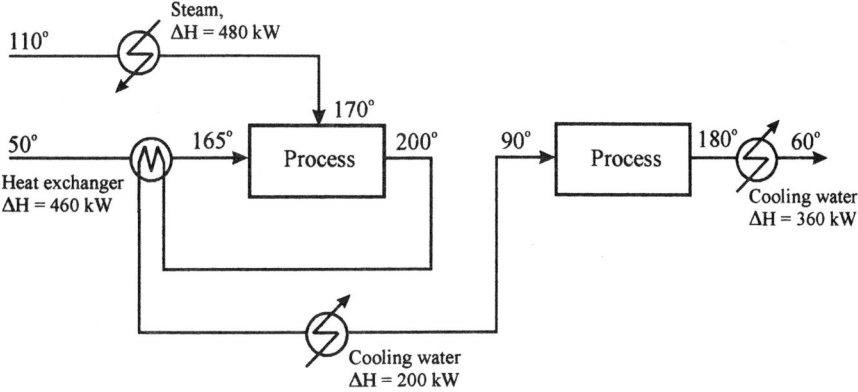

Fig. P3.5 Network for Q.5

6 Recalculate the problem in Q.5 using a $\Delta T_{min} = 10°C$. Comment on the effect of reducing the minimum temperature difference.

$$[(a)\ T_{C_{pinch}} = 110°C;\ T_{H_{pinch}} = 120°C;\ (b)\ Q_C = 560\ kW;\ Q_H = 480\ kW;$$
$$(c)\ Q_{C_{min}} = 120\ kW;\ Q_{H_{min}} = 40\ kW]$$

7 A network for a process plant is shown in Fig P3.7.

(a) Calculate the energy targets for $\Delta T_{min} = 10°C$ and show a design that achieves these targets.

(b) Explain why the existing network does not achieve the energy targets.

$$[(a)\ Q_{C_{min}} = 190\ kW;\ Q_{H_{min}}T = 90\ kW;\ (b)\ there\ is\ transfer\ across\ the\ pinch]$$

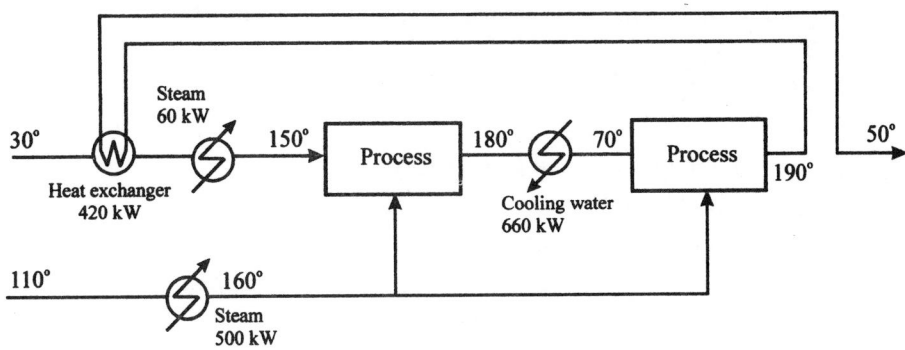

Fig. P3.7 Network for Q.7

4

Rational Efficiency of a Powerplant

4.1 The influence of fuel properties on thermal efficiency

The thermal efficiency of a cycle has been defined previously, in terms of specific quantities, as

$$\eta_{th} = \frac{w_{net}}{q_{in}} \tag{4.1}$$

where w_{net} = net work output from the cycle per unit mass of fluid,

and q_{in} = energy addition to the cycle per unit mass of fluid.

In this case q_{in} is the energy transfer to the working fluid, and does not take into account any losses in the boiler or heat transfer device. Equation (4.1) can be rewritten for the whole powerplant, including the boiler or heat transfer mechanism, as

$$\eta_0 = \eta_B \eta_{th} = \frac{w_{net}}{-\Delta h_0} \tag{4.2}$$

where η_0 = overall efficiency of powerplant,

η_B = efficiency of boiler,

η_{th} = thermal efficiency of cycle,

w_{net} = net work output from the cycle *per unit mass of fuel*,

and Δh_0 = specific enthalpy of reaction of fuel.

This might be considered to be an unfair, and possibly misleading, method of defining the efficiency because the energy addition cannot all be turned into work, as was shown when considering exergy and availability. Another definition of efficiency can be derived based on the Second Law, and this relates the work output from the cycle to the maximum work output obtainable.

The efficiency of the powerplant has been related, in eqn (4.2), to the amount of energy that has been added to the cycle by the combustion of the fuel. In the past this has been based on the enthalpy of reaction of the fuel, or usually its calorific value, Q_p'. It was shown previously (Chapter 2) that this is not the energy available for the production of work, and that the maximum available work that can be obtained from the fuel is based on the change of its *exergy* at the dead state conditions. Hence, the maximum available work *from unit mass of fuel* is

$$-\Delta g_0 = g_{R_0} - g_{P_0} \tag{4.3}$$

This is related to the enthalpy of formation by the equation

$$\Delta g_0 = \Delta h_0 - T_0(s_{R_0} - s_{P_0}) \tag{4.4}$$

It was shown in Chapter 2 that $|\Delta g_0|$ could be greater than or less than $|\Delta h_0|$, and the difference was dependent on the structure of the fuel and the composition of the exhaust products. The efficiency of the powerplant can then be redefined as

$$\eta_0 = \frac{w_{net}}{-\Delta g_0} \tag{4.5}$$

where w_{net} = actual net work output from the cycle *per unit mass of fuel*,

and Δg_0 = change of Gibbs function caused by combustion,

= maximum net work obtainable from unit mass of fuel.

Equation (4.5) is often referred to as the *Second Law Efficiency*, because the work output is related to the available energy in the fuel, rather than its enthalpy change. The actual effect on thermal efficiency of using the change of Gibbs function instead of the enthalpy of reaction is usually small (a few per cent).

4.2 Rational efficiency

When the efficiencies defined in eqns (4.2) and (4.5) are evaluated they contain terms which relate to the 'efficiency' of the energy transfer device (boiler) in transferring energy from the combustion gases to the working fluid. These effects are usually neglected when considering cycles, and the energy added is related to the change in enthalpy of the working fluid as it passes through the boiler, superheater, etc. Actual engine cycles will be considered later. First, a general heat engine will be considered (see Fig 4.1). For convenience the values will all be taken as specific values per unit mass flow of working fluid.

The engine shown in Fig 4.1 could be either a wholly reversible (i.e. internally and externally) one or an irreversible one. If it were internally reversible then it would follow the Carnot cycle 1–2–3s–4–1. If it were irreversible then it would follow the cycle 1–2–3–4–1. Consider first the reversible cycle. The efficiency of this cycle is

$$\eta_{th} = \frac{w_{net}}{q_{in}} = \frac{(T_1 - T_{3s})(s_2 - s_1)}{T_1(s_2 - s_1)} = \frac{(T'_H - T'_L)(s_2 - s_1)}{T'_H(s_2 - s_1)} \tag{4.6}$$

which is the efficiency of an internally reversible engine operating between the temperature limits T'_H and T'_L. This efficiency will always be less than unity unless $T'_L = 0$. However, the Second Law states that it is never possible to convert the full energy content of the energy supplied into work, and the maximum net work that can be achieved is

$$\hat{w}_{net} = b_2 - b_1 = h_2 - h_1 - T_0(s_2 - s_1) \tag{4.7}$$

where T_0 is the temperature of the dead state.

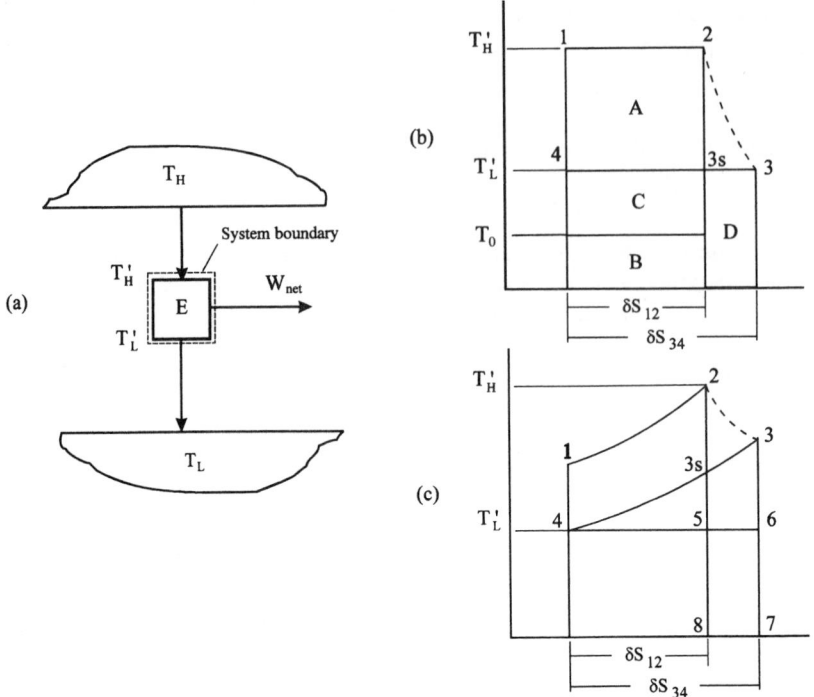

Fig. 4.1 A heat engine operating between two reservoirs: (a) schematic diagram of the engine; (b) $T-s$ diagram for the engine which receives and rejects energy isothermally; (c) $T-s$ diagram for the engine which receives and rejects energy non-isothermally

Hence, the Second Law efficiency of the heat engine is

$$\eta_{2,th} = \frac{W_{net}}{\hat{W}_{net}} = \frac{(T_1 - T_{3s})(s_2 - s_1)}{h_2 - h_1 - T_0(s_2 - s_1)} = \frac{T_1(s_2 - s_1) - T_{3s}(s_2 - s_1)}{h_2 - h_1 - T_0(s_2 - s_1)}$$

$$= \frac{h_2 - h_1 - T_{3s}(s_2 - s_1)}{h_2 - h_1 - T_0(s_2 - s_1)} = \frac{h_2 - h_1 - T_0(s_2 - s_1) - (T_{3s} - T_0)(s_2 - s_1)}{h_2 - h_1 - T_0(s_2 - s_1)}$$

$$= 1 - \frac{(T_{3s} - T_0)(s_2 - s_1)}{h_2 - h_1 - T_0(s_2 - s_1)} \tag{4.8}$$

where $\eta_{2,th}$ is called the Second Law efficiency, or the *rational efficiency*, η_R. This result can be interpreted in Fig 4.1(b) in the following way. Examination of eqn (4.8) shows that if the $T_{3s} = T_0$ then $\eta_R = 1.0$. This because when $T_{3s} = T_0$ then area C is zero (i.e. the line at $T'_L = T_0$ is coincident with that at T_{3s}). Then the difference in the enthalpies, $h_2 - h_1$, is depicted by areas A + B, and the unavailable energy $T_0 (s_2, s_1)$ by area B, and hence the energy available to produce work is area A. This shows that the cycle is as efficient as an internally reversible cycle operating between the same two temperature limits (T_{3s} and T_1). If T_0 is not equal to T_3, but is equal to a lower temperature T'_L, then the rational efficiency will be less than unity because energy which has the capacity to do work is being rejected. This is depicted by area C in Fig 4.1(b), and the rational efficiency becomes

$$\eta_R = \frac{\text{Areas}(A+B+C) - \text{Areas}(B+C)}{\text{Areas}(A+B+C) - \text{Areas}(B)} = \frac{\text{Area A}}{\text{Areas}(A+C)} < 1 \qquad (4.9)$$

Consideration of eqn (4.8) shows that it is made up of a number of different components which can be categorised as available energy and unavailable energy. This is similar to exergy, as shown below:

$$b = a - a_0 = \underbrace{h - h_0}_{\substack{\text{available} \\ \text{energy}}} - \underbrace{T_0(s - s_0)}_{\substack{\text{unavailable} \\ \text{energy}}} = \underbrace{E}_{\substack{\text{available} \\ \text{energy}}} - \underbrace{\Phi}_{\substack{\text{unavailable} \\ \text{energy}}} \qquad (4.10)$$

Thus eqn (4.7) may be rewritten

$$\hat{w}_{\text{net}} = b_2 - b_1 = h_2 - h_1 - T_0(s_2 - s_1) = \delta E_{12} - \delta \Phi_{12} \qquad (4.11)$$

giving eqn (4.8) as

$$\eta_R = \eta_{2,\text{th}} = \frac{w_{\text{net}}}{\hat{w}_{\text{net}}} = \frac{h_2 - h_1 - T_{3s}(s_2 - s_1)}{h_2 - h_1 - T_0(s_2 - s_1)} = \frac{\delta E_{12} - \delta \Phi_{12} - (T_{3s} - T_0)(s_2 - s_1)}{\delta E_{12} - \delta \Phi_{12}} \qquad (4.12)$$

which can be written as

$$\eta_R = \eta_{2,\text{th}} = 1 - \frac{(T_{3s} - T_0)(s_2 - s_1)}{\delta E_{12} - \delta \Phi_{12}} \qquad (4.13)$$

If $T_0 = T_{3s}$ then the rational efficiency, $\eta_R = 1.0$. If $T_0 < T_{3s}$, as would be the case if there were external irreversibilities, then $\eta_R < 1.0$ because the working fluid leaving the engine still contains the capacity to do work.

If the engine is not internally reversible then the T–s diagram becomes $1-2-3-4-1$, as depicted in Fig 4.1(b). The effect of the internal irreversibility is to cause the entropy at 3 to be bigger than that at 3s and hence the entropy difference $\delta S_{34} > \delta S_{12}$. The effect of this is that the net work becomes

$$
\begin{aligned}
w_{\text{net}} &= (h_2 - h_1) - (h_3 - h_4) = T_1(s_2 - s_1) - T_3(s_3 - s_4) \\
&= T_1(s_2 - s_1) - T_3(s_{3s} - s_4) - T_3(s_3 - s_{3s}) \\
&= h_2 - h_1 - T_0(s_2 - s_1) - (T_3 - T_0)(s_2 - s_1) - T_3(s_3 - s_{3s}) \\
&= \delta E_{12} - \delta \Phi_{12} - (T_3 - T_0)(s_2 - s_1) - T_3(s_3 - s_{3s}) \qquad (4.14)
\end{aligned}
$$

Hence, from eqn (4.14)

$$\eta_R = \eta_{2,\text{th}} = 1 - \frac{(T_3 - T_0)(s_2 - s_1) + T_3(s_3 - s_{3s})}{\delta E_{12} - \delta \Phi_{12}} \qquad (4.15)$$

In the case of the irreversible engine, even if $T_0 = T_3$ the rational efficiency is still less than unity because of the increase in entropy caused by the irreversible expansion from 2 to 3. The loss of available energy, in this case the irreversibility, is depicted by area D in Fig 4.1(b). In the case of $T_0 = T_3$,

$$\eta_R = 1 - \frac{T_3(s_3 - s_{3s})}{\delta E_{12} - \delta \Phi_{12}} \qquad (4.16)$$

If the dead state temperature is less than T_3 then the rational efficiency is even lower because of the loss of available energy shown as area C in Fig 4.1(b).

Up until now it has been assumed that the cycle is similar to a Carnot cycle, with isothermal heat supply and rejection. Such a cycle is typical of one in which the working fluid is a vapour which can change phase. However, many cycles use air as a working fluid (e.g. Otto, Diesel, Joule cycles, etc), and in this case it is not possible to supply and reject heat at constant temperature. A general cycle of this type is shown in Fig 4.1(c), and it can be seen that the heat is supplied over a range of temperatures from T_1 to T_2, and rejected over a range of temperatures from T_3 to T_4. If only the heat engine is considered then it is possible to neglect the temperature difference of the heat supply: this is an *external* irreversibility. Furthermore, to simplify the analysis it will be assumed that $T_L' = T_0 = T_4$.

However, it is not possible to neglect the varying temperature of heat rejection, because the engine is rejecting available energy to the surroundings. If the cycle is reversible, i.e. 1–2–3s–4–1, then the rational efficiency of the cycle is

$$\eta_R = \frac{W_{net}}{b_2 - b_1} \tag{4.17}$$

If the cycle shown is a Joule (gas turbine) cycle then

$$W_{net} = h_2 - h_1 - (h_{3s} - h_4) = b_2 - b_1 + T_0(s_2 - s_1) - \{ b_{3s} - b_4 + T_0(s_{3s} - s_4) \}$$
$$= b_2 - b_1 - (b_{3s} - b_4) \tag{4.18}$$

Hence, the net work is made up of the maximum net work supplied and the maximum net work rejected, i.e.

$$W_{net} = \{ \hat{W}_{net} \}_{supplied} - [\hat{W}_{net}]_{rejected} \tag{4.19}$$

These terms are defined by areas 1–2–3s–5–4–1 and 4–3s–5–4 respectively in Fig 4.1(c). Thus the rational efficiency of an engine operating on a Joule cycle is

$$\eta_R = \frac{b_2 - b_1 - (b_{3s} - b_4)}{b_2 - b_1} = 1 - \frac{b_{3s} - b_4}{b_2 - b_1} \tag{4.20}$$

Equation (4.20) shows that it is never possible for an engine operating on a cycle in which the temperature of energy rejection varies to achieve a rational efficiency of 100%. This is simply because there will always be energy available to produce work in the rejected heat.

If the cycle is not reversible, e.g. if the expansion is irreversible, then the cycle is defined by 1–2–3–4–1, and there is an increase in entropy from 2 to 3. Equation (4.18) then becomes

$$W_{net} = h_2 - h_1 - (h_3 - h_4) = b_2 - b_1 + T_0(s_2 - s_1) - \{ b_3 - b_4 + T_0(s_3 - s_4) \}$$
$$= \underbrace{b_2 - b_1}_{\substack{area \\ 1-2-3s-5-4-1}} - \underbrace{(b_3 - b_4)}_{\substack{area \\ 4-3-6-4}} - \underbrace{T_0(s_0 - s_{3s})}_{\substack{area \\ 5-6-7-8-5}} \tag{4.21}$$

Equation (4.21) shows that the irreversibility of the expansion process reduces the net work significantly by (i) increasing the amount of exergy rejected, and (ii) increasing the irreversibility of the cycle. The rational efficiency of this cycle is

$$\eta_R = \frac{(b_3 - b_4) - T_0(s_3 - s_{3s})}{b_2 - b_1} \tag{4.22}$$

The irreversible cycle can be seen to be less efficient than the reversible one by comparing eqns (4.20) and (4.22). In the case shown $b_2 - b_1$ is the same for both cycles, but $b_3 - b_4 > b_{3s} - b_4$, and in addition the irreversibility $T_0(s_3 - s_{3s})$ has been introduced.

4.3 Rankine cycle

It was stated above that the efficiency of a cycle is often evaluated neglecting the irreversibility of the heat transfer to the system. Such a situation can be seen on the steam plant shown in Fig 4.2(a), which can be represented by the simplified diagram shown in Fig 4.2(b), which shows the heat engine contained in system B of Fig 4.2(a).

Now, consider the passage of the working fluid through system A. It enters system A with a state defined by 3 on the Rankine cycle (see Fig 4.3), and leaves system A with a state defined by 2. The usual definition of thermal efficiency given in eqn (4.1) results in

$$\eta_{th} = \frac{w_{net}}{h_3 - h_2} \tag{4.23}$$

where $w_{net} =$ specific net work output from the cycle.

However, from the definition of exergy, the maximum work (per unit mass of steam) that is obtainable from the fluid is

$$\hat{w}_{net} = b_3 - b_2 = (h_3 - h_2) - T_0(s_3 - s_2) \tag{4.7}$$

Examination of eqn (4.7) shows that it consists of two terms: a difference of the enthalpies between states 3 and 2, and the product of the dead state temperature, T_0, and the difference of the entropies between states 3 and 2. The first term is obviously equal to the energy added between states 2 and 3, and is q_{in}. Considering the other term, the energy rejected to the cold reservoir is

$$q_{out} = h_4 - h_1 = T_1(s_4 - s_1) \tag{4.24}$$

If the condenser were of infinite size then the temperature at 1 could be equal to the environmental temperature, T_0. Also, if the feed pump and the turbine were both reversible then processes 3–4 and 1–2 would be isentropic and eqn (4.24) could be reduced to

$$q_{out} = h_4 - h_1 = T_0(s_3 - s_2) \tag{4.25}$$

Hence, for an internally reversible engine the maximum net work output is the sum of the heat supplied and the heat rejected, as shown previously simply from considerations of energy. If the cycle was not reversible due to inefficiencies in the turbine and feed pump, then the difference between the entropies at states 4 and 1 would be greater than the differences at states 2 and 3, and q_{out} would be greater than in the ideal case. Hence, the cycle would be less efficient.

The thermal efficiency of an ideal Rankine cycle (i.e. one in which the turbine and feed pump are both isentropic) is given by

$$\eta_{Rankine} = \frac{\hat{w}_{net}}{h_3 - h_2} = \frac{b_3 - b_2}{h_3 - h_2} \tag{4.26}$$

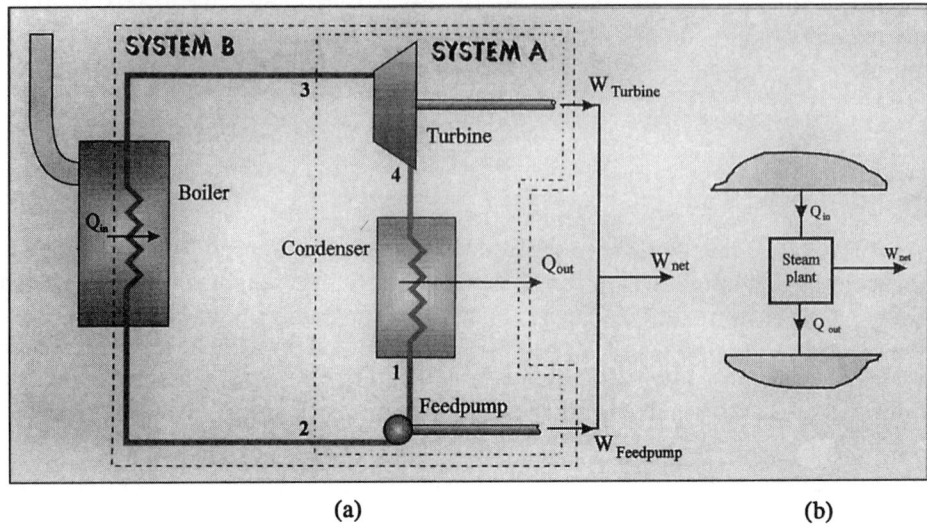

(a)   (b)

Fig. 4.2 Steam turbine power plant

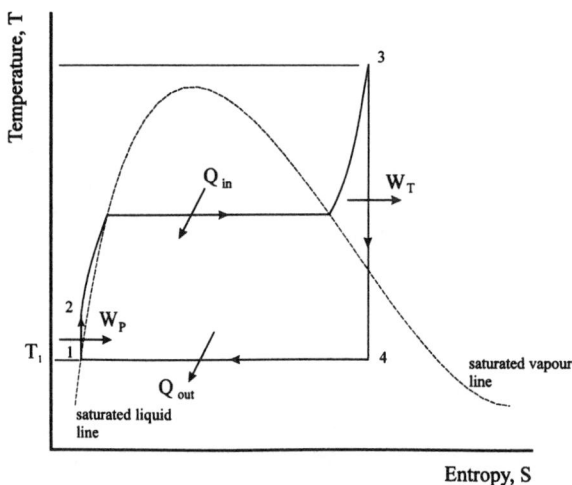

Fig. 4.3 Rankine cycle for steam turbine plant

If the turbine and feed pump of the cycle are not isentropic then the work output will be less than that of the ideal Rankine cycle and it is possible to define the *efficiency ratio* as

$$\text{Efficiency ratio} = \frac{\eta_{\text{cycle}}}{\eta_{\text{Rankine}}} \tag{4.27}$$

Substituting from eqns (4.23) and (4.26) gives

$$\text{Efficiency ratio} = \frac{W_{\text{net}}}{h_3 - h_2} \times \frac{h_3 - h_2}{b_3 - b_2} = \frac{W_{\text{net}}}{b_3 - b_2} \tag{4.28}$$

$$= \text{rational efficiency, } \eta_R$$

This means that a steam turbine which operates on a *reversible* Rankine cycle will have a rational efficiency of 100%. If there are any irreversibilities then the rational efficiency will be less than 100%. Rational efficiency shows the scope for improving the device within the constraints of, say, peak pressure and temperature. If the rational efficiency is low then the efficiency can be improved significantly, whereas if it is high then not much improvement is possible.

It has to be remembered that these definitions of rational efficiency for a steam plant cycle are based on the dead state temperature being made equal to the condenser temperature. If another temperature is chosen then the rational efficiency will be reduced as shown in eqn (4.22). It will be shown that this results in a rational efficiency of unity for an ideal (internally reversible) Rankine cycle, and such an approach can be used to indicate how far an actual (irreversible) steam plant cycle falls short of the yardstick set by the reversible one. However, this approach also masks the 'cost' of external irreversibilities between the working fluid in the condenser and the true dead state of environmental conditions. These effects are discussed in the examples.

When the term rational efficiency is applied to an air-standard cycle it is never possible to achieve a value of unity because the temperature at which energy is rejected is not constant. This will also be considered in the examples by reference to the Joule cycle for a gas turbine.

4.4 Examples

Q.1   Steam turbine cycles

A steam turbine operates on a basic Rankine cycle with a maximum pressure of 20 bar and a condenser pressure of 0.5 bar. Evaluate the thermal efficiency of the plant. Calculate the maximum net work available from the cycle, and evaluate the rational efficiency of the cycle.

Solution

The $T-s$ diagram for the Rankine cycle is shown in Fig 4.4. The parameters for the state points on cycle $1-2-3-4-6-1$ will now be evaluated.

Conditions at 4

$$p_4 = 20 \text{ bar}; \quad x = 1$$
$$t_s = 212.4°\text{C}; \quad h_g = 2799 \text{ kJ/kg}; \quad s_g = 6.340 \text{ kJ/kg K}$$

Conditions at 6

$$p_6 = 0.5 \text{ bar}; \quad s_6 = 6.340 \text{ kJ/kg K}$$
$$t_s = 81.3°\text{C}; \quad h_g = 2645 \text{ kJ/kg}; \quad h_f = 340 \text{ kJ/kg}$$
$$s_g = 7.593 \text{ kJ/kg K}; \quad s_f = 1.091 \text{ kJ/kg K}$$

Thus

$$x_6 = \frac{6.340 - 1.091}{7.593 - 1.091} = 0.8073$$

$$h_6 = x h_g + (1 - x) h_f = 0.8073 \times 2645 + (1 - 0.8073) \times 340 = 2200.8 \text{ kJ/kg}$$

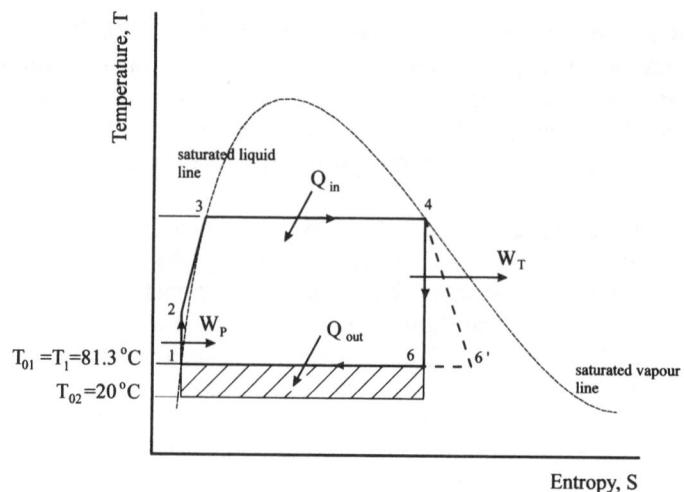

Fig. 4.4 Temperature–entropy diagram for a basic Rankine cycle

Conditions at 1

$$p_1 = 0.5 \text{ bar}; \ s_1 = 1.091 \text{ kJ/kg K}; \ h_1 = 340 \text{ kJ/kg}$$
$$v_1 = v_f = 0.001029 \text{ m}^3/\text{kg}$$

Conditions at 2

$$p_2 = 20 \text{ bar}; \ s_2 = 1.091 \text{ kJ/kg K}$$
$$v_2 = 0.001157 + 0.2 \times (-0.0009 \times 10^{-2}) = 0.001155 \text{ m}^3/\text{kg}$$

Hence feed pump work,

$$w_P = w_{12} = -v\,dp = -dh_{12}$$
$$= -(0.001029 + 0.001155)/2 \times 19.5 \times 100 = -2.11 \text{ kJ/kg}$$
$$h_2 = h_1 + dh_{12} = 340 + 2.11 = 342.1 \text{ kJ/kg}$$

Work done by turbine,

$$w_T = h_4 - h_6 = 2799 - 2200.8 = 598.2 \text{ kJ/kg K}$$

Thermal efficiency of the cycle is

$$\eta_{th} = \frac{w_{net}}{q_{in}} = \frac{w_T + w_P}{h_4 - h_2} = \frac{598.2 - 2.11}{2799 - 342.1} = 0.243$$

The maximum net work available can be evaluated from the change of exergy between state points 4 and 2. Hence

$$\hat{w}_{net} = b_4 - b_2$$

It is necessary to define a dead state condition. This is arbitrary, and in this case will be taken as the condition at 1. Hence, $p_0 = 0.5$ bar, $T_0 = 81.3 + 273 = 354.3$ K. Thus

$$b_4 = h_4 - T_0 s_4 - (h_0 - T_0 s_0) = 2799 - 354.3 \times 6.340 - a_0 = 552.7 - a_0 \text{ kJ/kg}$$
$$b_2 = h_2 - T_0 s_2 - (h_0 - T_0 s_0) = 342.1 - 354.3 \times 1.091 - a_0 = -44.4 - a_0 \text{ kJ/kg}$$

Hence

$$\hat{w}_{net} = 552.7 - (-44.4) = 597.1 \text{ kJ/kg}$$

Thus, the rational efficiency is

$$\eta_R = \frac{w_{net}}{b_4 - b_2} = \frac{w_{net}}{\hat{w}_{net}} = \frac{598.2 - 2.11}{597.1} = 1.00$$

In this case the rational efficiency is equal to unity because the turbine and feed pump have isentropic efficiencies of 100%, and it was assumed that the temperature of the working fluid in the condenser was equal to the dead state (ambient) temperature. Hence, although the cycle is not very efficient, at 24.3%, there is no scope for improving it *unless* the operating conditions are changed.

It is possible to evaluate the rational efficiency of a steam plant operating on a Rankine cycle in which the condenser temperature is above the ambient temperature. If the dead state temperature in the previous example was taken as 20°C, rather than 81.3°C, then the following values would be obtained:

$$b_4 = h_4 - T_0 s_4 - a_0 = 2799 - 293 \times 6.340 - a_0 = 941.4 - a_0 \text{ kJ/kg}$$
$$b_2 = h_2 - T_0 s_2 - a_0 = 342.1 - 293 \times 1.091 - a_0 = 22.4 - a_0 \text{ kJ/kg}$$

Hence

$$\hat{w}_{net} = 941.4 - 22.4 = 919.0 \text{ kJ/kg}$$

Thus, the rational efficiency is

$$\eta_R = \frac{w_{net}}{b_4 - b_2} = \frac{w_{net}}{\hat{w}_{net}} = \frac{598.2 - 2.11}{919.1} = 0.649$$

This result shows that irreversibilities in the condenser producing a temperature drop of 81.3°C to 20°C would reduce the potential efficiency of the powerplant significantly. Basically this irreversibility is equivalent to a loss of potential work equal to the area of the $T-s$ diagram bounded by the initial dead state temperature of 81.3°C and the final one of 20°C and the entropy difference, as shown in Fig 4.4, i.e. $(T_{0_1} - T_{0_2})(s_4 - s_2)$, which is equal to $(354.3 - 293) \times (6.340 - 1.09) = 321.8 \text{ kJ/kg}$.

Q.2   Steam turbine cycles

Re-evaluate the parameters for the steam plant in Q.1 based on $T_0 = 81.3$°C if (a) the isentropic efficiency of the turbine is 80%; (b) the isentropic efficiency of the feed pump is 70%; (c) the efficiency of the components is the combination of those given in (a) and (b).

Solution

(a)   Turbine efficiency, $\eta_T = 80$%. If the turbine efficiency is 80% then the work output of the turbine becomes

$$w_T = \eta_T (w_T)_{isen} = 0.80 \times 598.2 = 478.6 \text{ kJ/kg}$$

Hence, the thermal efficiency of the cycle is

$$\eta_{th} = \frac{w_{net}}{q_{in}} = \frac{w_T + w_P}{h_4 - h_2} = \frac{478.6 - 2.11}{2799 - 342.1} = 0.194$$

The rational efficiency is

$$\eta_R = \frac{[w_{net}]_{actual}}{b_4 - b_2} = \frac{[w_{net}]_{actual}}{\hat{w}_{net}} = \frac{478.6 - 2.11}{597.1} = 0.80$$

The rational efficiency has been significantly reduced by the inefficiency of the turbine, and the rational efficiency is approximately equal to the isentropic efficiency of the turbine.

(b) Feed pump efficiency, $\eta_P = 70\%$. The feed pump work becomes

$$w_P = \frac{(w_P)_{isen}}{\eta_P} = -\frac{2.11}{0.7} = -3.01 \text{ kJ/kg}$$

$$h_{2'} = h_1 + dh_{12'} = 340 + 3.01 = 343 \text{ kJ/kg}$$

The thermal efficiency of the cycle becomes

$$\eta_{th} = \frac{w_{net}}{q_{in}} = \frac{w_T + w_P}{h_4 - h_{2'}} = \frac{598.2 - 3.01}{2799 - 343} = 0.242$$

The exergy at 2' may be evaluated approximately by assuming that the entropy does not change significantly over the pumping process, i.e. $s_{2'} = s_1$. Hence

$$b_{2'} = h_{2'} - T_0 s_{2'} - a_0 = 343 - 354.3 \times 1.091 - a_0 = -43.5 - a_0 \text{ kJ/kg}$$

The rational efficiency is

$$\eta_R = \frac{[w_{net}]_{actual}}{b_4 - b_{2'}} = \frac{[w_{net}]_{actual}}{\hat{w}_{net}} = \frac{598.2 - 3.01}{596.2} = 0.998$$

The reduction in thermal efficiency is small, and the rational efficiency is almost 1. Hence, the Rankine cycle is not much affected by inefficiencies in the feed pump.

(c) Turbine efficiency $\eta_T = 80\%$, feed pump efficiency $\eta_P = 70\%$. The thermal efficiency is

$$\eta_{th} = \frac{w_{net}}{q_{in}} = \frac{w_T + w_P}{h_4 - h_{2'}} = \frac{478.6 - 3.01}{2799 - 343} = 0.194$$

The rational efficiency is

$$\eta_R = \frac{[w_{net}]_{actual}}{b_4 - b_{2'}} = \frac{[w_{net}]_{actual}}{\hat{w}_{net}} = \frac{478.6 - 3.01}{596.2} = 0.797$$

Q.3   Steam turbine cycles

The steam plant in Q.2 is modified so that the steam is superheated before entering the turbine so that the exit conditions from the turbine of the ideal cycle are dry saturated. Evaluate the parameters for the steam plant if (a) the isentropic efficiency of the turbine is 80%; (b) the isentropic efficiency of the feed pump is 70%; (c) the efficiency of the components is the combination of those given in (a) and (b). Assume that $t_0 = 81.3°C$. This cycle is shown in Fig 4.5.

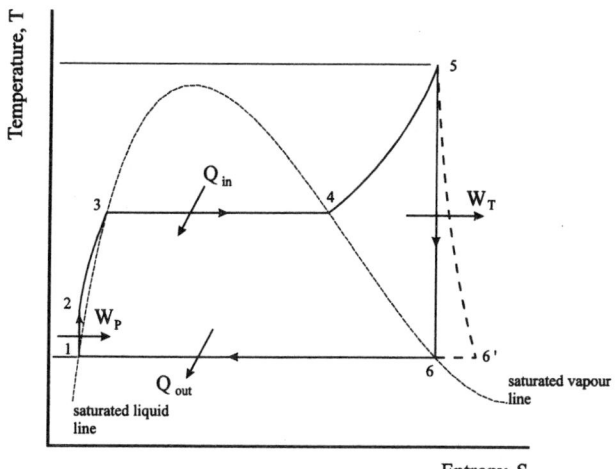

Fig. 4.5 Temperature–entropy diagram for a Rankine cycle with superheat

Conditions at 6

$p_6 = 0.5$ bar

$t_s = 81.3°C$; $h_g = 2645$ kJ/ kg; $h_f = 340$ kJ/kg

$s_g = 7.593$ kJ/kg K; $s_f = 1.091$ kJ/kg K

$h_6 = 2645$ kJ/kg; $s_6 = 7.593$ kJ/kg K

Conditions at 5

$p_5 = 20$ bar; $s_5 = s_6 = 7.593$ kJ/kg K

Hence

$$t_s = 500 + \frac{7.593 - 7.431}{7.701 - 7.431} \times 100 = 560°C$$

$$h_5 = 3467 + 0.6 \times (3690 - 3467) = 3601 \text{ kJ/kg}$$

Work done by turbine,

$$w_T = h_5 - h_6 = 3601 - 2645 = 956 \text{ kJ/kg K}$$

Thermal efficiency of the cycle is

$$\eta_{th} = \frac{w_{net}}{q_{in}} = \frac{w_T + w_P}{h_5 - h_2} = \frac{956 - 2.11}{3601 - 342.1} = 0.292$$

$$b_5 = h_5 - T_0 s_5 - a_0 = 3601 - 354.3 \times 7.593 - a_0 = 910.8 - a_0 \text{ kJ/kg}$$

$$b_2 = h_2 - T_0 s_2 - a_0 = 342.1 - 354.3 \times 1.091 - a_0 = -44.4 - a_0 \text{ kJ/kg}$$

The rational efficiency is

$$\eta_R = \frac{[w_{net}]_{actual}}{b_4 - b_2} = \frac{[w_{net}]_{actual}}{\hat{w}_{net}} = \frac{956 - 2.11}{955.2} \approx 1.00$$

(a)   Turbine efficiency, $\eta_T = 80\%$.

$$w_T = \eta_T (w_T)_{\text{isen}} = 0.80 \times 956 = 764.8 \text{ kJ/kg}$$

Hence, the thermal efficiency of the cycle is

$$\eta_{th} = \frac{w_{net}}{q_{in}} = \frac{w_T + w_P}{h_4 - h_2} = \frac{764.8 - 2.11}{3601 - 342.1} = 0.234$$

The rational efficiency is

$$\eta_R = \frac{[w_{net}]_{\text{actual}}}{b_5 - b_2} = \frac{[w_{net}]_{\text{actual}}}{\hat{w}_{net}} = \frac{764.8 - 2.11}{955.2} = 0.799$$

Hence, again, the effect of the turbine efficiency is to reduce the rational efficiency by an amount almost equal to its isentropic efficiency.

(b)   Feed pump efficiency, $\eta_P = 70\%$. The effect on the feed pump work is the same as above, and the thermal efficiency of the cycle becomes

$$\eta_{th} = \frac{w_{net}}{q_{in}} = \frac{w_T + w_P}{h_4 - h_{2'}} = \frac{956 - 3.01}{3601 - 343} = 0.295$$

The rational efficiency is

$$\eta_R = \frac{[w_{net}]_{\text{actual}}}{b_5 - b_{2'}} = \frac{[w_{net}]_{\text{actual}}}{\hat{w}_{net}} = \frac{956 - 3.01}{910.8 - (-44.4)} = 0.998$$

Again, the reduction in thermal efficiency is small, and the rational efficiency is almost 1.

(c)   Turbine efficiency, $\eta_T = 80\%$; feed pump efficiency, $\eta_P = 70\%$. The thermal efficiency is

$$\eta_{th} = \frac{w_{net}}{q_{in}} = \frac{w_T + w_P}{h_5 - h_{2'}} = \frac{764.8 - 3.01}{3601 - 343} = 0.234$$

The rational efficiency is

$$\eta_R = \frac{[w_{net}]_{\text{actual}}}{b_5 - b_{2'}} = \frac{[w_{net}]_{\text{actual}}}{\hat{w}_{net}} = \frac{764.8 - 3.01}{955.2} = 0.798$$

It can be seen that both the basic and superheated Rankine cycles are equally affected by inefficiencies in the individual components. The rational efficiency is an estimate of how close the cycle comes to the internally reversible cycle.

Q.4   Gas turbine cycle

A gas turbine operating on an ideal Joule cycle has a pressure ratio of 20 : 1 and a peak temperature of 1200 K. Calculate the net work output, the maximum work output, the thermal efficiency, and the rational efficiency of the cycle. Assume that the working fluid is air with a value of $\kappa = 1.4$ and a specific gas constant $R = 0.287$ kJ/kg K. The inlet conditions at 1 are 1 bar and 300 K, and these should be taken as the dead state conditions also.

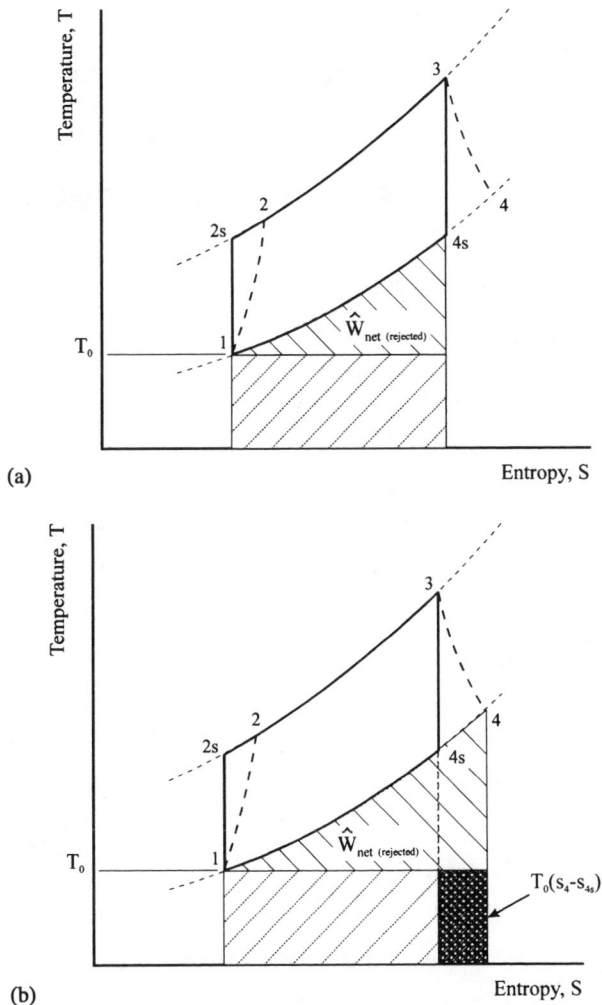

Fig. 4.6 (a) Temperature–entropy diagram for Joule cycle. (b) Temperature–entropy diagram for Joule cycle with irreversible compression and expansion

Solution

The ideal Joule cycle is depicted by 1–2s–3–4s–1 in Fig 4.6(a). The relationship between entropy and the primitive properties for an ideal gas is

$$s - s_0 = c_p \ln \frac{T}{T_0} - R \ln \frac{p}{p_0}$$

From the parameters given,

$$c_p = \frac{\kappa R}{\kappa - 1} = \frac{1.4 \times 0.287}{0.4} = 1.0045 \,\text{kJ/kg K}.$$

The compression process from 1 to 2s is isentropic, and hence

$$T_{2s} = T_1 \left(\frac{p_2}{p_1}\right)^{(\kappa-1)/\kappa} = 300 \times 20^{0.4/1.4} = 706.1 \text{ K}$$

The isentropic work done in the compressor is

$$w_{C_{isen}} = -dh_{12s} = c_P(T_1 - T_{2s}) = 1.0045 \times (706.1 - 300) = 407.9 \text{ kJ/kg}$$

The energy added to the cycle is

$$q_{2s3} = c_p(T_3 - T_{2s}) = 1.0045 \times (1200 - 706.1) = 496.1 \text{ kg}$$

For an isentropic expansion from 3 to 4s

$$T_{4s} = T_3 \left(\frac{p_{4s}}{p_3}\right)^{(\kappa-1)/\kappa} = 1200 \times \left(\frac{1}{20}\right)^{0.4/1.4} = 509.9 \text{ K}$$

Hence the isentropic turbine work is

$$w_{T_{isen}} = c_p (T_3 - T_{4s}) = 1.0045 \times (1200 - 509.9) = 693.2 \text{ kJ/kg}$$

The net work from the cycle is $w_{net} = w_T + w_C = 693.2 + (-407.9) = 285.3$ kJ/kg, and the thermal efficiency is

$$\eta_{th} = \frac{w_{net}}{q_{2s3}} = \frac{285.3}{496.1} = 0.575.$$

This value is equal to the standard expression for the efficiency of a Joule cycle,

$$\eta_{th} = 1 - \frac{1}{r_p^{(\kappa-1)/\kappa}}$$

The maximum net work output is defined by

$$\hat{w}_{net} = b_3 - b_2$$

Now, the values of exergy are defined, for a perfect gas, as

$$b = (h - T_0 s) - (h_0 - T_0 s_0) = (h - h_0) - T_0(s - s_0)$$

$$= c_p \left\{ (T - T_0) - T_0 \left(\ln\frac{T}{T_0} - \frac{\kappa-1}{\kappa} \ln\frac{p}{p_0} \right) \right\}$$

Thus, along an isobar

$$b = c_p \left\{ (T - T_0) - T_0 \ln\frac{T}{T_0} \right\}$$

Since both 2s and 3 are at the same pressure

$$b_3 - b_{2s} = c_p \left\{ (T_3 - T_0) - T_0 \ln\frac{T_3}{T_0} - \left[(T_2 - T_0) - T_0 \ln\frac{T_2}{T_0} \right] \right\} = c_p \left\{ (T_3 - T_2) - T_0 \ln\frac{T_3}{T_2} \right\}$$

Thus

$$\hat{w}_{net} = 1.0045 \times \left\{ (1200 - 706.1) - 300 \ln\frac{1200}{706.1} \right\} = 336.3 \text{ kJ/kg}$$

The rational efficiency of the cycle based on these dead state conditions is

$$\eta_R = \frac{w_{net}}{\hat{w}_{net}} = \frac{285.3}{336.3} = 0.848$$

This means that the ideal Joule cycle is only capable of extracting 84.8% of the maximum net work from the working fluid, whereas under similar conditions (with the dead state temperature defined as the minimum cycle temperature) the Rankine cycle had a rational efficiency of 100%. The reason for this is that the energy rejected by the Joule cycle, from 4s to 1, still has the potential to do work. The maximum net work obtainable from the rejected energy is $b_{4s} - b_1$. This is equal to

$$[\hat{w}_{net}]_{rejected} = b_{4s} - b_1 = c_p \left\{ (T_{4s} - T_1) - T_0 \ln \frac{T_{4s}}{T_1} \right\}$$

$$= 1.0045 \times \left\{ (509.9 - 300) - 300 \ln \frac{509.9}{300} \right\} = 50.99 \,\text{kJ/kg}$$

This term is shown in Fig 4.6(a), and is energy which is unavailable for the production of work.

If the efficiencies of the turbine and compressor are not 100% then the diagram is shown in Fig 4.6(b). It can be seen that the rejected energy is larger in this case.

Q.5   *Gas turbine cycle*

For the gas turbine cycle defined in Q.4, calculate the effect of (a) a turbine isentropic efficiency of 80%; (b) a compressor isentropic efficiency of 80%; and (c) the combined effect of both inefficiencies.

Solution

(a)   Turbine isentropic efficiency, $\eta_T = 80\%$. This will affect the work output of the turbine in the following way:

$$w_T = \eta_T w_{T_{isen}} = 0.8 \times 693.2 = 554.6 \,\text{kJ/kg}$$

Hence

$$w_{net} = w_T + w_C = 554.6 + (-407.9) = 146.7 \,\text{kJ/kg}$$

and

$$\eta_{th} = \frac{w_{net}}{q_{2s3}} = \frac{146.7}{496.1} = 0.296$$

The temperature after the turbine is $T_4 = T_3 - \eta_T \Delta T_{isen} = 1200 - 0.8 \times 690.1 = 647.9$ K, and the entropy at 4, related to 1, is

$$s_4 - s_1 = c_p \ln \frac{T_4}{T_1} - R \ln \frac{p_4}{p_1} = 1.0045 \ln \frac{647.9}{300} = 0.7734 \,\text{kJ/kg K}$$

Hence, the maximum net work rejected is

$$\hat{w}_{\text{net rejected}} = b_4 - b_1 = c_\text{p}\left\{(T_4 - T_1) - T_0 \ln \frac{T_4}{T_1}\right\}$$

$$= 1.0045 \times \left\{(647.9 - 300) - 300 \ln \frac{647.9}{300}\right\} = 117.4 \,\text{kJ/kg}$$

This is a significant increase over the rejected potential work for the ideal Joule cycle. However, in addition to the inefficiency of the turbine increasing the maximum net work rejected, it also increases the unavailable energy by the irreversibility, $T_0(s_4 - s_{4s})$, as shown in Fig 4.6(b). This is equal to

$$T_0(s_4 - s_{4s}) = 300 \times (0.7734 - 0.5328) = 72.2 \,\text{kJ/kG}$$

Since the maximum net work is the same for this case as the ideal one then the net work is given by

$$w_{\text{net}} = \hat{w}_{\text{net}} - [\hat{w}_{\text{net}}]_{\text{rejected}} - T_0(s_4 - s_{4s}) = 336.3 - 117.4 - 72.2 = 146.7 \,\text{kJ/kg}$$

This is the same value as found from the more basic calculation.
The rational efficiency of this cycle is

$$\eta_\text{R} = \frac{w_{\text{net}}}{\hat{w}_{\text{net}}} = \frac{146.7}{336.3} = 0.436$$

(b) Compressor isentropic efficiency, $\eta_\text{C} = 80\%$. The work done in the compressor is

$$w_\text{C} = \frac{w_{C_{\text{isen}}}}{\eta_\text{C}} = 509.9 \,\text{kJ/kg}$$

Temperature after compressor, at 2, is

$$T_2 = T_1 + \frac{w_\text{C}}{c_\text{p}} = 300 + \frac{509.9}{1.0045} = 807.6 \,\text{K}$$

The energy added to the cycle is

$$q_{23} = c_\text{p}(T_3 - T_2) = 1.0045 \times (1200 - 807.6) = 394.2 \,\text{kJ/kg}$$

The net work from the cycle is $w_{\text{net}} = w_\text{T} + w_\text{C} = 693.2 + (-509.9) = 183.3 \,\text{kJ/kg}$, and the thermal efficiency is

$$\eta_{\text{th}} = \frac{w_{\text{net}}}{q_{23}} = \frac{183.3}{394.2} = 0.465.$$

The maximum net work output is defined by

$$\hat{w}_{\text{net}} = b_3 - b_2$$

Since both 2 and 3 are at the same pressure

$$b_3 - b_2 = c_\text{p}\left\{(T_3 - T_2) - T_0 \ln \frac{T_3}{T_2}\right\}$$

Thus

$$\hat{w}_{net} = 1.0045 \times \left\{ (1200 - 807.6) - 300 \ln \frac{1200}{807.6} \right\} = 273.6 \, kJ/kg$$

The rational efficiency of the cycle based on these dead state conditions is

$$\eta_R = \frac{w_{net}}{\hat{w}_{net}} = \frac{183.3}{273.6} = 0.670$$

The net work output of the device is made up of

$$w_{net} = \hat{w}_{net} - [\hat{w}_{net}]_{rejected} - T_0(s_2 - s_{2s}) = 273.6 - 50.9 - 40.2 = 182.5 \, kJ/kg$$

In this case, the inefficiency of the compressor has introduced a quantity of unavailable energy $T_0(s_2 - s_{2s})$, which is depicted in Fig 4.7.

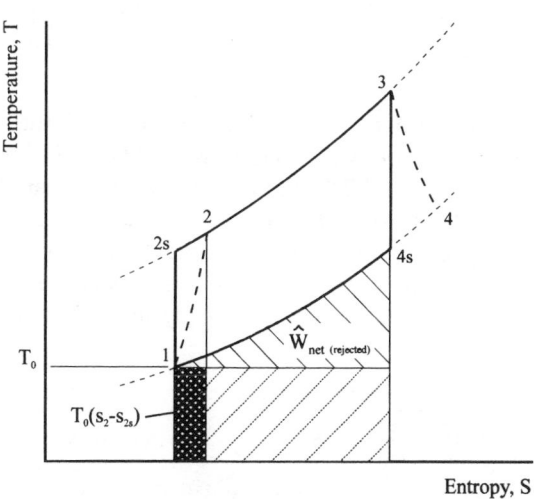

Fig. 4.7 The effect of inefficiency in the compressor

(c) Compressor isentropic efficiency $\eta_C = 80\%$, and turbine isentropic efficiency $\eta_T = 80\%$. This cycle combines the two inefficiencies considered above. Hence, the net work output of the cycle is

$$w_{net} = w_T + w_C = 554.6 + (-509.9) = 44.7 \, kJ/kg$$

Net heat addition $q_{23} = 394.2 \, kJ/kg$. Hence, thermal efficiency is

$$\eta_{th} = \frac{w_{net}}{q_{23}} = \frac{44.7}{394.2} = 0.113$$

The net work output of this cycle is made up in the following way:

$$w_{net} = \hat{w}_{net} - [\hat{w}_{net}]_{rejected} - T_0(s_4 - s_{4s}) - T_0(s_2 - s_{2s})$$
$$= 273.6 - 117.4 - 72.2 - 40.2 = 43.8 \, kJ/kg$$

This equation shows how the inefficiencies reduce the net work obtainable from the cycle. However, an additional quantity of net work has been lost which is not evident from the equation, and that is the reduction in maximum net work caused by the inefficiency of the compressor. The latter causes the temperature after compression to be higher than the isentropic value, and hence less fuel is necessary to reach the maximum cycle temperature.

4.5 Concluding remarks

The concept of a Second Law, or rational, efficiency has been introduced. This provides a better measure of how closely a particular power-producing plant approaches its maximum achievable efficiency than does the conventional 'First Law' thermal efficiency.

It has been shown that devices in which the working fluid changes phase (e.g. steam plant) can achieve rational efficiencies of 100 per cent, whereas those which rely on a single phase fluid (e.g. gas turbines) can never approach such a high value. The effect of irreversibility on rational efficiency has also been shown.

PROBLEMS

1 In a test of a steam powerplant (Fig P4.1), the measured rate of steam supply was 7.1 kg/s when the net rate of work output was 5000 kW. The condensate left the condenser as saturated liquid at 38°C and the superheated steam leaving the boiler was at 10 bar and 300°C. Neglecting the change in state of the feed water in passing through the feed pump, and taking the environment temperature as being equal to the saturation temperature of the condensate in the condenser, calculate the *rational efficiency* of the work-producing steam circuit (system A). Also calculate the *thermal efficiency* of the cyclic plant.

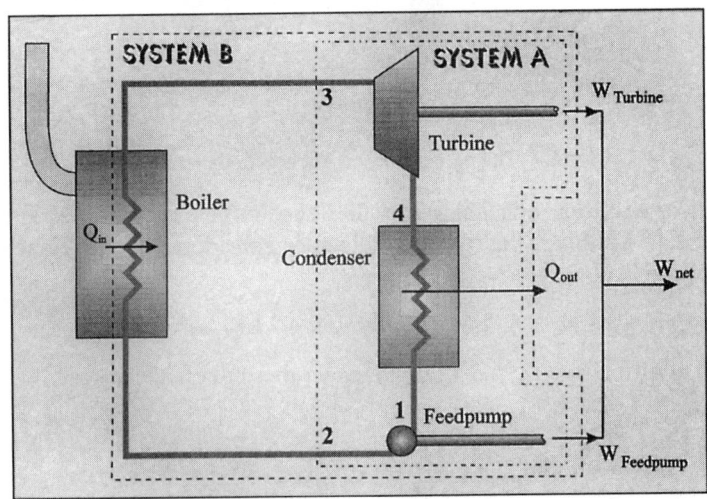

Fig. P4.1 Steam turbine powerplant

[83.3%; 24.3%]

2 Again neglecting the change in state of the feed water in passing through the feed pump, calculate the *thermal efficiency* of an ideal Rankine cycle in which the conditions of the fluid at inlet to and exit from the boiler are the same as those of the plant in Q.1. Hence confirm that the efficiency ratio of that plant is equal to the rational efficiency of the work-producing steam circuit calculated in Q.1.

[29.2%; 83.3%]

3 Neglecting the temperature rise of the feed water in passing through the feed pump, calculate the *mean temperature of heat reception* in the ideal Rankine cycle of Q.1. Hence make a second, alternative calculation of the *thermal efficiency* of this ideal Rankine cycle.

[440 K; 29.2%]

4 Air enters the compressor of a simple gas turbine at a pressure of 1 bar and a temperature of 25°C. The compressor has a pressure ratio of 15 and an isentropic efficiency of 85%, and delivers air to a combustion chamber which is supplied with methane (CH_4) at 25°C. The products of combustion leave the chamber at 1450 K and suffer a pressure loss of 5% in passing through it. They are then expanded in a turbine with an isentropic efficiency of 90% to a pressure of 1.05 bar. Calculate (a) the air–fuel ratio of the engine; (b) the power output per unit mass flow of air, and (c) the rational efficiency of the engine. What is the maximum work per unit mass of exhaust gas that could be obtained if the ambient conditions are 1 bar and 25°C?

The enthalpy of reaction based on 25°C is $-50\,000$ kJ/kg, and the Gibbs energy of reaction is $-51\,000$ kJ/kg. The products of combustion can be treated as a perfect gas with $c_p = 1.2$ kJ/kg, and $\kappa = 1.35$.

[0.02; 372.9 kJ/kg; 36.6%; 260.4 kJ/kg]

5 A steam turbine operates on a superheated Rankine cycle. The pressure and temperature of the steam leaving the boiler are 10 bar and 350°C respectively. The specific steam consumption of the plant is 4.55 kg/kWh. The pressure in the condenser is 0.05 bar.

If the feed pump work may be neglected, calculate the thermal efficiency of the plant, the turbine isentropic efficiency, and evaluate the rational efficiency. Also calculate the mean temperature of reception of heat in the boiler and use this in conjunction with the condenser temperature to evaluate the thermal efficiency. Explain why the value calculated by this method is higher than that obtained previously.

[26.5%; 86%; 84.84%; 30.85%]

6 Figure P4.6 depicts a closed cycle gas turbine operating on the Joule cycle (i.e. constant pressure heat addition and rejection, and isentropic compression and expansion). Energy is added to the working fluid (air) by a heat exchanger maintained at 1250 K, and rejected to another heat exchanger maintained at 300 K. The maximum temperature of the working fluid is 1150 K and its minimum temperature is 400 K. The pressure ratio of the compressor is 5 : 1.

Evaluate the irreversibilities introduced by the heat transfer processes and calculate

(a) the First Law efficiency
$$\eta_1 = \frac{\text{work ouput}}{\text{energy addition}}$$

(b) the Second Law efficiency $\eta_{\mathrm{II}} = \dfrac{\text{work output}}{\text{availability of energy addition}}$

Assume $c_p = 1.005$ kJ/kg K, $\kappa = 1.4$ and the specific gas constant, $R = 0.287$ kJ/kg K.

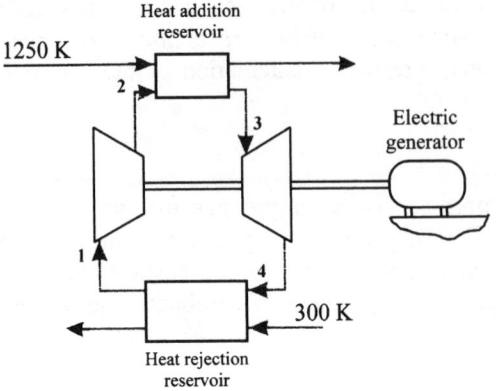

Fig. P4.6 Closed cycle gas turbine

Calculate the maximum efficiency that could be achieved from this system by modification of the heat exchangers.

[36.9%; 48.5%]

5

Efficiency of Heat Engines at Maximum Power

5.1 Efficiency of an internally reversible heat engine when producing maximum power output

The thermal efficiency of a Carnot cycle operating between high temperature (T_H) and low temperature (T_C) reservoirs is given by

$$\eta_{th} = 1 - \frac{T_C}{T_H} \tag{5.1}$$

This cycle is extremely idealised. It requires an ideal, reversible heat engine (internally reversible) but, in addition, the heat transfer from the reservoirs is also reversible (externally reversible). To achieve external reversibility it is necessary that the temperature difference between the reservoirs and the engine is infinitesimal, which means that the heat exchanger surface area must be very large or the time to transfer heat must be long. The former is limited by size and cost factors whilst the latter will limit the actual power output achieved for the engine. It is possible to evaluate the *maximum power output* achievable from an internally reversible (endoreversible) heat engine receiving heat irreversibly from two reservoirs at T_H and T_C. This will now be done, based on Bejan (1988).

Assume that the engine is a steady-flow one (e.g. like a steam turbine or closed cycle gas turbine): a similar analysis is possible for an intermittent device (e.g. like a Stirling engine). A schematic of such an engine is shown in Fig 5.1.

The reservoir at T_H transfers heat to the engine across a resistance and it is received by the engine at temperature T_1. In a similar manner, the engine rejects energy at T_2 but the cold reservoir is at T_C. It can be assumed that the engine itself is reversible and acts as a Carnot cycle device with

$$\eta_{th} = 1 - \frac{T_2}{T_1} \tag{5.2}$$

This thermal efficiency is less than the maximum achievable value given by eqn (5.1) because $T_2/T_1 > T_C/T_H$. The value can only approach that of eqn (5.1) if the temperature drops between the reservoirs and the engine approach zero.

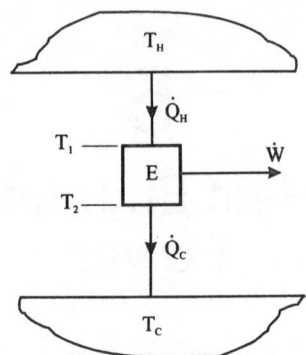

Fig. 5.1 Internally reversible heat engine operating between reservoirs at T_H and T_C

The heat transfer from the hot reservoir can be defined as

$$\dot{Q}_H = U_H A_H (T_H - T_1) \tag{5.3}$$

where U_H = heat transfer coefficient of hot reservoir (e.g. kW/m² K)
A_H = area of heat transfer surface of hot reservoir (e.g. m²)
and $\dot{Q}_H$ = rate of heat transfer (e.g. kW).

The heat transfer to the cold reservoir is similarly

$$\dot{Q}_C = U_C A_C (T_2 - T_C) \tag{5.4}$$

By the first law

$$\dot{W} = \dot{Q}_H - \dot{Q}_C \tag{5.5}$$

Now, the heat engine is internally reversible and hence the entropy entering and leaving it must be equal i.e.

$$\frac{\dot{Q}_H}{T_1} = \frac{\dot{Q}_C}{T_2} \tag{5.6}$$

This means that

$$\dot{W} = \dot{Q}_H \left(1 - \frac{\dot{Q}_C}{\dot{Q}_H} \right) = \dot{Q}_H \left(1 - \frac{T_2}{T_1} \right) \tag{5.7}$$

It is possible to manipulate these equations to give $\dot{W}$ in terms of T_H, T_C, $U_H A_H$, and the ratio $T_2/T_1 = \tau$. From eqn (5.4)

$$T_2 = \frac{\dot{Q}_C}{U_C A_C} + T_C \tag{5.8}$$

and, from eqn (5.6),

$$T_1 = \frac{\dot{Q}_H}{\dot{Q}_C} T_2 = \frac{\dot{Q}_H}{\dot{Q}_C} \left\{ \frac{\dot{Q}_C}{U_C A_C} + T_C \right\}$$

$$= \frac{1}{\tau} \left\{ \frac{\dot{Q}_C}{U_C A_C} + T_C \right\} \tag{5.9}$$

Hence

$$\dot{Q}_H = U_H A_H T_H - \frac{U_H A_H}{\tau}\left[\frac{\dot{Q}_C}{U_C A_C} + T_C\right] = U_H A_H T_H - \frac{U_H A_H \dot{Q}_H}{U_C A_C} - \frac{U_H A_H T_C}{\tau}$$

Rearranging gives

$$\frac{\dot{Q}_H}{U_H A_H T_H} = \left\{\frac{1 - \dfrac{T_C}{T_H}\dfrac{1}{\tau}}{1 + \dfrac{U_H A_H}{U_C A_C}}\right\} = \frac{\left(\tau - \dfrac{T_C}{T_H}\right)}{\tau\left(1 + \dfrac{U_H A_H}{U_C A_C}\right)} \tag{5.10}$$

and hence

$$\frac{\dot{W}}{U_H A_H T_H} = \frac{\left(\tau - \dfrac{T_C}{T_H}\right)}{\tau\left(1 + \dfrac{U_H A_H}{U_C A_C}\right)}(1 - \tau) \tag{5.11}$$

Thus the rate of work output is a function of the ratio of temperatures of the hot and cold reservoirs, the ratio of temperatures across the engine and the thermal resistances. The optimum temperature ratio across the engine (τ) to give maximum power output is obtained when

$$\frac{\partial \dot{W}}{\partial \tau} = 0$$

Differentiating eqn (5.11) with respect to τ gives

$$\frac{\partial \dot{W}}{\partial \tau} = \left(\frac{1}{1 + \dfrac{U_H A_H}{U_C A_C}}\right)\left[\frac{(1 - \tau)}{\tau} - \frac{(\tau - T_C/T_H)}{\tau} - \frac{(\tau - T_C/T_H)(1 - \tau)}{\tau^2}\right]$$

$$= \frac{1}{\tau}\left(\frac{1}{1 + \dfrac{U_H A_H}{U_C A_C}}\right)\left\{-\tau + \frac{1}{\tau}\frac{T_C}{T_H}\right\} \tag{5.12}$$

Hence $\partial \dot{W}/\partial \tau = 0$ when $\tau = \infty$ or $\tau^2 = T_C/T_H$.
   Considering only the non-trivial case, *for maximum work output*

$$\frac{T_2}{T_1} = \left(\frac{T_C}{T_H}\right)^{1/2}$$

This result has the effect of maximising the energy flow through the engine while maintaining the thermal efficiency ($\eta_{th} = 1 - T_2/T_1$) at a reasonable level. It compromises between the high efficiency ($\eta_{th} = 1 - T_C/T_H$) of the Carnot cycle (which produces zero energy flow rate) and the zero efficiency engine in which $T_2 = T_1$ (which produces high energy flow rates but no power).

Hence the efficiency of an internally reversible, ideal heat engine operating at maximum power output is

$$\eta_{th} = 1 - \left(\frac{T_C}{T_H}\right)^{1/2}$$

An example will be used to show the significance of this result.

Example

Consider a heat engine is connected to a hot reservoir at 1600 K and a cold one at 400 K. The heat transfer conductances (UA) are the same on both the hot and cold sides. Evaluate the high and low temperatures of the working fluid of the internally reversible heat engine for maximum power output; also calculate the maximum power.

Solution

Equation (5.11) gives the work rate (power) as

$$\frac{\dot{W}}{U_H A_H T_H} = \frac{\left(\tau - \dfrac{T_C}{T_H}\right)}{\tau\left(1 + \dfrac{U_H A_H}{U_C A_C}\right)}(1 - \tau)$$

For maximum power output $\tau = (T_C/T_H)^{1/2} = (400/1600)^{1/2}$. Then

$$\frac{\dot{W}}{U_H A_H} = 1600 \frac{\left(\dfrac{1}{2} - \dfrac{1}{4}\right)}{\dfrac{1}{2}(1 + 1)}(1 - \frac{1}{2}) = \frac{1600}{8} = 200 \text{ units}$$

and, from eqn (5.10)

$$\frac{\dot{Q}_H}{U_H A_H} = \frac{\dot{W}}{U_H A_H(1 - \tau)} = 400 \text{ units}$$

Then

$$\frac{\dot{Q}_C}{U_H A_H} = \frac{\dot{Q}_C}{U_C A_C} = \frac{1}{U_H A_H}(\dot{Q}_H - \dot{W}) = 200 \text{ units}$$

Thus

$$T_1 = \frac{1}{\tau}\left\{\frac{\dot{Q}_C}{U_C A_C} + T_C\right\} = 2 \times (200 + 400) = 1200 \text{ K}$$

and

$$T_2 = \frac{1200}{2} = 600 \text{ K}$$

This results in the temperature values shown in Fig 5.2 for the engine and reservoirs.

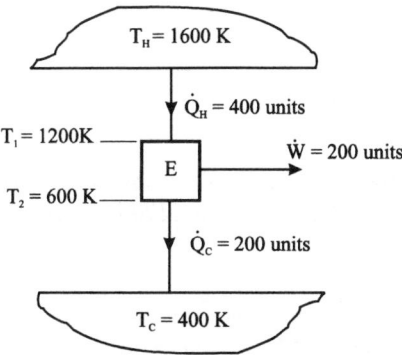

Fig. 5.2 Example of internally reversible heat engine operating between reservoirs at $T_H = 1600$ K and $T_C = 400$ K with $U_H A_H / U_C A_C = 1$

The efficiency of the Carnot cycle operating between the reservoirs would have been $\eta_{th} = 1 - 400/1600 = 0.75$ but the efficiency of this engine is $\eta_{th} = 1 - 600/1200 = 0.50$. Thus an engine which delivers maximum power is significantly less efficient than the Carnot engine.

It is possible to derive relationships for the intermediate temperatures. Equation (5.9) gives

$$T_1 = \frac{1}{\tau}\left[\frac{\dot{Q}_C}{U_C A_C} + T_C\right]$$

and eqn. (5.8) defines T_2 as

$$T_2 = \frac{\dot{Q}_C}{U_C A_C} + T_C \tag{5.8}$$

Also $\dot{Q}_C = \dot{Q}_H - \dot{W}$, and then

$$\frac{\dot{Q}_C}{U_H A_H T_H} = \frac{\dot{Q}_H}{U_H A_H T_H} - \frac{\dot{W}}{U_H A_H T_H}$$

Substituting from eqns (5.10) and (5.11) gives

$$\frac{\dot{Q}_C}{U_H A_H T_H} = \frac{(\tau - T_C/T_H)}{\left(1 + \dfrac{U_H A_H}{U_C A_C}\right)} \tag{5.13}$$

Substituting in eqn (5.8) gives

$$T_2 = \frac{U_H A_H T_H (\tau - T_C/T_H)}{U_C A_C \left(1 + \dfrac{U_H A_H}{U_C A_C}\right)} + T_C$$

$$= \frac{U_H A_H T_H (\tau - \tau^2)}{U_C A_C + U_H A_H} + T_C$$

$$= T_C^{1/2}\left\{\frac{U_H A_H T_H^{1/2} + U_C A_C T_C^{1/2}}{U_H A_H + U_C A_C}\right\} \tag{5.14}$$

Similarly

$$T_1 = T_H^{1/2} \left[\frac{U_H A_H T_H^{1/2} + U_C A_C T_C^{1/2}}{U_H A_H + U_C A_C} \right] \tag{5.15}$$

To be able to compare the effect of varying the resistances it is necessary to maintain the total resistance to heat transfer at the same value. For example, let

$$U_H A_H + U_C A_C = 2$$

Then, if $U_H A_H / U_C A_C = 1$ (as in the previous example), $U_H A_H = 1$.

Consider the effect of having a high resistance to the high temperature reservoir, e.g. $U_H A_H / U_C A_C = 1/2$. This gives

$$U_H A_H = \frac{2}{1 + \dfrac{U_C A_C}{U_H A_H}} = \frac{2}{3}.$$

Then

$$\frac{\dot{W}}{U_H A_H} = 1600 \, \frac{(1/2 - 1/4)}{(1 + 1/2)/2} (1 - 1/2) = 267 \text{ units, giving } \dot{W} = 177.8 \text{ units}$$

$$\frac{\dot{Q}_H}{U_H A_H} = \frac{\dot{W}}{U_H A_H (1 - \tau)} = 534 \text{ units and } \dot{Q}_H = 355.6 \text{ units}$$

Then

$$\frac{\dot{Q}_C}{U_C A_C} = \frac{\dot{Q}_C}{U_H A_H} \cdot \frac{1}{2} = \frac{1}{2 U_H A_H} (\dot{Q}_H - \dot{W}) = 133.5 \text{ units, giving } \dot{Q}_C = 177.8 \text{ units}$$

Hence

$$T_1 = \frac{1}{\tau} (133.5 + 400) = 2 \times 533.5 = 1067 \, K, \text{ and } T_2 = 533.5 \, K.$$

If the resistance to the low temperature reservoir is high, i.e. $U_H A_H / U_C A_C = 2$, the situation changes, as shown below. First,

$$U_H A_H = \frac{2}{1 + 1/2} = \frac{2}{3/2} = \frac{4}{3}$$

giving

$$\frac{\dot{W}}{U_H A_H} = 1600 \, \frac{(1/2 - 1/4)(1 - 1/2)}{(1 + 2)/2} = 133 \text{ units}$$

which results in $\dot{W} = 177.8$ kW and

$$\frac{\dot{Q}_H}{U_H A_H} = 267 \text{ units, or } \dot{Q} = 355.6 \, kW$$

Then

$$\frac{\dot{Q}_C}{U_C A_C} = \frac{2\dot{Q}_C}{U_H A_H} = \frac{2}{U_H A_H}(\dot{Q}_H - \dot{W}) = 267 \text{ units, giving } \dot{Q}_C = 200 \text{ units}$$

Hence

$$T_1 = \frac{1}{\tau}\left[\frac{\dot{Q}_C}{U_C A_C} + T_C\right] = 2(267 + 400) = 1334 \text{ K, and } T_2 = 667 \text{ K}$$

These results are shown graphically in Fig 5.3.

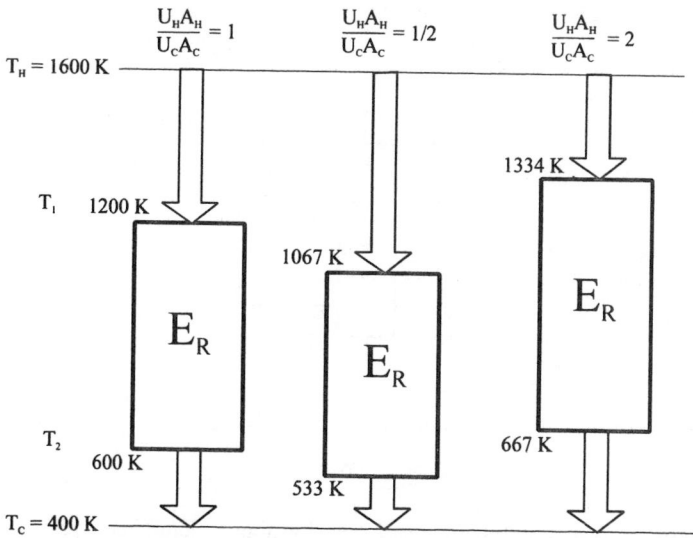

Fig. 5.3 The effect of heat transfer parameters on engine temperatures for a heat engine operating between reservoirs at $T_H = 1600$ K and $T_C = 400$ K

Comparing the power outputs based on the same total resistance, i.e. $U_H A_H + U_C A_C = 2$, gives the following table.

$U_H A_H / U_C A_C$	$\dot{Q}_H$	$\dot{W}$	$\dot{Q}_C$
1	400	200	200
1/2	355.6	177.7	177.7
2	355.6	177.7	177.7

It can be seen that the optimum system is one in which the high and low temperature resistances are equal. In this case the entropy generation (per unit of work) of the universe is minimised, as shown in the table below.

$U_H A_H / U_C A_C$	Q_H / T_H	Q_H / T_L	$\Delta S_H / W$	Q_C / T_2	Q_C / T_C	$\Delta S_C / W$	$\Sigma \Delta S / W$
1	-0.25	+0.333	$+4.17 \times 10^{-4}$	-0.333	+0.5	$+4.17 \times 10^{-4}$	$+8.34 \times 10^{-4}$
1/2	-0.225	+0.333	$+6.23 \times 10^{-4}$	-0.333	0.4443	$+6.24 \times 10^{-4}$	+0.00125
2	-0.225	+0.266	$+2.48 \times 10^{-4}$	-0.266	0.4443	$+10.00 \times 10^{-4}$	+0.00125

5.2   Efficiency of combined cycle internally reversible heat engines when producing maximum power output

One way of improving the overall efficiency of power production between two reservoirs is to use two engines. For example, a gas turbine and steam turbine can be used in series to make the most effective use of the available temperature drop. Such a power plant is referred to as a combined cycle gas turbine, and this type of generating system was introduced in Chapter 3 in connection with pinch technology. These plants can be examined in the following way, based on the two heat engines in series shown in Fig 5.4.

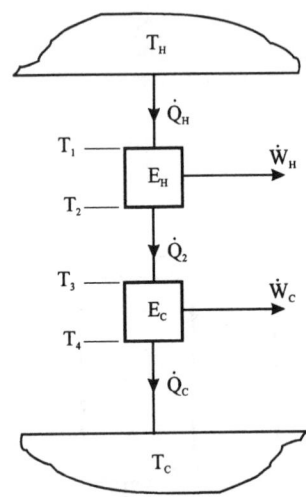

Fig. 5.4 Two internally reversible engines in series forming a combined cycle device operating between two reservoirs at T_H and T_C

In this case the product UA will be replaced by a 'conductivity' C to simplify the notation. Then

$$\dot{Q}_H = C_H(T_H - T_1) \tag{5.16}$$
$$\dot{Q}_2 = C_2(T_2 - T_3) \tag{5.17}$$
$$\dot{Q}_C = C_C(T_4 - T_C) \tag{5.18}$$

Also, for E_H

$$\frac{\dot{Q}_H}{T_1} = \frac{\dot{Q}_2}{T_2} \tag{5.19}$$

and for E_C

$$\frac{\dot{Q}_2}{T_3} = \frac{\dot{Q}_C}{T_4} \tag{5.20}$$

Let $\tau_1 = T_2/T_1$ and $\tau_2 = T_4/T_3$; then rearranging eqns (5.16), (5.17) and (5.18) gives

$$T_1 = T_H - \dot{Q}_H/C_H \tag{5.21}$$
$$T_2 = \dot{Q}_2/C_2 + T_3 \tag{5.22}$$

and

$$T_4 = \dot{Q}_C/C_C + T_C \qquad (5.23)$$

Also

$$T_3 = \frac{\dot{Q}_2}{\dot{Q}_C} T_4 = \frac{\dot{Q}_2}{\dot{Q}_C} \left[\frac{\dot{Q}_C}{C_C} + T_C \right] = \frac{1}{\tau_2} \left\{ \frac{\dot{Q}_C}{C_C} + T_C \right\} \qquad (5.24)$$

and

$$T_2 = \frac{\dot{Q}_2}{C_2} + \frac{1}{\tau_2} \left\{ \frac{\dot{Q}_C}{C_C} + T_C \right\} \qquad (5.25)$$

Similarly

$$T_1 = \frac{\dot{Q}_H}{\dot{Q}_2} T_2 = \frac{1}{\tau_1} \left[\frac{\dot{Q}_2}{C_2} + \frac{1}{\tau_2} \left\{ \frac{\dot{Q}_C}{C_C} + T_C \right\} \right] \qquad (5.26)$$

From eqns (5.16) and (5.26)

$$\dot{Q}_H = C_H \left[T_H - \frac{1}{\tau_1} \left(\frac{\dot{Q}_2}{C_2} + \frac{1}{\tau_2} \left\{ \frac{\dot{Q}_C}{T_C} + T_C \right\} \right) \right] \qquad (5.27)$$

Rearranging eqn (5.27) gives

$$\frac{\dot{Q}_H}{C_H T_H} = \left(1 - \frac{\dot{Q}_H}{C_2 T_H} - \frac{\dot{Q}_H}{C_C T_H} - \frac{1}{\tau_1 \tau_2} \frac{T_C}{T_H} \right) \qquad (5.28)$$

which can be written as

$$\frac{\dot{Q}_H}{C_H T_H} = \frac{(\tau_1 \tau_2 - T_C/T_H)}{\tau_1 \tau_2 \left(1 + \dfrac{C_H}{C_2} + \dfrac{C_H}{C_C} \right)} \qquad (5.29)$$

The power output of engine E_H can be obtained from eqn (5.29) as

$$\frac{\dot{W}_H}{C_H T_H} = \frac{\dot{Q}_H}{C_H T_H} (1 - \tau_1) = \frac{(\tau_1 \tau_2 - T_C/T_H)}{\tau_1 \tau_2 \left(1 + \dfrac{C_H}{C_2} + \dfrac{C_H}{C_C} \right)} (1 - \tau_1) \qquad (5.30)$$

In a similar manner $\dot{W}_C/C_H T_H$ can also be evaluated as

$$\frac{\dot{W}_C}{C_H T_H} = \frac{\dot{Q}_H}{C_H T_H} \tau_1 (1 - \tau_2) = \frac{(\tau_1 \tau_2 - T_C/T_H)(1 - \tau_2)\tau_1}{\tau_1 \tau_2 \left(1 + \dfrac{C_H}{C_2} + \dfrac{C_H}{C_C} \right)} \qquad (5.31)$$

$$= \tau_1 \frac{\dot{W}_H}{C_H T_H} \frac{(1 - \tau_2)}{(1 - \tau_1)} \qquad (5.32)$$

It can be seen from eqns (5.30) and (5.32) that it is possible to split the power output of the two engines in an arbitrary manner, dependent on the temperature drops across each engine. The ratio of work output of the two engines is

$$\frac{\dot{W}_H}{\dot{W}_C} = \frac{1}{\tau_1} \frac{(1-\tau_1)}{(1-\tau_2)} \tag{5.33}$$

This shows that if the temperature ratios across the high temperature and the low temperature engines are equal (i.e. $\tau_1 = \tau_2$), the work output of the low temperature engine will be

$$\dot{W}_C = \tau_1 \dot{W}_H \tag{5.34}$$

Hence the work output of the low temperature engine will be lower than that of the high temperature engine for the same temperature ratio. The reason comes directly from eqns (5.30) and (5.31), which show that the work output of an engine is directly proportional to the temperature of the 'heat' at entry.

The power output of a combined cycle power plant is the sum of the power of the individual engines, hence

$$\frac{\dot{W}_{CC}}{C_H T_H} = \frac{1}{C_H T_H} (\dot{W}_H + \dot{W}_C)$$

$$= \frac{(\tau_1 \tau_2 - T_C/T_H)}{\tau_1 \tau_2 \left(1 + \dfrac{C_H}{C_2} + \dfrac{C_H}{C_C}\right)} [(1 - \tau_1) + (1 - \tau_2)\tau_1]$$

This may be reduced to

$$\frac{\dot{W}_{CC}}{C_H T_H} = \frac{(\tau_1 \tau_2 - T_C/T_H)}{\tau_1 \tau_2 \left(1 + \dfrac{C_H}{C_2} + \dfrac{C_H}{C_C}\right)} [1 - \tau_1 \tau_2] \tag{5.35}$$

The efficiency of the combined cycle is defined by

$$\eta_{th} = \frac{\dot{W}_{CC}}{\dot{Q}_H} = 1 - \tau_1 \tau_2 \tag{5.36}$$

Equation (5.36) shows that the expression for the efficiency of the combined cycle engine at maximum power output is similar to that for the efficiency of a single heat engine, except that in this case the temperature ratio of the single engine is replaced by the product of the two temperature ratios. If the combined cycle device is considered to be two endoreversible heat engines connected by a perfect conductor (i.e. the resistance between the engines is zero; $C_2 = \infty$) then $T_3 = T_2$, and eqn (5.36) becomes

$$\eta_{th} = \frac{\dot{W}_{CC}}{\dot{Q}_H} = 1 - \tau_1 \tau_2 = 1 - \frac{T_1}{T_2} \frac{T_3}{T_4} = 1 - \frac{T_1}{T_4} \tag{5.37}$$

The efficiency given in eqn (5.37) is the same efficiency as would be achieved by a single endoreversible heat engine operating between the same two temperature limits, and what would be expected if there was no resistance between the two engines in the combined cycle plant.

To determine the efficiency of the combined cycle plant, composed of endoreversible heat engines, producing maximum power output requires the evaluation of the maxima of the surface $\dot{W}_{CC}$ plotted against the independent variables τ_1 and τ_2. It is difficult to obtain a mathematical expression for this and so the maximum work will be evaluated for some arbitrary conditions to demonstrate the necessary conditions. This will be done based on the following assumptions:

temperatures: $T_H = 1600$ K; $T_C = 400$ K

conductivities: $C_H = C_2 = C_C = 1$

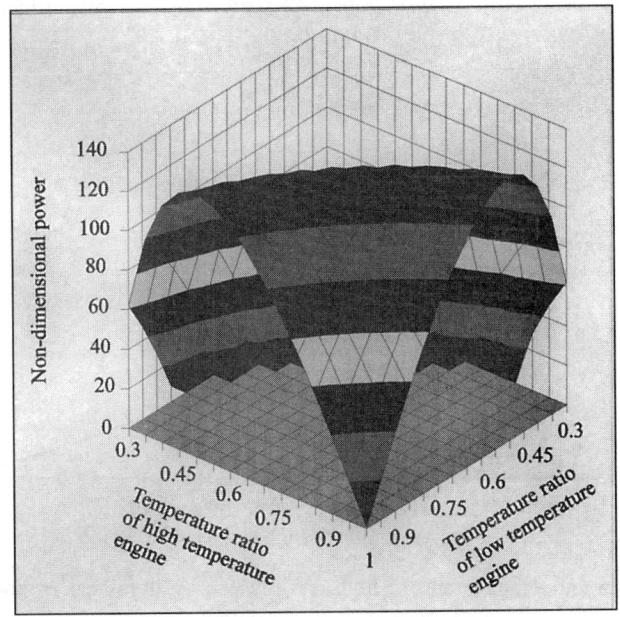

Fig. 5.5 Variation of maximum work output with temperature ratio across the two engines

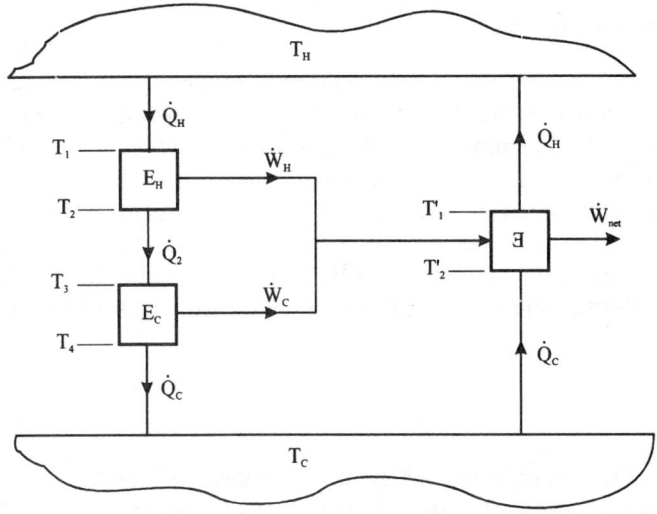

Fig. 5.6 Combined cycle heat engine driving a reversed heat engine

While these assumptions are arbitrary, it can be shown that the results obtained are logical and general. The variation of maximum power output with temperature ratios across the high and low temperature engines is shown in Fig 5.5. It can be seen that the maximum power occurs along a ridge which goes across the base plane. Examination shows that, in this case, this obeys the equation $\tau_1 \tau_2 = 0.5 = \sqrt{T_C/T_H}$. Hence, the efficiency of a combined cycle heat engine operating at maximum power output is the same as the efficiency obtainable from a single heat engine operating between the same two reservoirs. This solution is quite logical, otherwise it would be possible to arrange heat engines as in Fig 5.6, and produce net work output while transferring energy with a single reservoir.

Thus the variation of τ_1 with τ_2 to produce maximum power output is shown in Fig 5.7: all of these combinations result in the product being 0.5.

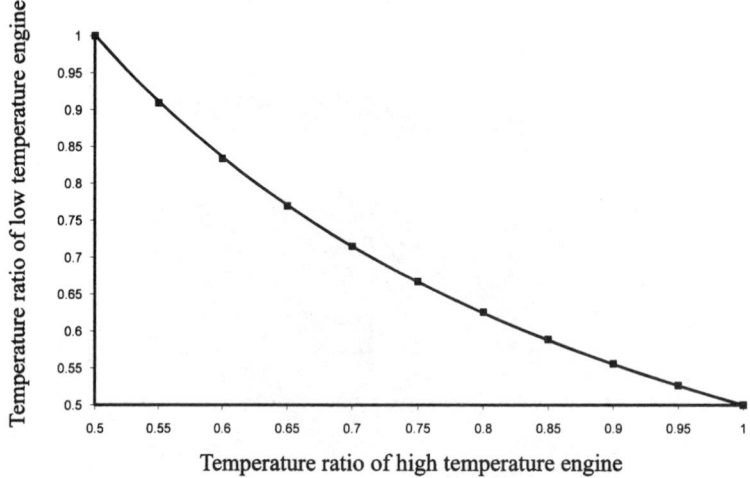

Fig. 5.7 Variations in temperature ratio of high and low temperature heat engines, which produce maximum power output

5.3   Concluding remarks

A new method for assessing the potential thermal efficiency of heat engines has been introduced. This is based on the engine operating at its maximum power output, and it is shown that the thermal efficiency is significantly lower than that of an engine operating on a Carnot cycle between the same temperature limits. The loss of efficiency is due to external irreversibilities which are present in all devices producing power output.

It has been shown that a combined cycle power plant cannot operate at a higher thermal efficiency than a single cycle plant *between the same temperature* limits. However, the use of two cycles enables the temperature limits to be widened, and thus better efficiencies are achieved.

PROBLEMS

1   Explain why the Carnot cycle efficiency is unrealistically high for a real engine. Introducing the concept of external irreversibility, evaluate the efficiency of an endoreversible engine at maximum power output.

Consider a heat engine connected to a high temperature reservoir at $T_H = 1200$ K and a low temperature one at $T_C = 300$ K. If the heat transfer conductances from the reservoirs to the engine are in the ratio $(UA)_H/(UA)_C = C_H/C_C = 2$, evaluate the following:

- the maximum Carnot cycle efficiency;
- the work output of the Carnot cycle;
- the engine efficiency at maximum power output;
- the maximum power output, $\dot{W}/C_H$;
- the maximum and minimum temperatures of the working fluid at maximum power output.

Derive all the necessary equations, but assume the Carnot efficiency, $\eta = 1 - T_C/T_H$.

[0.75; 0; 0.5; 100; 1000 K; 500 K]

2 A heat engine operates between two *finite* reservoirs, initially at 800 K and 200 K respectively. The temperature of the hot reservoir falls by 1 K for each kJ extracted from it, while the temperature of the cold reservoir rises by 1 K for each kJ added. What is the maximum work output from the engine as the reservoir temperatures equalise? Is the equalisation temperature for maximum work a higher or lower limit of the equalisation temperature?

[200 kJ; lower]

3 Closed cycle gas turbines operate on the *internally reversible* Joule cycle with an efficiency of

$$\eta_{joule} = 1 - \frac{1}{r_p^{(\kappa - 1)/\kappa}}$$

where r_p = pressure ratio of the turbine, and κ = ratio of specific heats, $c_p/c_v = 1.4$.

This equation significantly overestimates the efficiency of the cycle when *external irreversibilities* are taken into account. Shown in Fig P5.3 is a $T-s$ diagram for a closed

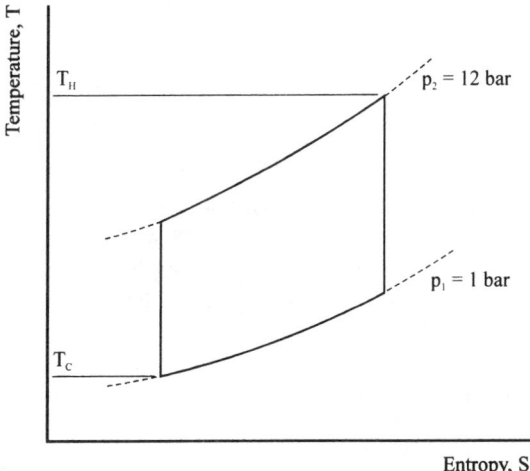

Fig. P5.3 Temperature–entropy diagram for Joule cycle

cycle gas turbine receiving energy from a high temperature reservoir at $T_H = 1200$ K and rejecting energy to a low temperature reservoir at $T_C = 400$ K.

(i)   Evaluate the Carnot efficiency and compare this with the Joule efficiency: explain why the Joule efficiency is the lower.

(ii)   Calculate the ratio of the gas turbine cycle work to the energy delivered from the high temperature reservoir (Q_H for the Carnot cycle). This ratio is less than the Joule efficiency – explain why in terms of unavailable energy.

(iii) Calculate the external irreversibilities and describe how these may be reduced to enable the Joule efficiency to be achieved.

(iv)   What are the mean temperatures of energy addition and rejection in the Joule cycle? What would be the thermal efficiency of a Carnot cycle based on these mean temperatures?

$$[0.667; 0.508; 0.4212;\ I_H/c_p = 79.9;\ 994\ \text{K};\ 489\ \text{K}]$$

4 Explain why the Carnot cycle overestimates the thermal efficiency achievable from an engine producing power output. Discuss why external irreversibility reduces the effective temperature ratio of an endoreversible engine.

Show that the thermal efficiency at maximum power output of an endoreversible engine executing an Otto cycle is

$$\eta_{th} = 1 - \left(\frac{T_C}{T_H}\right)^{1/2}$$

where $T_H = $ maximum temperature of the cycle and $T_C = $ minimum temperature of the cycle.

5 The operating processes of a spark-ignition engine can be represented by the Otto cycle, which is *internally reversible* and gives a thermal efficiency of

$$\eta_{Otto} = 1 - \frac{1}{r^{(\kappa - 1)}}$$

where $r = $ volumetric compression ratio; $\kappa = $ ratio of specific heats, c_p/c_v.

An Otto cycle is depicted in Fig P5.5, and the temperatures of the two reservoirs associated with the cycle are shown as T_H and T_C. The thermal efficiency of a Carnot cycle operating between these two reservoirs is $\eta = 1 - T_C/T_H$. This value is significantly higher than that of the Otto cycle operating between the same reservoirs. Show the ratio of net work output for the Otto cycle to the energy transferred from the high temperature reservoir for the Carnot cycle, Q_H, is

$$\eta = \left(1 - \frac{1}{r^{(\kappa - 1)}}\right) \frac{T_3 - T_2}{T_3 \ln(T_3/T_2)}$$

where T_2 and T_3 are defined in Fig P5.5.

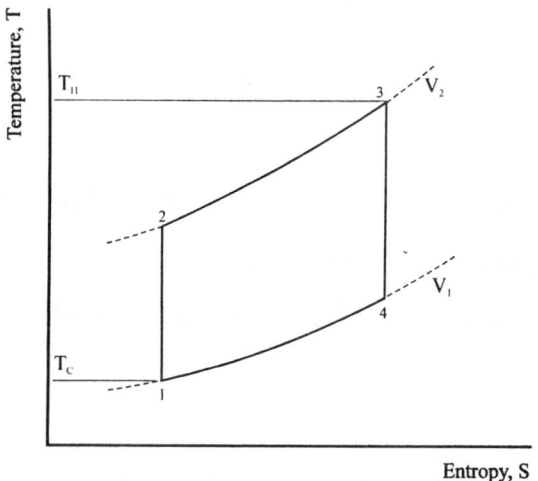

Fig. P5.5 Temperature–entropy diagram for Otto cycle

Explain why this value of η differs from that of the Otto cycle, and discuss the significance of the term

$$\frac{T_3 - T_2}{T_3 \ln(T_3/T_2)}$$

Evaluate the entropy change required at the high temperature reservoir to supply the Otto cycle in terms of the entropy span, $s_3 - s_2$, of the Otto cycle.

$$[\Delta s_{\mathrm{H}} = c_v T_2 (e^{\Delta s/c_v} - 1)/T_{\mathrm{H}}]$$

6 It is required to specify an ideal closed cycle gas turbine to produce electricity for a process plant. The first specification requires that the turbine produces the maximum work output possible between the peak temperature of 1200 K and the inlet temperature 300 K. The customer then feels that the efficiency of the turbine could be improved by

(i) incorporating a heat exchanger;
(ii) introducing reheat to the turbine by splitting the turbine pressure ratio such that the pressure ratio of each stage is the square root of the compressor pressure ratio.

Evaluate the basic cycle efficiency, and then evaluate separately the effects of the heat exchange and reheat. If these approaches have not increased the efficiency, propose another method by which the gas turbine performance might be improved (between the same temperature limits), without reducing the work output; evaluate the thermal efficiency of the proposed plant. The ratio of specific heats may be taken as $\kappa = 1.4$.

[0.5; 0.4233; 0.5731]

6

General Thermodynamic Relationships: single component systems, or systems of constant composition

The relationships which follow are based on single component systems, or systems of constant composition. These are a subset of the more general equations which can be derived, and which allow for changes in composition. It will be shown in Chapter 12 that if a change in composition occurs then another term defining the effect of this change is required.

6.1 The Maxwell relationships

The concept of functional relationships between properties was introduced previously. For example, the Second Law states that, for a reversible process, T, s, u, p and v are related in the following manner

$$T \, dS = dU + p \, dV \tag{6.1}$$

or, in specific (or molar) terms

$$T \, ds = du + p \, dv \tag{6.1a}$$

Rearranging eqn (6.1a) enables the change of internal energy, du, to be written

$$du = T \, ds - p \, dv \tag{6.2}$$

i.e.

$$u = f(s, v) \tag{6.2a}$$

It will be shown in Chapter 12 that, in the general case where the composition can change, eqn (6.1) should be written

$$T \, dS = dU + p \, dV - \sum_i \mu_i \, dn_i$$

where

μ_i = chemical potential of component i \tag{6.1b}
n_i = amount of component i

The chemical potential terms will be omitted in the following analysis, although similar equations to those below can be derived by taking them into account.

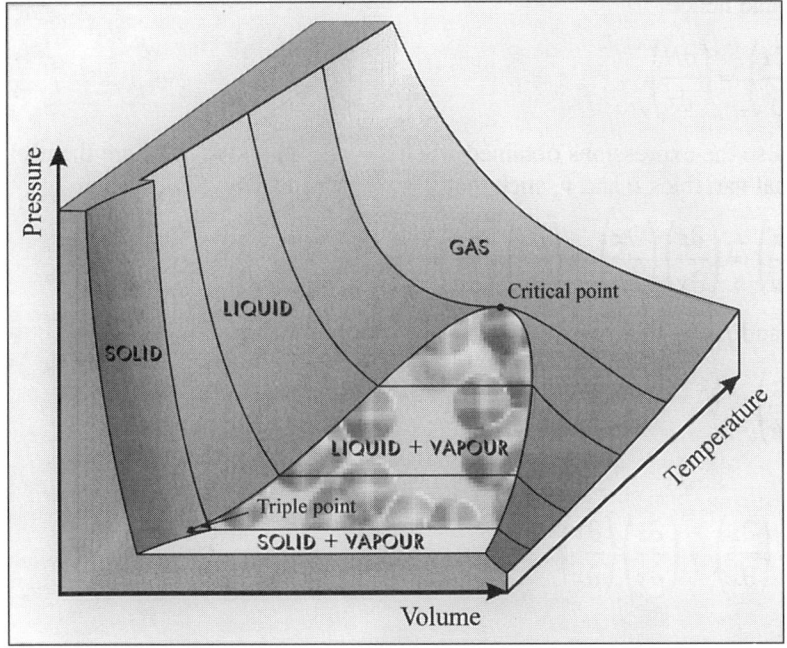

Fig. 6.1 $p-v-T$ surface for water

It can be seen from eqns (6.2) and (6.2a) that the specific internal energy can be represented by a three-dimensional surface based on the independent variables of entropy and specific volume. If this surface is *continuous* then the following relationships can be based on the mathematical properties of the surface. The restriction of a continuous surface means that it is 'smooth'. It can be seen from Fig 6.1 that the $p-v-T$ surface for water is continuous over most of the surface, but there are discontinuities at the saturated liquid and saturated vapour lines. Hence, the following relationships apply over the major regions of the surface, *but not across the boundaries*. For a continuous surface $z = z(x, y)$ where z is a continuous function. Then

$$dz = \left(\frac{\partial z}{\partial x}\right)_y dx + \left(\frac{\partial z}{\partial y}\right)_z dy \qquad (6.3)$$

Let

$$M = \left(\frac{\partial z}{\partial x}\right)_y \qquad \text{and} \qquad N = \left(\frac{\partial z}{\partial y}\right)_x \qquad (6.4)$$

Then

$$dz = M \, dx + N \, dy \qquad (6.5)$$

For continuous functions, the derivatives

$$\frac{\partial^2 z}{\partial x \, \partial y} \qquad \text{and} \qquad \frac{\partial^2 z}{\partial y \, \partial x}$$

are equal, and hence

$$\left(\frac{\partial M}{\partial y}\right)_x = \left(\frac{\partial N}{\partial x}\right)_y$$

Consider also the expressions obtained when $z = z(x, y)$ and x and y are themselves related to additional variables u and v, such that $x = x(u, v)$ and $y = y(u, v)$. Then

$$\left(\frac{\partial z}{\partial u}\right)_v = \left(\frac{\partial z}{\partial x}\right)_y\left(\frac{\partial x}{\partial u}\right)_v + \left(\frac{\partial z}{\partial y}\right)_x\left(\frac{\partial y}{\partial u}\right)_v \tag{6.6}$$

Let $z = v$, and $u = x$, then $x = x(z)$ and

$$\left(\frac{\partial z}{\partial u}\right)_v = 0, \quad \text{and} \quad \left(\frac{\partial x}{\partial u}\right)_v = 1 \tag{6.7}$$

Hence

$$0 = \left(\frac{\partial z}{\partial x}\right)_y + \left(\frac{\partial z}{\partial y}\right)_x\left(\frac{\partial y}{\partial x}\right)_z \tag{6.8}$$

giving

$$\left(\frac{\partial z}{\partial x}\right)_y\left(\frac{\partial y}{\partial z}\right)_x\left(\frac{\partial x}{\partial y}\right)_z = -1 \tag{6.9}$$

These expressions will now be used to consider relationships derived previously. The following functional relationships have already been obtained

$$du = T\,ds - p\,dv \tag{6.10}$$

$$dh = T\,ds + v\,dp \tag{6.11}$$

$$df = -p\,dv - s\,dT \tag{6.12}$$

$$dg = v\,dp - s\,dT \tag{6.13}$$

Consider the expression for du, given in eqn (6.10) then, by analogy with eqn (6.3)

$$T = \left(\frac{\partial u}{\partial s}\right)_v, \quad -p = \left(\frac{\partial u}{\partial v}\right)_s \quad \text{and} \quad \left(\frac{\partial T}{\partial v}\right)_s = -\left(\frac{\partial p}{\partial s}\right)_v \tag{6.14}$$

In a similar manner the following relationships can be obtained, *for constant composition or single component systems*

$$T = \left(\frac{\partial h}{\partial s}\right)_p, \quad v = \left(\frac{\partial h}{\partial p}\right)_s; \quad \left(\frac{\partial T}{\partial p}\right)_s = \left(\frac{\partial v}{\partial s}\right)_p \tag{6.15}$$

$$-p = \left(\frac{\partial f}{\partial v}\right)_T, \quad -s = \left(\frac{\partial f}{\partial T}\right)_v; \quad \left(\frac{\partial p}{\partial T}\right)_v = \left(\frac{\partial s}{\partial v}\right)_T \tag{6.16}$$

$$v = \left(\frac{\partial g}{\partial p}\right)_T, \quad -s = \left(\frac{\partial g}{\partial T}\right)_p; \quad \left(\frac{\partial v}{\partial T}\right)_p = \left(\frac{\partial s}{\partial p}\right)_T \tag{6.17}$$

If the pairs of relationships for T are equated then

$$\left(\frac{\partial u}{\partial s}\right)_v = \left(\frac{\partial h}{\partial s}\right)_p \qquad (6.18a)$$

and, similarly

$$\left(\frac{\partial g}{\partial p}\right)_T = \left(\frac{\partial h}{\partial p}\right)_s \qquad (6.18b)$$

$$\left(\frac{\partial u}{\partial v}\right)_s = \left(\frac{\partial f}{\partial v}\right)_T \qquad (6.18c)$$

$$\left(\frac{\partial g}{\partial T}\right)_p = \left(\frac{\partial f}{\partial T}\right)_v \qquad (6.18d)$$

In addition to these equivalences, eqns (6.14) to (6.17) also show that

$$\left(\frac{\partial T}{\partial v}\right)_s = -\left(\frac{\partial p}{\partial s}\right)_v \qquad (6.19a)$$

$$\left(\frac{\partial T}{\partial p}\right)_s = \left(\frac{\partial v}{\partial s}\right)_p \qquad (6.19b)$$

$$\left(\frac{\partial p}{\partial T}\right)_v = \left(\frac{\partial s}{\partial v}\right)_T \qquad (6.19c)$$

$$\left(\frac{\partial v}{\partial T}\right)_p = -\left(\frac{\partial s}{\partial p}\right)_T \qquad (6.19d)$$

Equations (6.19) are called the *Maxwell relationships*.

6.1.1   GRAPHICAL INTERPRETATION OF MAXWELL RELATIONS

Let a system comprising a pure substance execute a small reversible cycle 1–2–3–4–1 consisting of two isochors separated by dV and two isentropes separated by ds; these cycles are shown in Fig 6.2.

If the cycles are reversible, the area on the T–s diagram must equal the area on the p–v diagram. Now the difference in pressure between lines 1 and 2 is given by

$$\left(\frac{\partial p}{\partial s}\right)_v ds$$

and to the first order the difference in pressure between lines 3 and 4 is the same. Hence

$$\text{the area on the } p\text{–}v \text{ diagram} = \left(\frac{\partial p}{\partial s}\right)_v ds\, dv \qquad (6.20a)$$

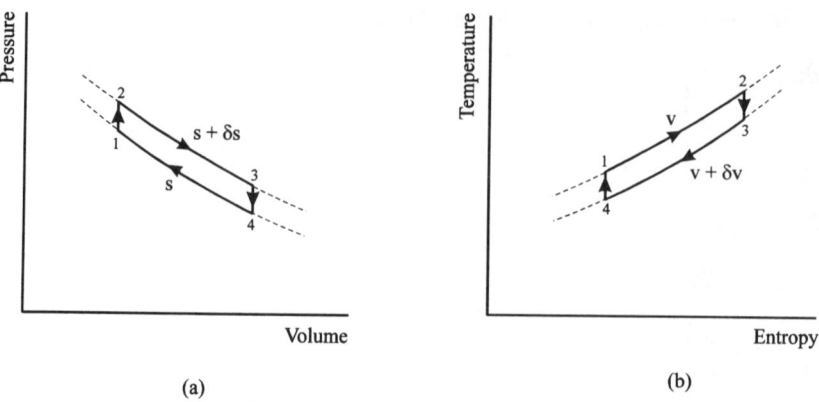

Fig. 6.2 Elemental cycles illustrating the Maxwell relationships

Applying the same approach to the T–s diagram, the difference in temperature between lines 1 and 4 is

$$-\left(\frac{\partial T}{\partial v}\right)_s dv \tag{6.20b}$$

which is negative because the temperature decreases as the volume increases, and the area of the diagram is

$$-\left(\frac{\partial T}{\partial v}\right)_s dv\, ds$$

Thus, equating the two areas gives

$$-\left(\frac{\partial T}{\partial v}\right)_s dv\, ds = \left(\frac{\partial p}{\partial s}\right)_v ds\, dv \tag{6.21}$$

and hence

$$\left(\frac{\partial p}{\partial s}\right)_v = -\left(\frac{\partial T}{\partial v}\right)_s$$

6.2   Uses of the thermodynamic relationships

When performing certain types of calculation it is useful to have data on the values of the specific heat capacities of the substance under consideration and also the variation of these specific heat capacities with the other properties. The specific heat capacities themselves are first derivatives of the internal energy (u) and the enthalpy (h), and hence the variations of specific heat capacities are second derivatives of the basic properties.

By definition, the specific heat capacity at constant volume is

$$c_v = \left(\frac{\partial u}{\partial T}\right)_v = \left(\frac{\partial u}{\partial s}\right)_v \left(\frac{\partial s}{\partial T}\right)_v \tag{6.22}$$

But

$$\left(\frac{\partial u}{\partial s}\right)_v = T, \text{ and thus } c_v = T\left(\frac{\partial s}{\partial T}\right)_v \tag{6.23}$$

Similarly

$$c_p = \left(\frac{\partial h}{\partial T}\right)_p = \left(\frac{\partial h}{\partial s}\right)_p\left(\frac{\partial s}{\partial T}\right)_p = T\left(\frac{\partial s}{\partial T}\right)_p \tag{6.24}$$

Consider the variation of the specific heat capacity at constant volume, c_v, with specific volume, v, if the temperature is maintained constant. This can be derived in the following way:

$$\left(\frac{\partial c_v}{\partial v}\right)_T = \frac{\partial}{\partial v}\left[T\left(\frac{\partial s}{\partial T}\right)_v\right] = T\frac{\partial^2 s}{\partial v\, \partial T} = T\frac{\partial}{\partial T}\left(\frac{\partial s}{\partial v}\right)_T$$

Now, from eqn (6.19c)

$$\left(\frac{\partial s}{\partial v}\right)_T = \left(\frac{\partial p}{\partial T}\right)_v$$

giving

$$\left(\frac{\partial c_v}{\partial v}\right)_T = T\left(\frac{\partial^2 s}{\partial T\, \partial v}\right) = T\left(\frac{\partial^2 p}{\partial T^2}\right)_v \tag{6.25}$$

Hence, if data are available for a substance in the form of a $p-v-T$ surface, or a mathematical relationship, it is possible to evaluate $(\partial^2 p/\partial T^2)_v$ and then $(\partial c_v/\partial v)_T$. By integration it is then possible to obtain the values of c_v at different volumes. For example, consider whether the specific heat capacity at constant volume is a function of the volume for both an ideal gas and a van der Waals' gas.

Ideal gas

The gas relationship for an ideal gas is

$$pv = RT \tag{6.26}$$

and

$$\left(\frac{\partial p}{\partial T}\right)_v = \frac{R}{v} \quad \text{and} \quad \left(\frac{\partial^2 p}{\partial T^2}\right)_v = 0 \tag{6.27}$$

Hence $c_v \ne f(v)$ for an ideal gas. This is in agreement with Joule's experiment for assessing the change of internal energy, u, with volume.

van der Waals' gas

The equation of state of a van der Waals' gas is

$$p = \frac{RT}{v-b} - \frac{a}{v^2} \tag{6.28}$$

where a and b are constants. Hence

$$\left(\frac{\partial p}{\partial T}\right)_v = \frac{R}{v-b} \quad \text{and} \quad \left(\frac{\partial^2 p}{\partial T^2}\right)_v = 0 \tag{6.29}$$

Again $c_v \neq f(v)$ for a van der Waals' gas.

So, for 'gases' obeying these state equations $c_v \neq f(v)$, but in certain cases and under certain conditions c_v could be a function of the volume of the system and it would be calculated in this way. Of course, if these equations are to be integrated to evaluate the internal energy it is necessary to know the value of c_v at a datum volume and temperature.

Similarly, the variation of the specific heat capacity at constant pressure with pressure can be investigated by differentiating the specific heat capacity with respect to pressure, giving

$$\left(\frac{\partial c_p}{\partial p}\right)_T = \frac{\partial}{\partial p}\left(T\left(\frac{\partial s}{\partial T}\right)_p\right)_T = T\frac{\partial}{\partial T}\left(\frac{\partial s}{\partial p}\right)_T = -T\left(\frac{\partial^2 v}{\partial T^2}\right)_p \tag{6.30}$$

This equation can be used to see if the specific heat capacity at constant pressure of gases obeying the ideal gas law, and those obeying van der Waals' equation are functions of pressure. This is done below.

Ideal gas

$$pv = RT$$

Hence

$$\left(\frac{\partial v}{\partial T}\right)_p = \frac{R}{p} \quad \text{and} \quad \left(\frac{\partial^2 v}{\partial T^2}\right)_p = 0 \tag{6.31}$$

This means that the specific heat capacity at constant pressure for a gas obeying the ideal gas law is not a function of pressure, i.e. $c_p \neq f(p)$ for an ideal gas. This conclusion is in agreement with the Joule–Thomson experiment for superheated gases.

Van der Waals' gas

$$p = \frac{RT}{v-b} - \frac{a}{v^2}$$

Equation (6.28) can be rewritten as

$$\left(p + \frac{a}{v^2}\right)(v - b) = RT \tag{6.32}$$

which expands to

$$pv + \frac{a}{v} - pb - \frac{ab}{v^2} = RT$$

Differentiating implicitly gives

$$p\left(\frac{\partial v}{\partial T}\right)_p + (-)\frac{a}{v^2}\left(\frac{\partial v}{\partial T}\right)_p - \frac{(-2)ab}{v^3}\left(\frac{\partial v}{\partial T}\right)_p = R$$

which can be rearranged to give

$$\left(\frac{\partial v}{\partial T}\right)_p\left\{p - \frac{a}{v^2} + \frac{2ab}{v^3}\right\} = R$$

and hence

$$\left(\frac{\partial v}{\partial T}\right)_p = \frac{R}{p - \frac{a}{v^2}\left(1 - \frac{2b}{v}\right)} \qquad (6.33)$$

This can be differentiated again to give

$$\left(\frac{\partial^2 v}{\partial T^2}\right)_p = \frac{\partial}{\partial T}\left(\frac{\partial v}{\partial T}\right)_p$$

which results in

$$\left(\frac{\partial^2 v}{\partial T^2}\right)_p\left\{p - \frac{a}{v^2} + \frac{2ab}{v^3}\right\} + \left(\frac{\partial v}{\partial T}\right)_p^2\left\{\frac{2a}{v^3} - \frac{6ab}{v^4}\right\} = 0$$

giving

$$\left(\frac{\partial^2 v}{\partial T^2}\right)_p = \frac{-\left(\frac{\partial v}{\partial T}\right)_p^2\left[\frac{2a}{v^3} - \frac{6ab}{v^4}\right]}{\left\{p - \frac{a}{v^2} + \frac{2ab}{v^3}\right\}}, \text{ which, on substituting for }\left(\frac{\partial v}{\partial T}\right)_p \text{ becomes}$$

$$\left(\frac{\partial^2 v}{\partial T^2}\right)_p = -\frac{R^2\left(\frac{2a}{v^3} - \frac{6ab}{v^4}\right)}{\left(p - \frac{a}{v^2}\left(1 - \frac{2b}{v}\right)\right)^3} \qquad (6.34)$$

Hence, for a van der Waals' gas, $c_p = f(p)$.

This means that allowance would have to be taken of this when evaluating the change of specific heat capacity at constant pressure for a process in which the pressure changed. This can be evaluated from

$$c_p = \int_{p_1}^{p_2} -T\left(\frac{\partial^2 v}{\partial T^2}\right)_p dp \qquad (6.35)$$

where $(\partial^2 v/\partial T^2)_p$ can be calculated from eqn (6.34).

6.3   *Tds relationships*

Two approaches will have been used previously in performing cycle calculations. When evaluating the performance of steam plant and refrigeration equipment, much emphasis was placed on the use of tables, and if the work done between two states was required, enthalpy values at these states were evaluated and suitable subtractions were performed. When doing cycle calculations for gas turbines and internal combustion engines the air was assumed to be an ideal gas and the specific heat capacity was used. The most accurate way of performing such calculations is, in fact, to use the enthalpy or internal energy values because these implicitly integrate the specific heat capacity over the range involved. The following analyses will demonstrate how these values of internal energy and enthalpy are obtained.

The two-property rule can be used to define entropy as the following functional relationship

$$s = s(v, T) \qquad\qquad (6.36)$$

Hence

$$ds = \left(\frac{\partial s}{\partial T}\right)_v dT + \left(\frac{\partial s}{\partial v}\right)_T dv \qquad\qquad (6.37)$$

Now, by definition

$$c_v = T\left(\frac{\partial s}{\partial T}\right)_v$$

and from the Maxwell relationships, eqn (6.19c),

$$\left(\frac{\partial s}{\partial v}\right)_T = \left(\frac{\partial p}{\partial T}\right)_v$$

Hence

$$T\,ds = c_v\,dT + T\left(\frac{\partial p}{\partial T}\right)_v dv \qquad\qquad (6.38)$$

This is called the *first Tds relationship*.

The second *Tds* relationship can be derived in the following way. If it is assumed that entropy is a function of temperature and pressure, then

$$s = s(T, p) \qquad\qquad (6.39)$$

and

$$ds = \left(\frac{\partial s}{\partial T}\right)_p dT + \left(\frac{\partial s}{\partial p}\right)_T dp \qquad\qquad (6.40)$$

By definition,

$$c_p = T\left(\frac{\partial s}{\partial T}\right)_p$$

and, from eqn (6.19d)

$$\left(\frac{\partial s}{\partial p}\right)_T = -\left(\frac{\partial v}{\partial T}\right)_p$$

Thus, the *second Tds relationship* is

$$T\,ds = c_p\,dT - T\left(\frac{\partial v}{\partial T}\right)_p dp \qquad (6.41)$$

The *third Tds relationship* is derived by assuming that

$$s = s(p, v) \qquad (6.42)$$

$$ds = \left(\frac{\partial s}{\partial p}\right)_v dp + \left(\frac{\partial s}{\partial v}\right)_p dv \qquad (6.43)$$

Now, by definition,

$$c_p = T\left(\frac{\partial s}{\partial T}\right)_p = T\left(\frac{\partial s}{\partial v}\right)_p\left(\frac{\partial v}{\partial T}\right)_p \quad \Rightarrow \quad \left(\frac{\partial s}{\partial v}\right)_p = \frac{1}{T}c_p\left(\frac{\partial T}{\partial v}\right)_p$$

A similar expression can be derived for c_v, and this enables the equation

$$T\,ds = c_v\left(\frac{\partial T}{\partial p}\right)_v dp + c_p\left(\frac{\partial T}{\partial v}\right)_p dv \qquad (6.44)$$

to be obtained. This is the *third Tds relationship*.

It is now possible to investigate the variation of internal energy and enthalpy with independent properties for gases obeying various gas laws. The internal energy of a substance can be expressed as

$$u = u(T, v) \qquad (6.45)$$

Hence, the change in internal energy is

$$du = \left(\frac{\partial u}{\partial T}\right)_v dT + \left(\frac{\partial u}{\partial v}\right)_T dv \qquad (6.46)$$

Now

$$c_v = \left(\frac{\partial u}{\partial T}\right)$$

and from the second law

$$du = T\,ds - p\,dv \qquad (6.47)$$

Substituting for Tds in eqn (6.47) from the first Tds relation (eqn (6.38)) gives

$$du = c_v\,dT + T\left(\frac{\partial p}{\partial T}\right)_v dv - p\,dv \qquad (6.48)$$

and hence eqn (6.46) becomes

$$c_v \, dT + T\left(\frac{\partial p}{\partial T}\right)_v dv - p \, dv = c_v \, dT + \left(\frac{\partial u}{\partial v}\right)_T dv \tag{6.49}$$

or

$$\left(\frac{\partial u}{\partial v}\right)_T = \left[T\left(\frac{\partial p}{\partial T}\right)_v - p\right] \tag{6.50}$$

Equation (6.50) can be used to evaluate the variation in internal energy with volume for both ideal and van der Waals' gases.

Ideal gas

The equation of state of an ideal gas is $pv = RT$ and hence

$$\left(\frac{\partial p}{\partial T}\right)_v = \frac{R}{v} \tag{6.51}$$

Thus

$$\left(\frac{\partial u}{\partial v}\right)_T = \left[\frac{RT}{v} - p\right] = 0 \tag{6.52}$$

Hence, the specific internal energy of an ideal gas is not a function of its specific volume (or density). This is in agreement with Joule's experiment that $u \neq f(v)$ at constant temperature.

Van der Waals' gas

The equation of state of a van der Waals' gas is

$$p = \frac{RT}{v - b} - \frac{a}{v^2}$$

and hence

$$\left(\frac{\partial p}{\partial T}\right)_v = \frac{R}{v - b} \tag{6.53}$$

which gives the change of internal energy with volume as

$$\left(\frac{\partial u}{\partial v}\right)_T = \left[\frac{RT}{v - b} - p\right] = \frac{a}{v^2} \tag{6.54}$$

This means that the internal energy of a van der Waals' gas is a function of its specific volume, or density. This is not surprising because density is a measure of the closeness of the molecules of the substance, and the internal energy variation is related to the force of attraction between the molecules. This means that some of the internal energy in a van der

Waals' gas is stored in the attraction forces between the molecules, and not all of the thermal energy is due to molecular motion, as was the case for the ideal gas.

Now, if it is required to calculate the total change in internal energy, u, for a change in volume and temperature it is necessary to use the following:

$$du = \left(\frac{\partial u}{\partial T}\right)_v dT + \left(\frac{\partial u}{\partial v}\right)_T dv$$

For an *ideal gas*

$$du = \left(\frac{\partial u}{\partial T}\right)_v dT = c_v\, dT \tag{6.55}$$

i.e. for an ideal gas $u = f(T)$.

It is also possible to use the following approach to show whether $u = f(p)$. This may be done in the following way:

$$\left(\frac{\partial u}{\partial v}\right)_T = \left(\frac{\partial u}{\partial p}\right)_T \left(\frac{\partial p}{\partial v}\right)_T = 0$$

Now

$$\left(\frac{\partial p}{\partial v}\right)_T = -\frac{p}{v},\ \text{and since } p \neq 0, \left(\frac{\partial u}{\partial p}\right)_T = 0$$

Hence, for an ideal gas $u \neq f(p)$.

For a *van der Waals' gas*

$$du = c_v\, dT + \frac{a}{v^2}\, dv \tag{6.56}$$

Hence for a van der Waals' gas $u = f(T, v)$.

6.4 Relationships between specific heat capacities

These relationships enable the value of one specific heat capacity to be calculated if the other is known. This is useful because it is much easier to measure the specific heat capacity at constant pressure, c_p, than that at constant volume, c_v.

Using the two-property rule it is possible to write

$$s = s(T, v) \tag{6.57}$$

and, if s is a continuous function of temperature and volume,

$$ds = \left(\frac{\partial s}{\partial T}\right)_v dT + \left(\frac{\partial s}{\partial v}\right)_T dv \tag{6.58}$$

If eqn (6.58) is differentiated with respect to temperature, T, with pressure maintained constant, then

$$\left(\frac{\partial s}{\partial T}\right)_p = \left(\frac{\partial s}{\partial T}\right)_v + \left(\frac{\partial s}{\partial v}\right)_T \left(\frac{\partial v}{\partial T}\right)_p \tag{6.59}$$

The definitions of the specific heat capacities are

$$c_v = T\left(\frac{\partial s}{\partial T}\right)_v \qquad \text{and} \qquad c_p = T\left(\frac{\partial s}{\partial T}\right)_p$$

Hence

$$\frac{c_p - c_v}{T} = \left(\frac{\partial s}{\partial v}\right)_T\left(\frac{\partial v}{\partial T}\right)_p \tag{6.60}$$

From Maxwell relations (eqn (6.19c)),

$$\left(\frac{\partial s}{\partial v}\right)_T = \left(\frac{\partial p}{\partial T}\right)_v$$

Thus

$$\frac{c_p - c_v}{T} = \left(\frac{\partial p}{\partial T}\right)_v\left(\frac{\partial v}{\partial T}\right)_p \tag{6.61}$$

The mathematical relationship in eqn (6.9)

$$\left(\frac{\partial p}{\partial T}\right)_v\left(\frac{\partial T}{\partial v}\right)_p\left(\frac{\partial v}{\partial p}\right)_T = -1 \tag{6.62}$$

can be rearranged to give

$$\left(\frac{\partial p}{\partial T}\right)_v = -\left(\frac{\partial p}{\partial v}\right)_T\left(\frac{\partial v}{\partial T}\right)_p \tag{6.63}$$

Thus

$$c_p - c_v = -T\left(\frac{\partial p}{\partial v}\right)_T\left(\frac{\partial v}{\partial T}\right)_p^2 \tag{6.64}$$

Examination of eqn (6.64) can be used to define which specific heat capacity is the larger. First, it should be noted that T and $(\partial v/\partial T)_p^2$ are both positive, and hence the sign of $c_p - c_v$ is controlled by the sign of $(\partial p/\partial v)_T$. Now, for all known substances $(\partial p/\partial v)_T$ is negative. If it were not negative then the substance would be completely unstable because a positive value means that as the pressure is increased the volume increases, and vice versa. Hence if the pressure on such a substance were decreased its volume would decrease until it ceased to exist.

Thus $c_p - c_v$ must always be positive or zero, i.e. $c_p \geqslant c_v$. The circumstances when $c_p = c_v$ are when $T = 0$ or when $(\partial p/\partial T)_v = 0$, e.g. at 4°C for water. It can also be shown that $c_p \geqslant c_v$ by considering the terms in eqn (6.61) in relation to the state diagrams for substances, as shown in Fig 6.3. The term $(\partial p/\partial T)_v$, which can be evaluated along a line at constant volume, v, can be seen to be positive, because as the temperature increases, the pressure increases over the whole of the section. Similarly, $(\partial v/\partial T)_p$, which can be evaluated along a line at constant pressure, p, is also positive. If both these terms are positive then $c_p \geqslant c_v$.

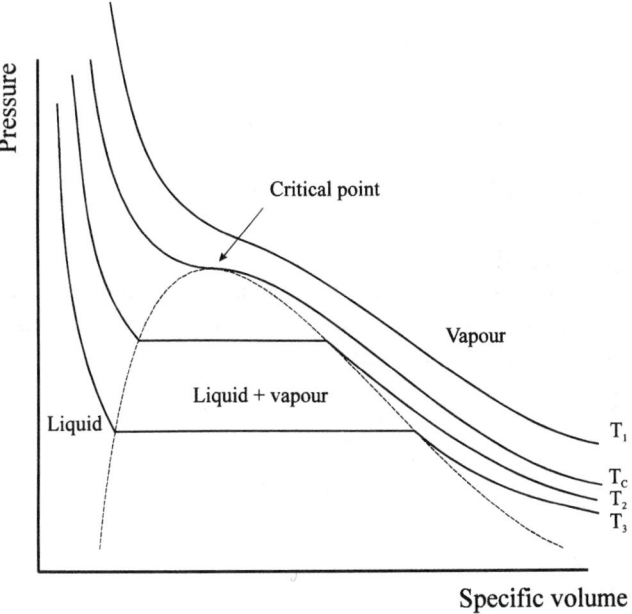

Fig. 6.3 *p–v* section of state diagram showing isotherms

Now consider $c_p - c_v$ for an ideal gas, which is depicted by the superheated region in Fig 6.3. The state equation for an ideal gas is

$$pv = RT$$

Differentiating gives

$$\left(\frac{\partial p}{\partial v}\right)_T = -\frac{p}{v} \tag{6.65}$$

and

$$\left(\frac{\partial v}{\partial T}\right)_p = \frac{R}{p} \tag{6.66}$$

Thus

$$c_p - c_v = -T\left(\frac{-p}{v}\right)\frac{R^2}{p^2} = R \tag{6.67}$$

The definition of an ideal gas is one which obeys the ideal gas equation (eqn (6.26)), and in which the specific heat capacity at constant volume (or pressure) is a *function of temperature alone*, i.e. $c_v = f(T)$. Hence, from eqn (6.67), $c_p = c_v + R = f(T) + R = f'(T)$ if R is a function of T alone. Hence, the difference of specific heat capacities for an ideal gas is the gas constant R. Also

$$\left(\frac{\partial p}{\partial v}\right)_T = -\frac{1}{kv} \quad \text{and} \quad \left(\frac{\partial v}{\partial T}\right)_p = \beta v$$

and thus

$$c_p - c_v = -T\left(\frac{-1}{kv}\right)(\beta v)^2 = \frac{T\beta^2 v}{k} \tag{6.68}$$

where

$\beta =$ coefficient of expansion (isobaric expansivity)
$k =$ isothermal compressibility.

These parameters are defined in Chapter 7.

Expressions for the difference between the specific heat capacities, $c_p - c_v$, have been derived above. It is also interesting to examine the ratio of specific heat capacities, $\kappa = c_p/c_v$. The definitions of c_p and c_v are

$$c_v = T\left(\frac{\partial s}{\partial T}\right)_v \quad \text{and} \quad c_p = T\left(\frac{\partial s}{\partial T}\right)_p$$

and thus the ratio of specific heat capacities is

$$\frac{c_p}{c_v} = \frac{(\partial s/\partial T)_p}{(\partial s/\partial T)_v} \tag{6.69}$$

Now, from the mathematical relationship (eqn (6.9)) for the differentials,

$$\left(\frac{\partial s}{\partial T}\right)_p\left(\frac{\partial T}{\partial p}\right)_s\left(\frac{\partial p}{\partial s}\right)_T = -1 = \left(\frac{\partial s}{\partial T}\right)_v\left(\frac{\partial T}{\partial v}\right)_s\left(\frac{\partial v}{\partial s}\right)_T \tag{6.70}$$

giving

$$\frac{c_p}{c_v} = \left(\frac{\partial T}{\partial v}\right)_s\left(\frac{\partial v}{\partial s}\right)_T\left(\frac{\partial p}{\partial T}\right)_s\left(\frac{\partial s}{\partial p}\right)_T \tag{6.71}$$

From the Maxwell relationships

$$\left(\frac{\partial s}{\partial v}\right)_T = \left(\frac{\partial p}{\partial T}\right)_v \quad \text{and} \quad \left(\frac{\partial s}{\partial p}\right)_T = -\left(\frac{\partial v}{\partial T}\right)_p$$

giving

$$\frac{c_p}{c_v} = -\left(\frac{\partial T}{\partial v}\right)_s\left(\frac{\partial T}{\partial p}\right)_v\left(\frac{\partial p}{\partial T}\right)_s\left(\frac{\partial v}{\partial T}\right)_p$$

$$= -\left(\frac{\partial p}{\partial v}\right)_s\left(\frac{\partial T}{\partial p}\right)_v\left(\frac{\partial v}{\partial T}\right)_p \tag{6.72}$$

Now, from the two-property rule, $T = T(p, v)$ and hence

$$\left(\frac{\partial p}{\partial v}\right)_T = -\left(\frac{\partial p}{\partial T}\right)_v\left(\frac{\partial T}{\partial v}\right)_p \tag{6.73}$$

Thus

$$\frac{c_p}{c_v} = \left(\frac{\partial p}{\partial v}\right)_s \bigg/ \left(\frac{\partial p}{\partial v}\right)_T = v\left(\frac{\partial p}{\partial v}\right)_s \bigg/ v\left(\frac{\partial p}{\partial v}\right)_T \tag{6.74}$$

The denominator of eqn (6.74), $v(\partial p/\partial v)_T$, is the reciprocal of the *isothermal compressibility*, k. By analogy, the numerator can be written as $1/k_s$, where k_s = *adiabatic, or isentropic, compressibility*. Thus

$$\frac{c_p}{c_v} = \frac{k}{k_s} = \kappa \tag{6.75}$$

If the slopes of isentropes in the $p-v$ plane are compared with the slopes of isotherms (see Fig 6.4), it can be seen that $c_p/c_v > 1$ for a gas in the superheat region.

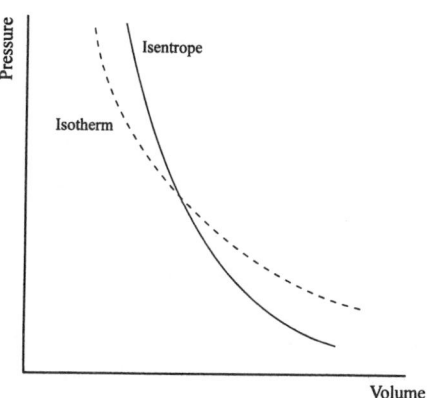

Fig. 6.4 Isentropic and isothermal lines for a perfect gas

6.5 The Clausius–Clapeyron equation

From the Maxwell relationships (eqn 6.19c),

$$\left(\frac{\partial s}{\partial v}\right)_T = \left(\frac{\partial p}{\partial T}\right)_v$$

The left-hand side of this equation is the rate of change of entropy with volume at constant temperature.

If the change of phase of water from, say, liquid to steam is considered, then this takes place at constant pressure and constant temperature, known as *the saturation values*. Hence, if the change of phase between points 1 and 2 on Fig 6.5 is considered,

$$\left(\frac{\partial s}{\partial v}\right)_T = \frac{s_2 - s_1}{v_2 - v_1} \tag{6.76}$$

which, on substituting from eqn (6.19c), becomes

$$\left(\frac{\partial p}{\partial T}\right)_v = \frac{s_2 - s_1}{v_2 - v_1} \tag{6.77}$$

If there is a mixture of two phases, say steam and water, then the pressure is a function of temperature alone, as shown in Chapter 7. Thus

$$\frac{dp}{dT} = \frac{s_2 - s_1}{v_2 - v_1} \tag{6.78}$$

Now $(\partial h/\partial s)_p = T$, and during a change of phase both T and p are constant, giving $h_2 - h_1 = T(s_2 - s_1)$. Thus

$$\frac{h_2 - h_1}{v_2 - v_1} = T\frac{\mathrm{d}p}{\mathrm{d}T} \tag{6.79}$$

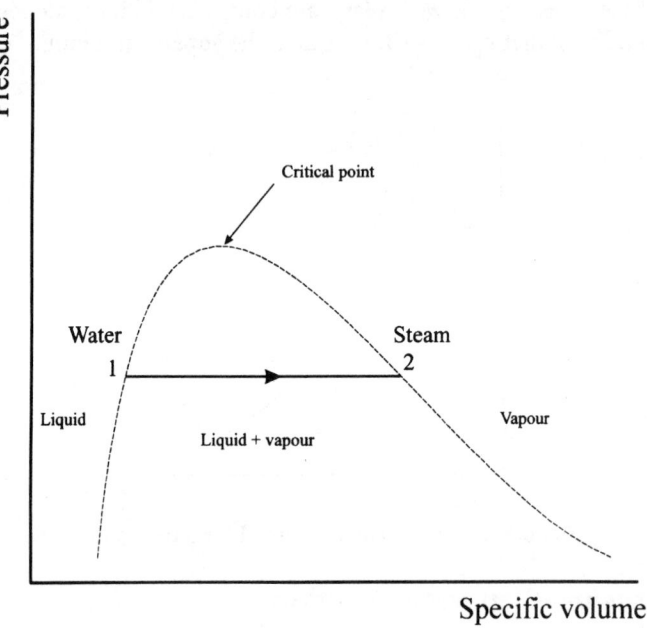

Fig. 6.5 Change of phase depicted on a $p-v$ diagram

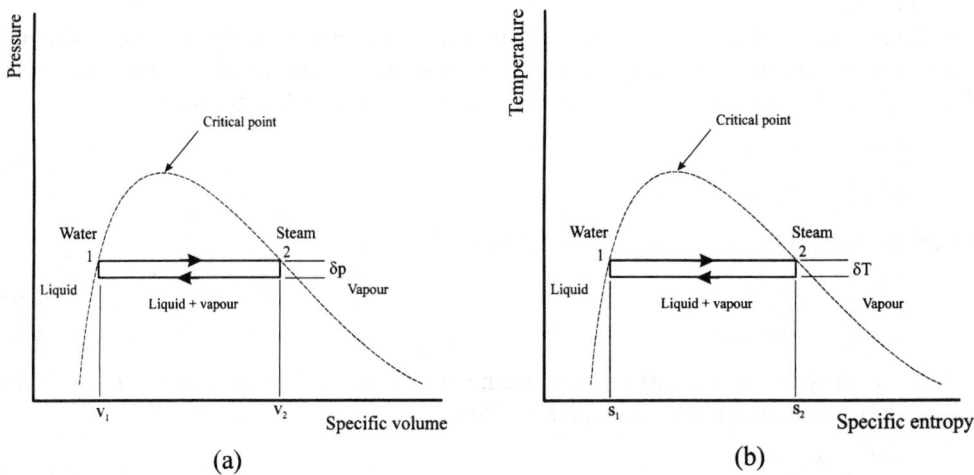

Fig. 6.6 Evaporation processes shown on state diagrams: (a) $p-v$ diagram; (b) $T-s$ diagram

This equation is known as the *Clapeyron equation*. Now $h_2 - h_1$, is the latent heat, h_{fg}. If point 1 is on the saturated liquid line and point 2 is on the saturated vapour line then eqn (6.79) can be rewritten as

$$T \frac{dp}{dT} = \frac{h_{fg}}{v_{fg}} \tag{6.80}$$

This can be depicted graphically, and is shown in Fig 6.6.

If the processes shown in Figs 6.6(a) and (b) are reversible, then the areas are equivalent to the work done. If the two diagrams depict the same processes on different state diagrams then the areas of the 'cycles' must be equal. Hence

$$(v_2 - v_1)\, dp = (s_2 - s_1)\, dT \tag{6.81}$$

Now, the change of entropy is $s_{fg} = s_2 - s_1 = h_{fg}/T$, again giving (eqn (6.80))

$$T \frac{dp}{dT} = \frac{h_{fg}}{v_{fg}}$$

Equation (6.80) is known as the *Clausius–Clapeyron equation*.

6.5.1 THE USE OF THE CLAUSIUS–CLAPEYRON EQUATION

If an empirical expression is known for the saturation pressure and temperature, then it is possible to calculate the change of entropy and specific volume due to the change in phase. For example, the change of enthalpy and entropy during the evaporation of water at 120°C can be evaluated from the slope of the saturation line defined as a function of pressure and temperature. The values of entropy and enthalpy for dry saturated steam can also be evaluated if their values in the liquid state are known. An extract of the properties of water is given in Table 6.1, and values of pressure and temperature on the saturation line have been taken at adjacent temperatures.

Table 6.1 Excerpt from steam tables, showing values on the saturated liquid and vapour lines

Temperature /(°C)	Pressure /(bar)	Specific volume /(m³/kg)	Change of specific volume, v_{fg}	Entropy of liquid, s_f /(kJ/kg K)	Enthalpy of liquid, h_f /(kJ/kg K)
115	1.691				
120	1.985	0.001060	0.8845	1.530	505
125	2.321				

At 120°C the slope of the saturation line is

$$\frac{dp}{dT} \approx \frac{2.321 - 1.691}{10} \times 10^5 = 6.30 \times 10^3 \, \text{N/m}^2 \, \text{K}$$

From the Clausius–Clapeyron equation, eqn (6.80),

$$\frac{s_2 - s_1}{v_{fg}} = \frac{dp}{dT} = \frac{s_g - s_f}{v_{fg}}$$

giving

$$s_g = s_f + v_{fg} \frac{dp}{dT} = 1.530 + 0.8845 \times 6.30 \times 10^3/10^3$$

$$= 1.530 + 5.57235 = 7.102 \text{ kJ/kgK}$$

The specific enthalpy on the saturated vapour line is

$$h_g = h_f + Ts_{fg}$$
$$= 505 + (120 + 273) \times 5.57235$$
$$= 2694.9 \text{ kJ/kg K}$$

These values are in good agreement with the tables of properties; the value of $s_g = 7.127$ kJ/kg K, while $h_g = 2707$ kJ/kg at a temperature of 120.2°C.

6.6 Concluding remarks

Thermodynamic relationships between properties have been developed which are independent of the particular fluid. These can be used to evaluate derived properties from primitive ones and to extend empirical data.

PROBLEMS

1 For a van der Waals' gas, which obeys the state equation

$$p = \frac{RT}{v - b} - \frac{a}{v^2}$$

show that the coefficient of thermal expansion, β, is given by

$$\beta = \frac{Rv^2(v - b)}{RTv^3 - 2a(v - b)^2}$$

and the isothermal compressibility, k, is given by

$$k = \frac{v^2(v - b)^2}{RTv^3 - 2a(v - b)^2}$$

Also evaluate

$$T\left(\frac{\partial p}{\partial v}\right)_T \left(\frac{\partial v}{\partial T}\right)_p^2$$

for the van der Waals' gas. From this find $c_p - c_v$ for an ideal gas, stating any assumptions made in arriving at the solution.

2 Prove the general thermodynamic relationship

$$\left(\frac{\partial c_p}{\partial p}\right)_T = -T\left(\frac{\partial^2 v}{\partial T^2}\right)_p$$

and evaluate an expression for the variation of c_p with pressure for a van der Waals' gas which obeys a law

$$p = \frac{\Re T}{v_m - b} - \frac{a}{v_m^2}$$

where the suffix 'm' indicates molar quantities and $\Re$ is the universal gas constant.
It is possible to represent the properties of water by such an equation, where

$$a = 5.525 \; \frac{(m^3)^2 \, bar}{(kmol)^2}; \qquad b = 0.03042 \; \frac{m^3}{kmol}; \qquad \Re = 8.3143 \; kJ/kmol \, K$$

Evaluate the change in c_p as the pressure of superheated water vapour is increased
from 175 bar to 200 bar at a constant temperature of 425°C. Compare this with the
value you would expect from the steam tables abstracted below:

	p = 175 bar		p = 200 bar	
t/(°C)	v/(m³/kg)	h/(kJ/kg)	v/(m³/kg)	h/(kJ/kg)
425	0.013914	3014.9	0.011458	2952.9
450	0.015174	3109.7	0.012695	3060.1

What is the value of $(\partial c_p / \partial p)_T$ of a gas which obeys the Clausius equation of state,
$p = \Re T / (v_m - b)$?

3 Show that if the ratio of the specific heats is 1.4 then

$$\left(\frac{\partial p}{\partial T} \right)_s = \frac{7}{2} \left(\frac{\partial p}{\partial T} \right)_v$$

4 Show that

(a) $\quad h - u = T^2 \left(\left(\frac{\partial (f/T)}{\partial T} \right)_v - \left(\frac{\partial (g/T)}{\partial T} \right)_p \right)$

(b) $\quad \dfrac{c_p}{c_v} = \left(\dfrac{\partial^2 g}{\partial T^2} \right)_p \bigg/ \left(\dfrac{\partial^2 f}{\partial T^2} \right)_v$

5 Show

(a) $\quad T \, ds = c_p \, dT - T \left(\frac{\partial v}{\partial T} \right)_p dp$

(b) $\quad T \, ds = c_v \left(\frac{\partial T}{\partial p} \right)_v dp + c_p \left(\frac{\partial T}{\partial v} \right)_p dv$

(c) $\quad \left(\frac{\partial h}{\partial p} \right)_T = v - T \left(\frac{\partial v}{\partial T} \right)_p$

6 Show that, for a pure substance,

$$\left(\frac{\partial s}{\partial T}\right)_p = \left(\frac{\partial s}{\partial T}\right)_v + \left(\frac{\partial s}{\partial v}\right)_T\left(\frac{\partial v}{\partial T}\right)_p$$

and from this prove that

$$c_p - c_v = T\left(\frac{\partial p}{\partial T}\right)_v\left(\frac{\partial v}{\partial T}\right)_p$$

The following table gives values of the specific volume of water, in m³/kg. A quantity of water is initially at 30°C, 20 bar and occupies a volume of 0.2 m³. It is heated at constant volume to 50°C and then cooled at constant pressure to 30°C. Calculate the net heat transfer to the water.

	Specific volume/(m³/kg)	
Pressure/(bar)	20	200
Temperature/(°C)		
30	0.0010034	0.0009956
50	0.0010112	0.0010034

[438 kJ]

7 Assuming that entropy is a continuous function, $s = s(T, v)$, derive the expression for entropy change

$$ds = c_v\frac{dT}{T} + \left(\frac{\partial p}{\partial T}\right)_v dv$$

and similarly, for $s = s(T, p)$ derive

$$ds = c_p\frac{dT}{T} - \left(\frac{\partial v}{\partial T}\right)_p dp$$

Apply these relationships to a gas obeying van der Waals' equation

$$p = \frac{RT}{v - b} - \frac{a}{v^2}$$

and derive an equation for the change of entropy during a process in terms of the basic properties and gas parameters. Also derive an expression for the change of internal energy as such a gas undergoes a process.

$$\left[s_2 - s_1 = c_p\ln\frac{T_2}{T_1} - R\ln\frac{\left(p_2 + \dfrac{a}{v^2} - \dfrac{2a(v - b)}{v^3}\right)}{\left(p_1 + \dfrac{a}{v^2} - \dfrac{2a(v - b)}{v^3}\right)}; \quad u_2 - u_1 = \int_1^2 c_v\,dT - a\left(\frac{v_1 - v_2}{v_1 v_2}\right) \right]$$

7
Equations of State

The properties of fluids can be defined in two ways, either by the use of tabulated data (e.g. steam tables) or by state equations (e.g. perfect gas law). Both of these approaches have been developed by observation of the behaviour of fluids when they undergo simple processes. It has also been possible to model the behaviour of such fluids from 'molecular' models, e.g. the kinetic theory of gases. A number of models which describe the relationships between properties for single component fluids, or constant composition mixtures, will be developed here.

7.1 Ideal gas law

The ideal and perfect gas laws can be developed from a number of simple experiments, or a simple molecular model. First the experimental approach will be considered.

If a fixed mass of a single component fluid is contained in a closed system then two processes can be proposed:

(i)   the volume of the gas can be changed by varying the pressure, while maintaining the temperature constant;

(ii) the volume of the system can be changed by varying the temperature, while maintaining the pressure constant.

The first process is an isothermal one, and is the experiment proposed by Boyle to define Boyle's law (also known as Mariotte's law in France). The second process is an isobaric one and is the one used to define Charles' law (also known as Gay-Lussac's law in France).

The process executed in (i) can be described mathematically as

$$v = v(p)_T \tag{7.1}$$

while the second one, process (ii), can be written

$$v = v(T)_p \tag{7.2}$$

Since these processes can be undergone independently then the relationship between the three properties is

$$v = v(p, T) \tag{7.3}$$

Equation (7.3) is a functional form of the *equation of state* of a single component fluid. It can be seen to obey the two-property rule, which states that any property of a single component fluid or constant composition mixture can be defined as a function of two independent properties. The actual mathematical relationship has to be found from experiment (or a simulation of the molecular properties of the gas molecules), and this can be derived by knowing that, if the property, v, is a continuous function of the other two properties, p and T, as discussed in Chapter 6, then

$$dv = \left(\frac{\partial v}{\partial p}\right)_T dp + \left(\frac{\partial v}{\partial T}\right)_P dT \qquad (7.4)$$

Hence, if the partial derivatives $(\partial v/\partial p)_T$ and $(\partial v/\partial T)_p$ can be evaluated then the gas law will be defined. It is possible to evaluate the first derivative by a Boyle's law experiment, and the second one by a Charles' law experiment. It is found from Boyle's law that

$$pv = \text{constant} \qquad (7.5)$$

giving

$$\left(\frac{\partial p}{\partial v}\right)_T = -\frac{p}{v} \qquad (7.6)$$

Similarly, it is found from Charles' law that

$$\frac{v}{T} = \text{constant} \qquad (7.7)$$

giving

$$\left(\frac{\partial v}{\partial T}\right)_p = \frac{v}{T} \qquad (7.8)$$

Substituting eqns (7.6) and (7.8) into eqn (7.4) gives

$$dv = -\frac{v}{p} dp + \frac{v}{T} dT \qquad (7.9)$$

which may be integrated to give

$$\frac{pv}{T} = \text{constant} \qquad (7.10)$$

Equation (7.10) is known as the *ideal gas law*. This equation contains no information about the internal energy of the fluid, and does not define the specific heat capacities. If the specific heat capacities are not functions of temperature then the gas is said to obey the *perfect gas law*: if the specific heat capacities are functions of temperature (i.e. the internal energy and enthalpy do not vary linearly with temperature) then the gas is called an ideal gas.

It is possible to define two coefficients from eqn (7.4), which are analogous to concepts used for describing the properties of materials. The first is called the *isothermal compressibility*, or *isothermal bulk modulus*, k. This is defined as the 'volumetric strain' produced by a change in pressure, giving

$$k = -\frac{1}{v}\left(\frac{\partial v}{\partial p}\right)_T = -\frac{1}{V}\left(\frac{\partial V}{\partial p}\right)_T \tag{7.11}$$

The negative sign is introduced because an increase in pressure produces a decrease in volume and hence $(\partial v/\partial p) < 0$; it is more convenient to have a coefficient with a positive value and the negative sign achieves this. The isothermal compressibility of a fluid is analogous to the Young's modulus of a solid.

The other coefficient that can be defined is the *coefficient of expansion, β*. This is defined as the 'volumetric strain' produced by a change in temperature, giving

$$\beta = \frac{1}{v}\left(\frac{\partial v}{\partial T}\right)_p = \frac{1}{V}\left(\frac{\partial V}{\partial T}\right)_p \tag{7.12}$$

The coefficient of expansion is analogous to the coefficient of thermal expansion of a solid material.

Hence, eqn (7.4) may be written

$$\frac{dv}{v} = \frac{dV}{V} = -kdp + \beta dT \tag{7.13}$$

The ideal gas law applies to gases in the superheat phase. This is because when gases are superheated they obey the kinetic theory of gases in which the following assumptions are made:

- molecules are solid spheres;
- molecules occupy a negligible proportion of the total volume of the gas;
- there are no forces of attraction between the molecules, but there are infinite forces of repulsion on contact.

If a gas is not superheated the molecules are closer together and the assumptions are less valid. This has led to the development of other models which take into account the interactions between the molecules.

7.2   Van der Waals' equation of state

In an attempt to overcome the limitations of the perfect gas equation, a number of modifications have been made to it. The two most obvious modifications are to assume:

- the diameters of the molecules are an appreciable fraction of the mean distances between them, and the mean free path between collisions – this basically means that the volume occupied by the molecules is not negligible;
- the molecules exert forces of attraction which vary with the distance between them, while still exerting infinite forces of repulsion on contact.

First, if it is assumed that the molecules occupy a significant part of the volume occupied by the gas then eqn (7.10) can be modified to

$$p = \frac{\Re T}{v - b} \tag{7.14}$$

where b is the volume occupied by the molecules. This is called the *Clausius equation of state*.

Second, allowing for the forces of attraction between molecules gives the *van der Waals' equation of state*, which is written

$$p = \frac{\mathfrak{R}T}{v - b} - \frac{a}{v^2} \tag{7.15}$$

Figure 7.1 shows five isotherms for water calculated using van der Waals' equation (the derivation of the constants in the eqn (7.15) is described below). It can be seen that eqn (7.15) is a significant improvement over the perfect gas law as the state of the water approaches the saturated liquid and vapour lines. The line at 374°C is the isotherm at the critical temperature. This line passes through the critical point, and follows closely the saturated vapour line (in fact, it lies just in the two-phase region, which indicates an inaccuracy in the method). At the critical temperature the isotherm exhibits a point of inflection at the critical point. At temperatures above the critical temperature the isotherms exhibit monotonic behaviour, and by 500°C the isotherm is close to a rectangular hyperbola, which would be predicted for a perfect gas. At temperatures below the critical isotherm, the isotherms are no longer monotonic, but exhibit the characteristics of a cubic equation. While this characteristic is not in agreement with empirical experience it does result in the correct form of function in the saturated liquid region – which could never be achieved by a perfect gas law. It is possible to resolve the problem of the correct pressure to use for an isotherm in the two-phase region by considering the Gibbs function, which must remain constant during the evaporation process (see Chapter 1). This results in a constant pressure (horizontal) line which obeys an equilibrium relationship described in Section 7.4. It can also be seen from Fig 7.1 that the behaviour of a substance obeying the van der Waals' equation of state approaches that of a perfect gas when

- the temperature is above the critical temperature
- the pressure is low compared with the critical pressure.

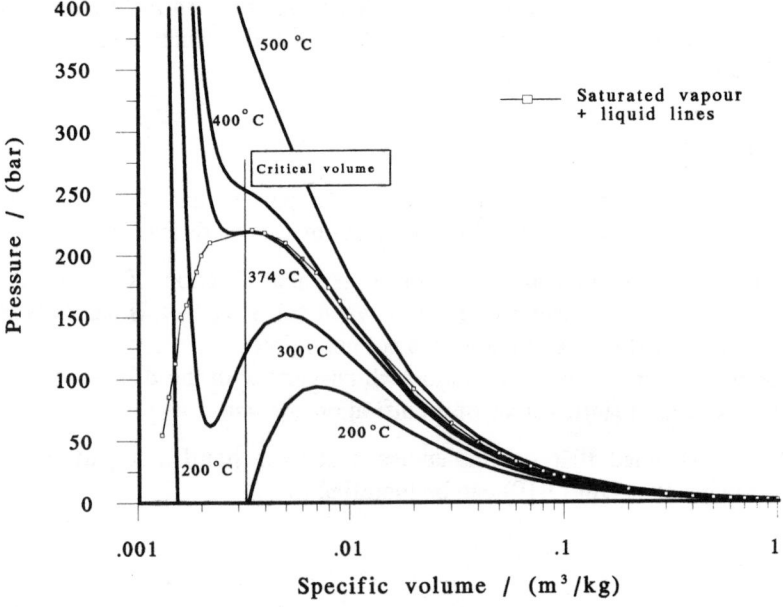

Fig. 7.1 Isotherms on a $p-v$ diagram calculated using van der Waals' equation

Exercise

Evaluate the isothermal compressibility, k, and coefficient of expansion, β, of a van der Waals' gas.

7.3 Law of corresponding states

If it is assumed that all substances obey an equation of the form defined by the van der Waals' equation of state then all state diagrams will be *geometrically similar*. This means that if the diagrams were normalised by dividing the properties defining a particular state point (p, v, T) by the properties at the critical point (p_c, v_c, T_c) then all state diagrams based on these *reduced properties* will be identical. Thus there would be a general relationship

$$v_R = f(p_R, T_R) \tag{7.16}$$

where $v_R = v/v_c = $ reduced volume;

$p_R = p/p_c = $ reduced pressure;

$T_R = T/T_c = $ reduced temperature.

To be able to define the general equation that might represent all substances it is necessary to evaluate the values at the critical point in relation to the other parameters. It has been seen from Fig 7.1 that the critical point is the boundary between two different forms of behaviour of the van der Waals' equation. At temperatures above the critical point the curves are monotonic, whereas below the critical point they exhibit maxima and minima. Hence, the critical isotherm has the following characteristics: $(\partial p/\partial v)_{T_c}$ and $(\partial^2 p/\partial v^2)_{T_c} = 0$.

Consider van der Waals' equation

$$p = \frac{RT}{v - b} - \frac{a}{v^2}$$

This can be differentiated to give the two differentials, giving

$$\left(\frac{\partial p}{\partial v}\right)_T = -\frac{RT}{(v - b)^2} + \frac{2a}{v^3} \tag{7.17}$$

$$\left(\frac{\partial^2 p}{\partial^2 v}\right)_T = \frac{2RT}{(v - b)^3} - \frac{6a}{v^4} \tag{7.18}$$

and at the critical point

$$-\frac{RT_c}{(v_c - b)^2} + \frac{2a}{v_c^3} = 0 \tag{7.19}$$

$$\frac{2RT_c}{(v_c - b)^3} - \frac{6a}{v_c^4} = 0 \tag{7.20}$$

giving

$$v_c = 3b; \qquad T_c = \frac{8a}{27bR}; \qquad p_c = \frac{a}{27b^2} \tag{7.21}$$

Hence

$$a = 3p_c v_c^2; \qquad b = \frac{v_c}{3}; \qquad R = \frac{8}{3} \frac{p_c v_c}{T_c} \qquad (7.22)$$

The equation of state for a gas obeying the law of corresponding states can be written

$$p_R = \frac{8T_R}{(3v_R - 1)} - \frac{3}{v_R^2} \qquad (7.23)$$

or

$$\left(p_R + \frac{3}{v_R^2} \right)(3v_R - 1) = 8T_R \qquad (7.24)$$

Equation (7.24) is the equation of state for a substance obeying van der Waals' equation: it should be noted that it does not explicitly contain values of a and b. It is possible to obtain a similar solution which omits a and b for any two-parameter state equation, but such a solution has not been found for state equations with more than two parameters. The law of corresponding states based on van der Waals' equation does not give a very accurate prediction of the properties of substances over their whole range, but it does demonstrate some of the important differences between real substances and perfect gases. Figure 7.1 shows the state diagram for water evaluated using parameters in van der Waals' equation based on the law of corresponding states. The parameters used for the calculation of Fig 7.1 will now be evaluated.

The values of the relevant parameters defining the critical point of water are

$$v_c = 0.00317 \text{ m}^3/\text{kg}; \qquad p_c = 221.2 \text{ bar}; \qquad T_c = 374.15°C$$

which gives

$$b = \frac{v_c}{3} = \frac{0.00317}{3} = 0.0010567 \text{ m}^3/\text{kg}$$

$$a = 3p_c v_c^2 = 3 \times 221.2 \times 0.00317^2 = 0.0066684 \text{ bar m}^6/\text{kg}^2$$

$$R = \frac{8 \times 221.2 \times 0.00317}{3 \times (374.15 + 273)} = 0.002889 \text{ bar m}^3/\text{kg K}$$

Thus, the van der Waals' equation for water is

$$p = \frac{0.002889T}{(v - 0.0010567)} - \frac{0.0066684}{v^2} \qquad (7.25)$$

Note that the value for 'R' in this equation is not the same as the specific gas constant for the substance behaving as a perfect gas. This is because it has been evaluated using values of parameters at the critical point, which is not in the region where the substance performs as a perfect gas.

It is possible to manipulate the coefficients of van der Waals' equation to use the correct value of R for the substance involved. This does not give a very good approximation for the critical isotherm, but is reasonable elsewhere. Considering eqn (7.22), the term for v_c

can be eliminated to give

$$a = \frac{27R^2T_c^2}{64p_c} \quad \text{and} \quad b = \frac{RT_c}{8p_c}$$

(7.26)

Substituting the values of p_c and T_c into these terms, using $R = \mathfrak{R}/18$, gives

$$a = \frac{27 \times (8.3143 \times 10^3/18)^2 \times (374.1 + 273)^2}{64 \times 221.2 \times 10^5}$$

$$= 1703.4 \, \text{Pa} \, \text{m}^6/\text{kg}^2 = 0.017034 \, \text{bar} \, \text{m}^6/\text{kg}^2$$

$$b = \frac{8.3143 \times 10^3 \times (374.1 + 273)}{18 \times 8 \times 221.2 \times 10^5} = 0.0016891 \, \text{m}^3/\text{kg}$$

which results in the following van der Waals' equation for water:

$$p = \frac{0.004619T}{(v - 0.0016891)} - \frac{0.017034}{v^2}$$

(7.27)

It can be readily seen that eqn (7.27) does not accurately predict the critical isotherm at low specific volumes, because the value of b is too big. However, it gives a reasonable prediction of the saturated vapour region, as will be demonstrated for the isotherms at 200°C, 300°C and the critical isotherm at 374°C. These are shown in Fig 7.2, where the predictions are compared with those from eqn (7.25).

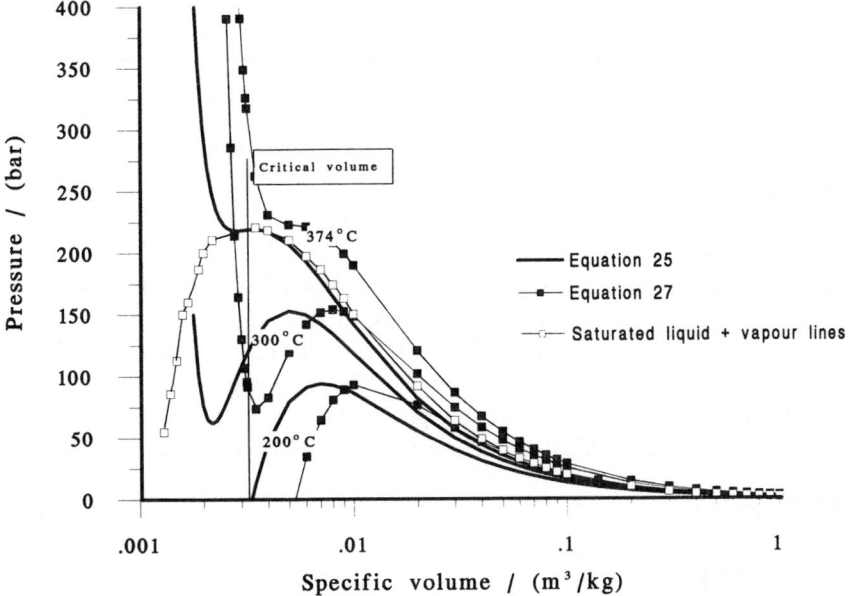

Fig. 7.2 Comparison of isotherms calculated by van der Waals' equation based on eqns (7.25) and (7.27)

It is interesting to compare the values calculated using van der Waals' equation with those in tables. This has been done for the isotherm at 200°C, and a pressure of 15 bar (see Fig 7.3), and the results are shown in Table 7.1.

Table 7.1 Specific volume at 200°C and 15 bar

Region	Specific volume from eqn (7.25)/(m³/kg)	Specific volume from eqn (7.27)/(m³/kg)	Specific volume from tables/(m³/kg)
Saturated liquid	0.0016	0.0017	0.001157
Saturated vapour	0.09	0.14	0.1324

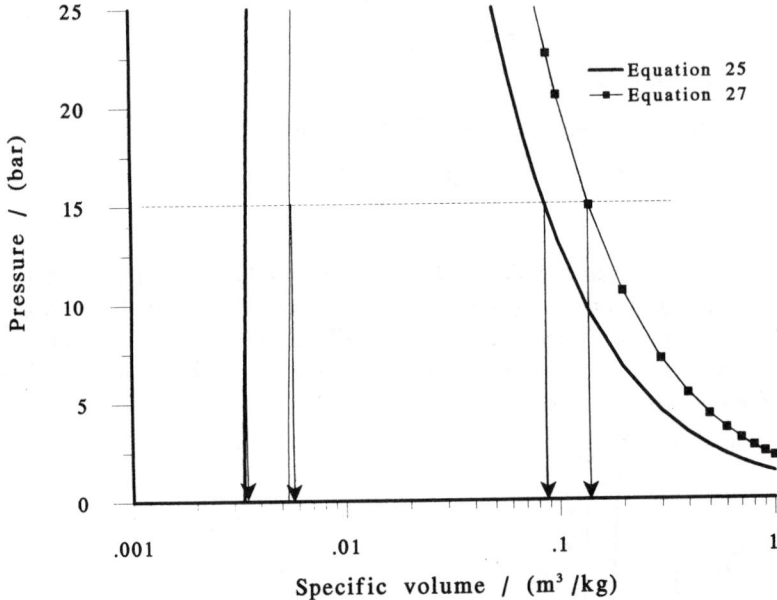

Fig. 7.3 Expanded diagram of 200°C isotherms, showing specific volumes at 15 bar

The above diagrams and tables show that van der Waals' equation does not give a good overall representation of the behaviour of a gas in the liquid and mixed state regions. However, it is a great improvement on the perfect, or ideal, gas equation in regions away from superheat. It will be shown later that van der Waals' equation is capable of demonstrating certain characteristics of gases in the two-phase region, e.g. liquefaction. There are other more accurate equations for evaluating the properties of substances, and these include

- the virial equation:

$$pv = RT\left(1 + \frac{b}{v} + \frac{c}{v^2} + \frac{d}{v^3} + \cdots\right) \qquad (7.28)$$

- the Beattie–Bridgeman equation:

$$p = \frac{RT}{v^2}(1 - e)(v + B) - \frac{A}{v^2}$$

where

$$A = A_0\left(1 - \frac{a}{v}\right), \qquad B = B_0\left(1 - \frac{b}{v}\right), \qquad e = \frac{c}{vT^3}$$

(7.29)

- the Bertholet equation:

$$p = \frac{RT}{v-b} - \frac{a}{Tv^2}$$

(7.30)

- the Dieterici equation:

$$p = \frac{RT}{v-b} e^{-a/RTv}$$

(7.31)

Returning to the van der Waals' gas, examination of eqns (7.22) shows that

$$z_c = \frac{p_c v_c}{RT_c} = \frac{3}{8} = 0.375$$

(7.32)

This value z_c (sometimes denoted μ_c) is the compression coefficient, or compressibility factor, of a substance at the critical point. In general, the compression coefficient, z, is defined as

$$z = \frac{pv}{RT}$$

(7.33)

It is obvious that, for a real gas, z is not constant, because $z = 1$ for a perfect gas (i.e. in the superheat region), but it is 0.375 (ideally) at the critical point. There are tables and graphs which show the variation of z with state point, and these can be used to calculate the properties of real gases.

7.4   Isotherms or isobars in the two-phase region

Equation (7.15) predicts that along an isotherm the pressure is related to the specific volume by a cubic equation, and this means that in some regions there are three values of volume which satisfy a particular pressure. Since the critical point has been defined as an inversion point then the multi-valued region lies below the critical point, and experience indicates that this is the two-phase (liquid + vapour) region. It is known that pressure and temperature are not independent variables in this region, and that fluids evaporate at constant temperature in a constant pressure chamber; hence these isotherms should be at constant pressure. This anomaly can be resolved by considering the Gibbs energy of the fluid in the two-phase region. Since the evaporation must be an equilibrium process then it must obey the conditions of equilibrium. Considering the isotherm shown in Fig 7.4, which has been calculated using eqn (7.27) for water at 300°C, it can be seen that in the regions from 3 to 1 and from 7 to 5, a decrease in pressure results in an increase in the specific volume. However, in the region between 5 and 3 an increase in pressure results in an increase in specific volume: this situation is obviously unstable (see also Chapter 6). It was shown in Chapter 1 that equilibrium was defined by $dG|_{p,T} \geq 0$. Now

$$dg = v\,dp - s\,dT$$

(7.34)

and hence along an isotherm the variation of Gibbs function from an initial point, say, 1 to another point is

$$\Delta g = g - g_1 = \int_{p_1}^{p} v \, dp \qquad (7.35)$$

This variation has been calculated for the region between points 1 and 7 in Fig 7.4, and is shown in Fig 7.5.

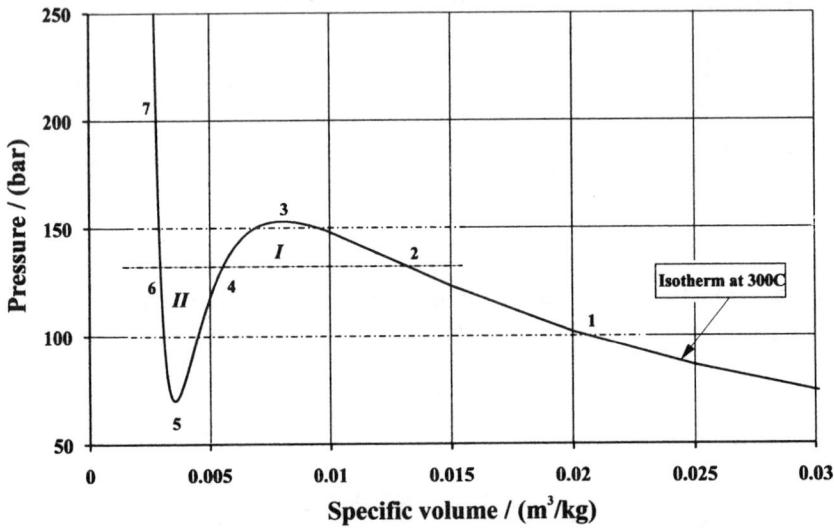

Fig. 7.4 Variation of pressure and specific volume for water along the 300°C isotherm

It can be seen from Fig 7.5 that the Gibbs function increases along the isotherm from point 1 to point 3. Between point 3 and point 5 the Gibbs function decreases, but then begins to increase again as the fluid passes from 5 to 7. The regions between 1 and 3 and 5 and 7 can be seen to be stable in terms of variation of Gibbs function, whereas that between 3 and 5 is unstable: this supports the argument introduced earlier based on the variation of pressure and volume. Figure 7.5 also shows that the values of Gibbs energy and pressure are all equal at the points 2, 4 and 6, which is a line of constant pressure in Fig 7.4 This indicates that the Gibbs function of the liquid and vapour phases can be equal if the two ends of the evaporation process are at equal pressures for an isotherm. Now considering the region between the saturated liquid and saturated vapour lines: the Gibbs function along the 'isotherm' shown in Fig 7.4 will increase between points 2 and 3, and the substance would attempt to change spontaneously back from 3 to 2, or from 3 to 4, and hence point 3 is obviously unstable. In a similar manner the stability of point 5 can be considered. An isobar at 100 bar is shown in Fig 7.5, and it can be seen that it crosses the Gibbs function line at three points – the lowest Gibbs function is at state 1. Hence, state 1 is more stable than the other points. If that isobar were moved down further until it passed through point 5 then there would be other points of higher stability than point 5 and the system would tend to move to these points. This argument has still not defined the equilibrium line

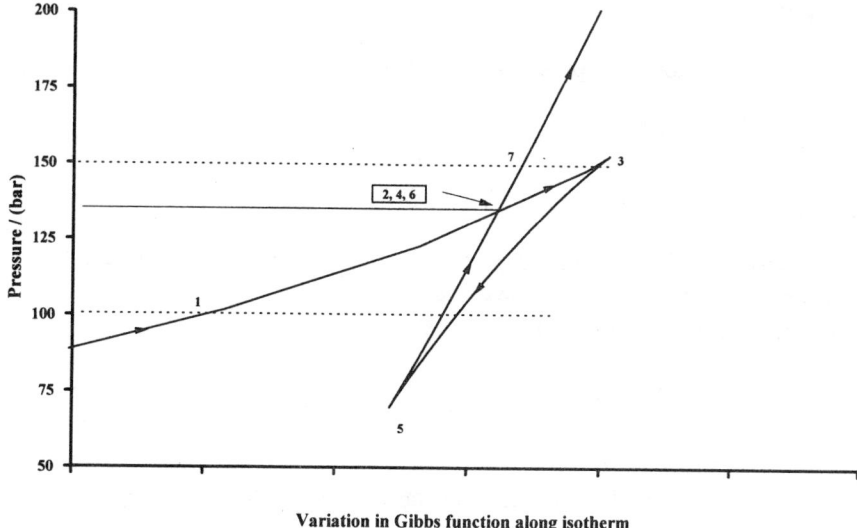

Variation in Gibbs function along isotherm

Fig. 7.5 Variation of Gibbs function for water along the 300°C isotherm

between the liquid and vapour lines, but this can be defined by ensuring that the Gibbs function remains constant in the two-phase region, and this means the equilibrium state must be defined by constant temperature and constant Gibbs function. Three points on that line are defined by states 2, 4 and 6. The pressure which enables constant Gibbs function to be achieved is such that

$$\int_{p_2}^{p_6} v\,dp = 0 \tag{7.36}$$

and this means that

$$\underbrace{\int_{p_2}^{p_4} v\,dp}_{\text{Region I}} + \underbrace{\int_{p_4}^{p_6} v\,dp}_{\text{Region II}} = 0 \tag{7.37}$$

which means that the area of the region between the equilibrium isobar and line 2-3-4 (Region I) must be equal to that between the isobar and line 4-5-6 (Region II). This is referred to as *Maxwell's equal area rule*, and the areas are labelled I and II in Fig 7.4.

7.5 Concluding remarks

A number of different equations of state have been introduced which can describe the behaviour of substances over a broader range than the common perfect gas equation. Van der Waal's equation has been analysed quite extensively, and it has been shown to be capable of defining the behaviour of gases close to the saturated vapour line. The law of corresponding states has been developed, and enables a general equation of state to be considered for all substances. The region between the saturated liquid and vapour lines has been analysed, using Gibbs energy to define the equilibrium state.

PROBLEMS

1   The Dieterici equation for a pure substance is given by

$$p = \frac{\Re T}{v - b} e^{-(a/\Re Tv)}$$

Determine

(a) the constants a and b in terms of the critical pressure and temperature;
(b) the compressibility factor at the critical condition;
(c) the law of corresponding states.

$$\left[a = \frac{4\Re^2 T_c^2}{p_c e^2}; b = \frac{\Re T_c}{p_c e^2}; z_c = 0.2707; p_R = \frac{T_R}{(2v_R - 1)} \exp\left[2\left(1 - \frac{1}{v_R T_R}\right)\right] \right]$$

2   Derive expressions for $(\partial c_v / \partial v)_T$ for substances obeying the following laws:

(a) $$p = \frac{\Re T}{v - b} e^{-(a/\Re Tv)}$$

(b) $$p = \frac{\Re T}{v - b} - \frac{a}{Tv^2}$$

(c) $$p = \frac{\Re T}{v - b} - \frac{a}{v(v - b)} + \frac{c}{v^3}.$$

Discuss the physical implication of the results.

$$\left[\frac{pa^2}{\Re^2 v^2 T^3}; -\frac{2a}{v^2 T^2}; 0 \right]$$

3   The difference of specific heats for an ideal gas, $c_{p,m} - c_{v,m} = \Re$. Evaluate the difference in specific heats for gases obeying (a) the van der Waals', and (b) the Dieterici equations of state. Comment on the results for the difference in specific heat for these gases compared with the ideal gas.

$$\left[\Re \bigg/ \left(1 - \frac{2a(v - b)^2}{\Re Tv^3}\right) \right]$$

4   Derive an expression for the law of corresponding states for a gas represented by the following expression:

$$p = \frac{\Re T}{v - b} - \frac{a}{Tv^2}$$

$$\left[p_R = \frac{8 T_R}{3 v_R - 1} - \frac{3}{T_R v_R^2} \right]$$

5 Show, for a gas obeying the state equation

$$pv = (1 + \alpha)\Re T$$

where α is a function of temperature alone, that the specific heat at constant pressure is given by

$$c_p = -\Re T \frac{\mathrm{d}^2 (\alpha T)}{\mathrm{d}T^2} \ln p + c_{p_0}$$

where c_{p_0} is the specific heat at unit pressure.

6 The virial equation of state is

$$pv = \Re T \left(b_1 + \frac{b_2}{v} + \frac{b_3}{v^2} + \cdots \right)$$

Compare this equation with van der Waals' equation of state and determine the first two virial coefficients, b_1 and b_2, as a function temperature and the van der Waals' constants.

Determine the critical temperature and volume (T_c, v_c) for the van der Waals' gas, and show that

$$b_2 = \frac{v_c}{3} \left(1 - \frac{27T_c}{8v_c} \right)$$

$$[b_1 = 1; \; b_2 = (b - a/\Re T)]$$

7 The equation of state for a certain gas is

$$\frac{pv}{\Re T} = 1 + pe^{-AT}$$

where A is constant.

Show that if the specific heat at constant pressure at some datum pressure p_0 is c_{p_0} then the value of the specific heat at constant pressure at the state (T, p) is given by

$$c_p - c_{p_0} = \Re T A \, e^{-AT} (2 - AT)(p - p_0)$$

8 How can the equation of state in the form of a relationship between pressure, volume and temperature be used to extend limited data on the entropy of a substance.

A certain gas, A, has the equation of state

$$pv = \Re T (1 + \alpha p)$$

where α is a function of temperature alone. Show that

$$\left(\frac{\partial s}{\partial p} \right)_T = -\Re \left(\frac{1}{p} + \alpha + T \frac{\mathrm{d}\alpha}{\mathrm{d}T} \right)$$

Another gas B behaves as an ideal gas. If the molar entropy of gas A is equal to that of gas B when both are at pressure p_0 and the same temperature T, show that if the pressure is increased to p with the temperature maintained constant at T, the molar entropy of gas B exceeds that of gas A by an amount

$$\Re(p - p_0)\left(a + T\frac{da}{dT}\right)$$

9 A gas has the equation of state

$$\frac{pv}{\Re T} = a - bT$$

where a and b are constants. If the gas is compressed reversibly and isothermally at the temperature T' show that the compression will also be adiabatic if

$$T' = \frac{a}{2b}$$

8

Liquefaction of Gases

If the temperature and pressure of a gas can be brought into the region between the saturated liquid and saturated vapour lines then the gas will become 'wet' and this 'wetness' will condense giving a liquid. Most gases existing in the atmosphere are extremely superheated, but are at pressures well below their critical pressures. Critical point data for common gases and some hydrocarbons are given in Table 8.1.

Table 8.1 Critical temperatures and pressures for common substances

Substance	Critical temperature $T_c°C$ [K]	Critical pressure, p_c (bar)
Water (H_2O)	374 [647]	221.2
Methane (CH_4)	−82 [191]	46.4
Ethane (C_2H_6)	32 [305]	49.4
Propane (C_3H_8)	96 [369]	43.6
Butane (C_4H_{10})	153 [426]	36.5
Carbon dioxide (CO_2)	31 [304]	89
Oxygen (O_2)	−130 [143]	51
Hydrogen (H_2)	−243 [30]	13
Nitrogen (N_2)	−147 [126]	34

Figure 8.1 depicts qualitatively the state point of oxygen at ambient conditions and shows that it is a superheated gas at this pressure and temperature, existing at well above the critical temperature but below the critical pressure. If it is desired to liquefy the gas it is necessary to take its state point into the saturated liquid–saturated vapour region. This can be achieved in a number of ways. First, experience indicates that 'heat' has to be taken out of the gas. This can be done by two means:

(i)   cooling the gas by heat transfer to a cold reservoir, i.e. refrigeration;
(ii) expanding the gas in a reversible manner, so that it does work.

8.1   Liquefaction by cooling – method (i)

This method is satisfactory if the liquefaction process does not require very low temperatures. A number of common gases can be obtained in liquid form by cooling.

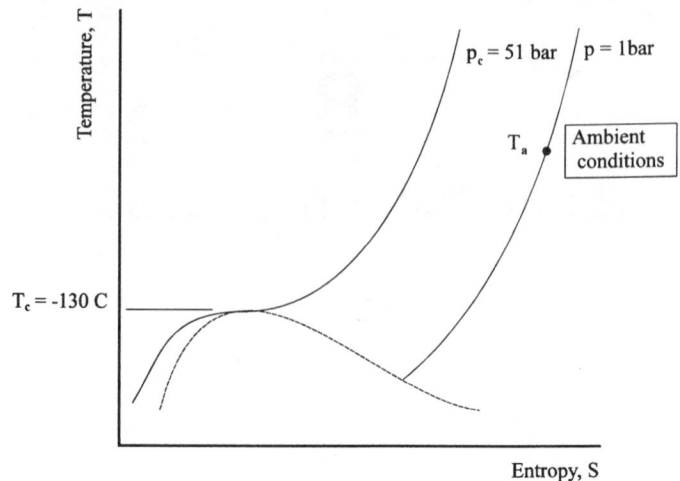

Fig. 8.1 State point of oxygen at ambient temperature and pressure

Examples of these are the hydrocarbons butane and propane, which can both exist as liquids at room temperature if they are contained at elevated pressures. Mixtures of hydrocarbons can also be obtained as liquids and these include liquefied petroleum gas (LPG) and liquefied natural gas (LNG).

A simple refrigerator and the refrigeration cycle are shown in Figs 8.2(a) and (b) respectively.

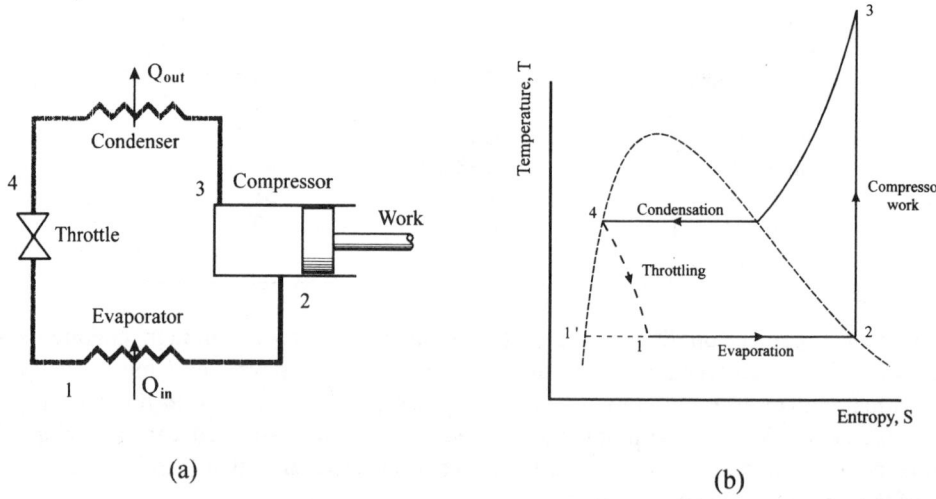

(a)   (b)

Fig. 8.2 A simple refrigerator: (a) schematic of plant; (b) temperature–entropy diagram

Consider the throttling process in Fig 8.2, which is between 4 and 1. The working fluid enters the throttle at a high pressure in a liquid state, and leaves it at a lower pressure and temperature as a wet vapour. If the mass of working fluid entering the throttle is 1 kg then the mass of liquid leaving the throttle is $(1 - x_1)$ kg. If this liquid were then withdrawn to a vessel, and the mass of fluid in the system were made up by adding $(1 - x_1)$ kg of gas at state 3 then it would be possible to liquefy that gas. The liquefaction has effectively taken place because, in passing through the throttle, the quality (dryness fraction) of the fluid has been increased, and the energy to form the vapour phase has been obtained from the latent heat of the liquid thus formed: the throttling process is an isenthalpic one, and hence energy has been conserved. If the working fluid for this cycle was a gas, which it was desired to liquefy, then the liquid could be withdrawn at state $1'$.

The simple refrigeration system shown in Fig 8.2 might be used to liquefy substances which boil at close to the ambient temperature, but it is more common to have to use refrigeration plants in series, referred to as a *cascade*, to achieve reasonable levels of cooling. The cascade plant is also more cost-effective because it splits the temperature drop between two working fluids. Consider Fig 8.2: the difference between the top and bottom temperatures is related to the difference between the pressures. To achieve a low temperature it is necessary to reduce the evaporator pressure, and this will increase the specific volume of the working fluid and hence the size of the evaporator. The large temperature difference will also decrease the coefficient of performance of the plant. The use of two plants in cascade enables the working fluid in each section to be optimised for the temperature range encountered. Two plants in cascade are shown in Fig 8.3(a), and the $T-s$ diagrams are depicted in Fig 8.3(b).

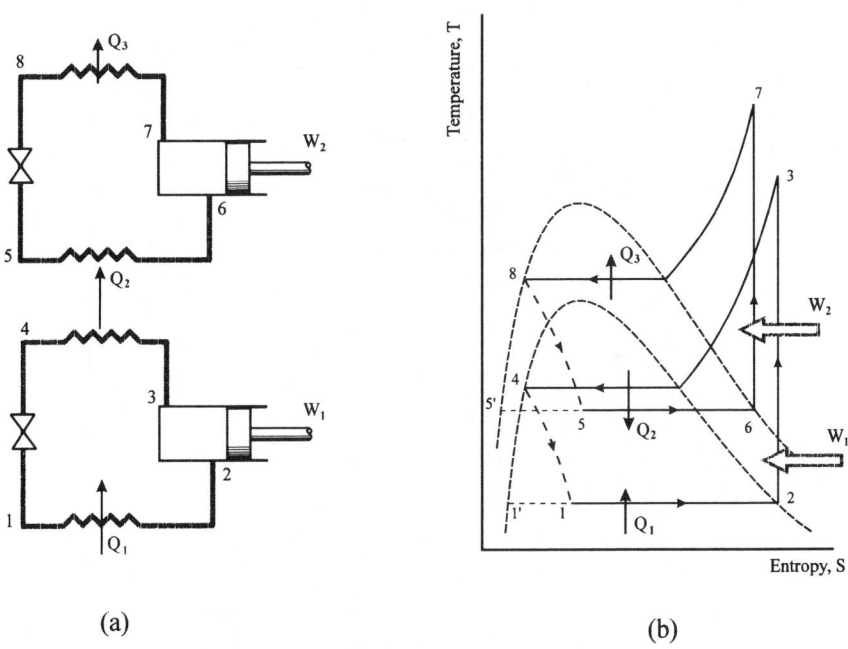

(a) (b)

Fig. 8.3 Cascade refrigeration cycle

In the cascade arrangement the cooling process takes place in two stages: it is referred to as a binary cascade cycle. This arrangement is the refrigeration equivalent of the combined cycle power station. The substance to be liquefied follows cycle $1-2-3-4-1$, and the liquid is taken out at state $1'$. However, instead of transferring its waste energy, Q_2, to the environment it transfers it to another refrigeration cycle which operates at a higher temperature level. The working fluids will be different in each cycle, and that in the high temperature cycle, $5-6-7-8-5$, will have a higher boiling point than the substance being liquefied. The two cycles can also be used to liquefy both of the working fluids, in which case liquid will also be taken out at state 5'.

The overall coefficient of performance of the two plants working in cascade can be evaluated by considering the heat flow through each section and is given by

$$\left(1 + \frac{1}{\beta'}\right) = \left(1 + \frac{1}{\beta_1'}\right)\left(1 + \frac{1}{\beta_2'}\right) \qquad (8.1)$$

where β' is the overall coefficient of the combined plant, and β'_1 and β'_2 are the coefficients of performance of the separate parts of the cascade. The coefficient of performance of the overall plant is less than that of either individual part.

The liquefaction of natural gas is achieved by means of a ternary cascade refrigeration cycle, and the components are shown in Fig 8.4. In this system a range of hydrocarbons are used to cool each other. The longer chain hydrocarbons have higher boiling points than the shorter ones, and hence the alkanes can be used as the working fluids in the plant.

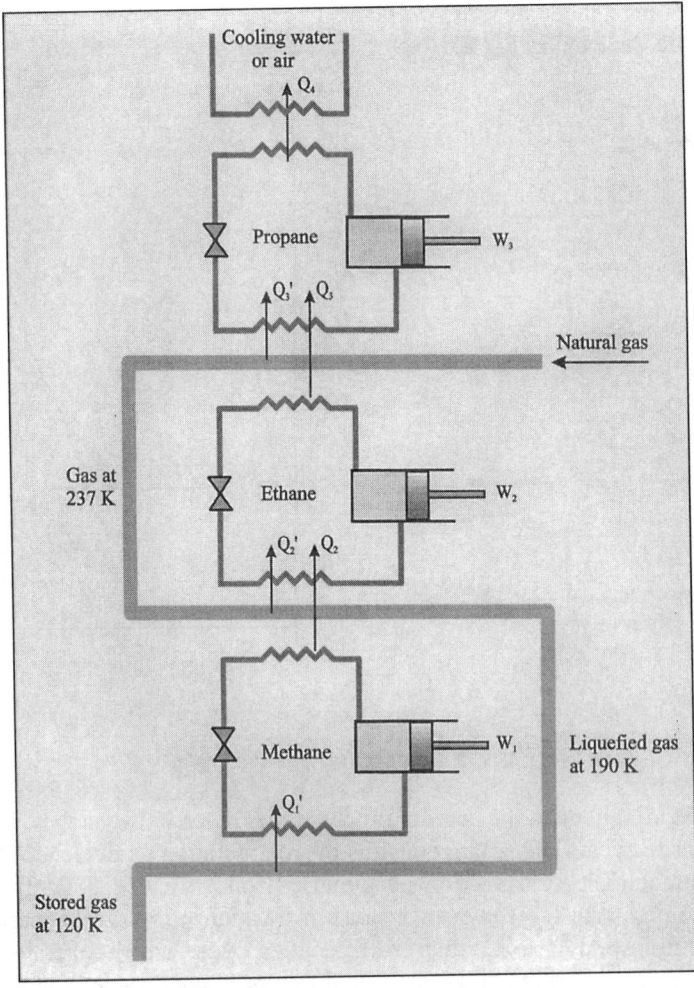

Fig. 8.4 Plant for liquefaction of natural gas

Another common gas that can be obtained in a non-gaseous form is carbon dioxide (CO_2). This is usually supplied nowadays in a solid form called *dry ice*, although it was originally provided as a pressurised liquid. Dry ice is obtained by a modification of the refrigeration process, and the plant and relevant $T-s$ diagrams are shown in Figs 8.5(a) and (b) respectively. This is like a ternary cascade system, but in this case the working fluid passes through all the stages of the plant. The carbon dioxide gas enters the plant at state 'a', and passes through the final stage, referred to as the snow chamber, because by this stage the gas has been converted to dry ice. Heat is transferred from the carbon dioxide to the final product causing some evaporation: this evaporated product is passed back to the first compressor. The fluid then passes to the first compressor, where it is compressed to state 'c'. It is then passed into a *flash tank* where it is cooled by transferring heat with liquid already in there; this again causes some evaporation of the product, which is passed back into the second compressor, along with the working fluid. In the example shown the process has three stages, and the final stage must be at a pressure high enough

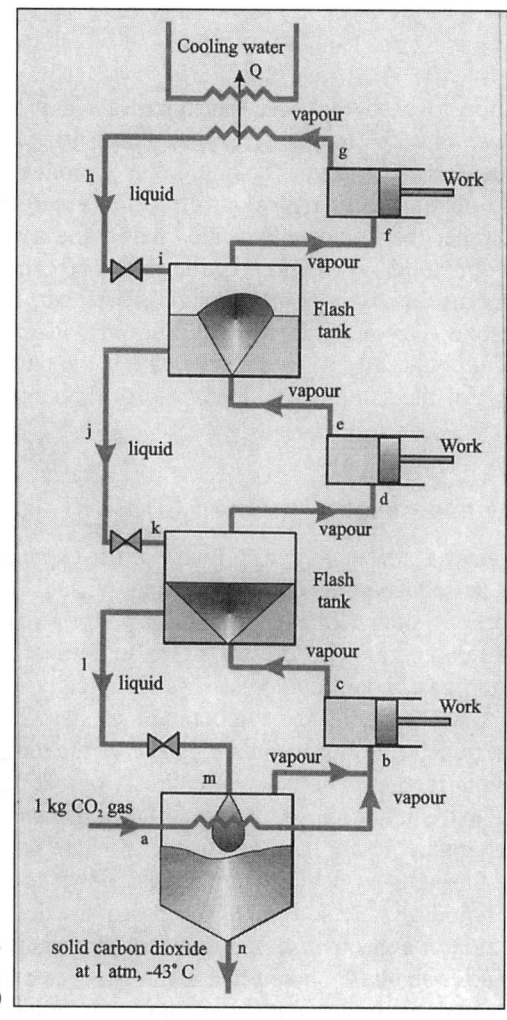

(a)

Fig. 8.5 Plant for manufacture of dry ice (solid carbon dioxide)

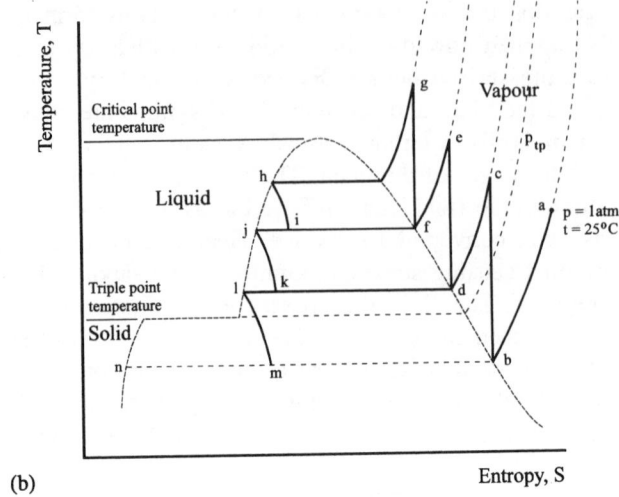

(b)

Fig. 8.5 *Continued*

to enable the condensation to occur by heat transfer to a fluid at close to atmospheric temperature, depicted by point 'h' in Fig 8.5(b); the pressure must also be below the critical pressure for condensation to occur. The liquid that is produced at 'j' passes through the series of throttles to obtain the refrigeration effect necessary to liquefy the carbon dioxide. In the final chamber the liquid is expanded below the triple point pressure, and the phase change enters the solid–gas region. Carbon dioxide cannot exist as a liquid at atmospheric pressure, because its triple point pressure is 5.17 bar; hence it can be either a pressurised liquid or a solid. Dry ice is not in equilibrium with its surroundings, and is continually evaporating. It could only be in a stable state if the ambient temperature were dropped to the equivalent of the saturated temperature for the solid state at a pressure of 1 bar.

8.2 Liquefaction by expansion – method (ii)

If the gas does work against a device (e.g. a turbine) whilst expanding adiabatically then the internal energy will be reduced and liquefaction may ensue. A cycle which includes such an expansion process is shown in Fig 8.6. The gas, for example oxygen, exists at state 1 when at ambient temperature and pressure: it is in a superheated state, but below the critical pressure. If the gas is then compressed isentropically to a pressure above the critical pressure it will reach state 2. The temperature at state 2 is above the ambient temperature, T_a, but heat transfer to the surroundings allows the temperature to be reduced to state point 2a. If the liquefaction plant is a continuously operating plant, then there will be available a supply of extremely cold gas or liquid and this can be used, by a suitable arrangement of heat exchangers, to cool the gas further to state point 3. If the gas is now expanded isentropically through the device from state 3 down to its original pressure it will condense out as a liquid at state 4. Hence, the processes defined in Fig 8.6 can be used to achieve liquefaction of a gas by use of a device taking energy out of the substance by producing work output. It should be noted that if the gas was a *supercritical vapour* at ambient conditions it would be impossible to obtain it in liquid form at atmospheric pressure.

In the liquefaction process described above, the gas may be expanded from state 3 to state 4 either by a reciprocating machine or against an expansion turbine. Both these machines suffer the problem that the work done in the expansion process is a function of the initial temperature because the temperature drop across the turbine is

$$\Delta T = T_3 \eta_T \left\{ 1 - \left(\frac{p_4}{p_3} \right)^{(\kappa - 1)/\kappa} \right\} \tag{8.2}$$

It can be seen from eqn (8.2) that as T_3 becomes very small, the temperature drop achieved by the expansion gets smaller, which means that the pressure ratio to obtain the same temperature drop has to be increased for very low critical temperatures. Another major problem that occurs at very low temperatures is that lubrication becomes extremely difficult. For this reason the turbine is a better alternative than a reciprocating device because it may have air bearings, or gas bearings of the same substance as that being liquefied, thus reducing contamination.

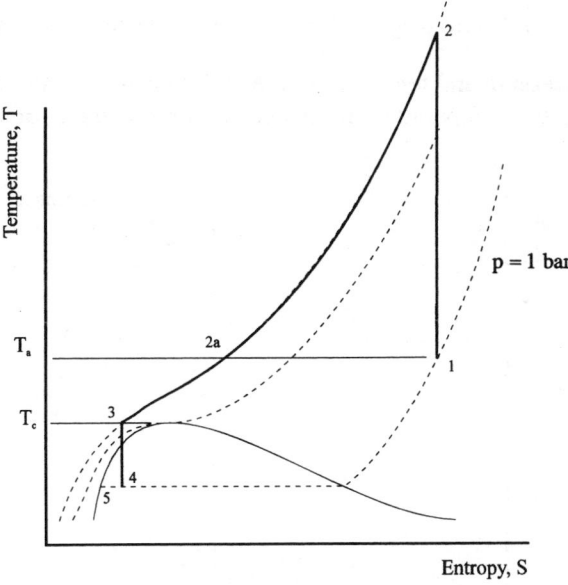

Fig. 8.6 A method for liquefying a gas

Fortunately another method of liquefaction is available which overcomes many of the problems described above. This is known as the Joule–Thomson effect and it can be evaluated analytically. The Joule–Thomson effect is the result of relationships between the properties of the gas in question.

8.3 The Joule–Thomson effect

The Joule–Thomson effect was discovered in the mid-19th century when experiments were being undertaken to define the First Law of Thermodynamics. Joule had shown that the specific heat at constant volume was not a function of volume, and a similar experiment was developed to ascertain the change of enthalpy with pressure. The experiment consisted of forcing a gas through a porous plug by means of a pressure drop. It was found that, for some gases, at a certain entry temperature there was a temperature drop in the gas after it had passed through the plug. This showed that, for these gases, *the enthalpy of the gas was a function of both temperature and pressure.* A suitable apparatus for conducting the experiment is shown in Fig 8.7.

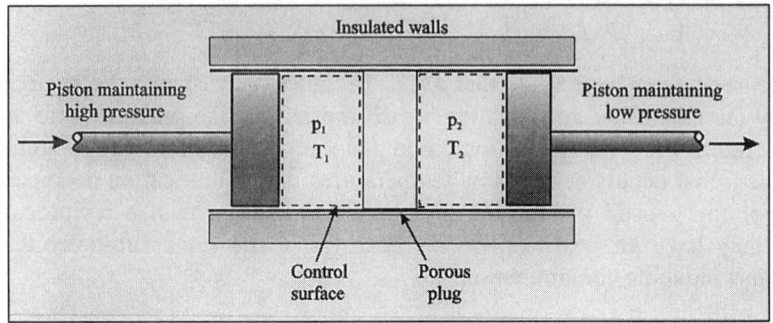

Fig. 8.7 Porous plug device for the Joule–Thomson experiment

If the upstream pressure and temperature are maintained constant, and the downstream pressure is varied, the temperature measured follows a trace of the form shown in Fig 8.8.

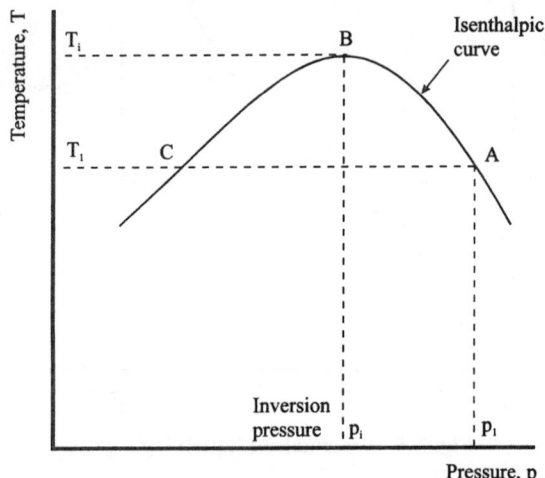

Fig. 8.8 Isenthalpic curve for flow through a porous plug

Line ABC in Fig 8.8 *does not show the change of temperature as the gas flows through the porous plug*: it is made up of results from a series of experiments and shows the effect of lowering the back pressure on the downstream temperature. It is very difficult to evaluate the variation of temperature along the plug, and this will not be attempted here. However, as the pressure p_2 is decreased below p_1 the temperature increases until point B is reached; after this the temperature decreases with decreasing pressure, p_2. Eventually a point is reached, denoted C on the diagram, where $T_2 = T_1$, and after this $T_2 < T_1$. The maximum temperature on this isenthalpic curve is referred to as the *inversion temperature*, and point B is called the *inversion point*.

The process shown in Fig 8.8 may be analysed by applying the steady flow energy equation (SFEE) across control volume, when

$$\dot{Q} - \dot{W} = \dot{m}\left(h_2 - h_1 + \frac{V_2^2}{2} - \frac{V_1^2}{2}\right)$$

(8.3)

where

$\dot{Q}$ = rate of heat transfer,

$\dot{W}$ = rate of work transfer,

$\dot{m}$ = mass flow rate,

h_1, h_2 = enthalpy upstream and downstream of plug,

V_1, V_2 = velocity upstream and downstream of plug.

Now $\dot{Q} = 0$, $\dot{W} = 0$, $V_2 \approx V_1$. Hence $h_2 = h_1$, and the enthalpy along line ABC is constant; thus ABC is an *isenthalpic line*. The most efficient situation from the viewpoint of obtaining cooling is achieved if p_1 and T_1 are at B, the *inversion point*. If state (p_1, T_1) is to the right of B then T_2 depends on the pressure drop, and it could be greater than, equal to or less than T_1. If the upstream state (p_1, T_1) is at B then the downstream temperature, T_2, will always be less than T_1. It is possible to analyse whether heating or cooling will occur by evaluating the sign of the derivative $(\partial T/\partial p)_h$. This term is called the *Joule–Thomson coefficient*, μ.

If $\mu < 0$, then there can be either heating or cooling, depending on whether the downstream pressure is between A and C, or to the left of C. If $\mu > 0$, then the gas will be cooled on passing through the plug (i.e. the upstream state point defined by (p_1, T_1) is to the left of B on the isenthalpic line).

This situation can be analysed in the following way. From the Second Law of Thermodynamics

$$dh = T\,ds + v\,dp$$

(8.4)

But, for this process $dh = 0$, and thus

$$0 = T\left(\frac{\partial s}{\partial p}\right)_h + v$$

(8.5)

If it is assumed that entropy is a continuous function of pressure and temperature, i.e. $s = s(p, T)$, then

$$ds = \left(\frac{\partial s}{\partial p}\right)_T dp + \left(\frac{\partial s}{\partial T}\right)_p dT$$

(8.6)

which can be rearranged to give

$$\left(\frac{\partial s}{\partial p}\right)_h = \left(\frac{\partial s}{\partial p}\right)_T + \left(\frac{\partial s}{\partial T}\right)_p\left(\frac{\partial T}{\partial p}\right)_h$$

(8.7)

Hence, substituting this expression into eqn (8.5) gives

$$0 = T\left\{\left(\frac{\partial s}{\partial p}\right)_T + \left(\frac{\partial s}{\partial T}\right)_p\left(\frac{\partial T}{\partial p}\right)_h\right\} + v$$

(8.8)

Now, from the thermodynamic relationships (eqn (6.24)),

$$T\left(\frac{\partial s}{\partial T}\right)_p = c_p$$

and, from the Maxwell relationships (eqn (6.19d)),

$$\left(\frac{\partial s}{\partial p}\right)_T = -\left(\frac{\partial v}{\partial T}\right)_p.$$

Thus

$$0 = -T\left(\frac{\partial v}{\partial T}\right)_p + c_p\left(\frac{\partial T}{\partial p}\right)_h + v \tag{8.9}$$

which may be rearranged to give

$$\left(\frac{\partial T}{\partial p}\right)_h = \frac{1}{c_p}\left[T\left(\frac{\partial v}{\partial T}\right)_p - v\right] = \mu \tag{8.10}$$

This can be written in terms of the coefficient of expansion

$$\beta = \frac{1}{v}\left(\frac{\partial v}{\partial T}\right)_p$$

and eqn (8.10) becomes

$$\mu = \frac{v}{c_p}[\beta T - 1] \tag{8.11}$$

At the inversion temperature,

$$\left(\frac{\partial T}{\partial p}\right)_h = 0 \quad \Rightarrow \quad T_i = \frac{1}{\beta} \tag{8.12}$$

It is possible to evaluate the Joule–Thomson coefficient for various gases from their state equations. For example, the Joule–Thomson coefficient for a perfect gas can be evaluated by evaluating μ from eqn (8.10) by differentiating the perfect gas equation

$$pv = RT \tag{8.13}$$

Now

$$\mu = \frac{1}{c_p}\left[T\left(\frac{\partial v}{\partial T}\right)_p - v\right]$$

and, from eqn (8.13), the derivative

$$\left(\frac{\partial v}{\partial T}\right)_p = \frac{R}{p} \tag{8.14}$$

Hence

$$\mu = \frac{1}{c_p}\left[T\frac{R}{p} - v\right] = 0$$

This means that it is not possible to cool a perfect gas by the Joule–Thomson effect. This is what would be expected, because the enthalpy of an ideal gas is not a function of pressure. However, it does not mean that gases which obey the ideal gas law at normal atmospheric conditions (e.g. oxygen, nitrogen, etc) cannot be liquefied using the Joule–Thomson effect, because they cease to obey this law close to the saturated vapour line. The possibility of liquefying a gas obeying the van der Waals' equation is considered below.

Van der Waals' equation for air may be written, using a and b for air as 1.358 bar $(m^3/kmol)^2$ and 0.0364 $m^3/kmol$ respectively, as

$$p = \frac{0.083143T}{(v - 0.0364)} - \frac{1.358}{v^2} \tag{8.15}$$

To assess whether a van der Waals' gas can be liquefied using the Joule–Thomson effect it is necessary to evaluate the Joule–Thomson coefficient, μ. This is related to $(\partial v/\partial T)_p$, which can be evaluated in the following way.

From eqn (8.15), in general form

$$T = \left(p + \frac{a}{v^2}\right)\frac{(v - b)}{\mathfrak{R}} \tag{8.16}$$

and hence,

$$\left(\frac{\partial T}{\partial v}\right)_p = \frac{1}{\mathfrak{R}}\left(p + \frac{a}{v^2}\right) - \left(\frac{v - b}{\mathfrak{R}}\right)\left(\frac{2a}{v^3}\right)$$

giving

$$\left(\frac{\partial v}{\partial T}\right)_p = \frac{\mathfrak{R}}{\left(p + \frac{a}{v^2}\right) - (v - b)\left(\frac{2a}{v^3}\right)} \tag{8.17}$$

Thus, from eqn (8.17), the Joule–Thomson coefficient for a van der Waals' gas is not zero at all points, being given by

$$\mu = \frac{1}{c_p}\left[\frac{\mathfrak{R}T}{\left(p + \frac{a}{v^2}\right) - (v - b)\left(\frac{2a}{v^3}\right)} - v\right] \tag{8.18}$$

8.3.1 MAXIMUM AND MINIMUM INVERSION TEMPERATURES

The general form of van der Waals' equation, eqn (8.15), can be rewritten as

$$T = \left(p + \frac{a}{v^2}\right)\frac{(v - b)}{\mathfrak{R}}$$

The inversion temperature is defined as

$$T_i = v\left(\frac{\partial T}{\partial v}\right)_p \tag{8.19}$$

which can be evaluated from eqn (8.16) in the following way:

$$\left(\frac{\partial T}{\partial v}\right)_p = \frac{1}{\Re}\left(p + \frac{a}{v^2}\right) - \left(\frac{(v-b)}{\Re}\right)\left(\frac{2a}{v^3}\right)$$

$$\Rightarrow v\left(\frac{\partial T}{\partial v}\right)_p = \frac{1}{\Re}\left\{v\left(p + \frac{a}{v^2}\right) - (v-b)\left(\frac{2a}{v^2}\right)\right\}$$

(8.20)

It is now necessary to solve for v in terms of T_i alone, and this can be achieved by multiplying eqn (8.16) by v and eqn (8.20) by $(v-b)$. This gives

$$vT_i = \left(p + \frac{a}{v^2}\right)\frac{v(v-b)}{\Re}$$

(8.21)

and

$$(v-b)T_i = \frac{v(v-b)}{\Re}\left(p + \frac{a}{v^2}\right) - \frac{(v-b)^2}{\Re}\frac{2a}{v^2}$$

(8.22)

Subtracting eqns (8.21) and (8.22) gives

$$\left(\frac{v-b}{v}\right) = \sqrt{\frac{\Re bT_i}{2a}} = x = 1 - \frac{b}{v}$$

(8.23)

Substituting eqn (8.23) in eqn (8.15) gives

$$p = \frac{a}{b^2}\left(1 - \sqrt{\frac{\Re bT_i}{2a}}\right)\left(3\sqrt{\frac{\Re bT_i}{2a}} - 1\right)$$

(8.24)

The maximum and minimum inversion temperatures are achieved when the pressure, p, is zero. This gives

$$\hat{T}_i = \frac{2a}{\Re b} \quad \text{and} \quad \check{T}_i = \frac{2a}{9\Re b} = \frac{\hat{T}_i}{9}$$

(8.25)

The value of the critical temperature for a van der Waals' gas is

$$T_c = \frac{8a}{27\Re b}$$

(8.26)

and hence

$$\frac{\hat{T}_i}{T_c} = \frac{\text{Maximum inversion temperature}}{\text{Critical temperature}} = 6.75$$

$$\frac{\check{T}_i}{T_c} = \frac{\text{Minimum inversion temperature}}{\text{Critical temperature}} = 0.75$$

(8.27a, b)

These equations will now be applied to air. The values of a and b for air are 1.358 bar $(m^3/kmol)^2$ and 0.0364 $m^3/kmol$ respectively, and van der Waals' equation for air was given as eqn (8.15). Substituting these values into eqn (8.24), which can be solved as a quadratic equation in T_i, gives the diagram shown in Fig 8.9. This shows that it is not possible to liquefy air if it is above a pressure of about 340 bar.

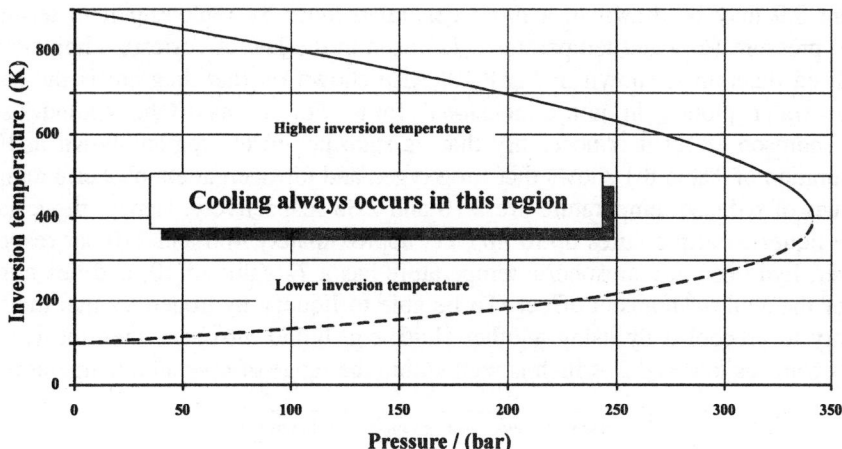

Fig. 8.9 Higher and lower inversion temperatures for air

The maximum pressure for which inversion can occur is when $dp/dT = 0$. This can be related to van der Waals' equation in the following way, using the expression for x introduced in eqn (8.23):

$$\frac{dp}{dT} = \frac{dp}{dx} \frac{dx}{dT}$$

and since dx/dT is a simple single term expression

$$\frac{dp}{dT} = 0 \quad \text{when} \quad \frac{dp}{dx} = 0$$

$$\frac{dp}{dx} = \frac{a}{b^2} \{3(1-x) - (3x-1)\} \tag{8.28}$$

Equation (8.28) is zero when $3(1-x) - (3x-1) = 0$, i.e. when $x = 2/3$.

Hence the maximum pressure at which inversion can occur is

$$p = \frac{a}{b^2} \left(1 - \frac{2}{3}\right)\left(3 \times \frac{2}{3} - 1\right) = \frac{a}{3b^2} = 9p_c \tag{8.29}$$

Thus the maximum pressure at which a Van der Waals' gas can be liquefied using the Joule–Thomson effect is nine times the critical pressure. Substituting for a and b in eqn (8.29) gives the maximum pressure as 341.6 bar.

Figure 8.9 indicates that it is not possible to cool air using the Joule–Thomson effect if it is at a temperature of greater than 900 K, or less than about 100 K. Similar calculations for hydrogen give 224 K and 24.9 K, respectively, for the maximum and minimum inversion temperatures. The maximum pressure at which inversion can be achieved for hydrogen is 117 bar.

The inversion curve goes through the lines of constant enthalpy at the point where $(\partial T/\partial p)_h = 0$. To the left of the inversion curve, cooling always occurs; to the right, heating or cooling occurs depending on the values of the pressure.

Figure 8.9 may be drawn in a more general manner by replotting it in terms of the reduced pressure (p_R) and temperature (T_R) used in the law of corresponding states. This generalised diagram is shown in Fig 8.10. Also shown on that diagram is the saturation line for water plotted in non-dimensional form. For a gas to be cooled using the Joule–Thomson effect it is necessary that its state lies in the region shown in Fig 8.10. Consideration of Table 8.1 shows that for oxygen and nitrogen at atmospheric temperature the values of reduced temperature are 2.10 and 2.38 respectively. Hence, these gases can both be liquefied at pressures up to $8p_c$, i.e. approximately 400 and 270 bar respectively. However, hydrogen at atmospheric temperature has a T_R value of 10, and lies outside the range of the Joule–Thomson effect. To be able to liquefy hydrogen by this method it is necessary to precool it by using another fluid, e.g. liquid nitrogen when the value of T_R could become as low as 4.2, which is well within the range of inversion temperatures.

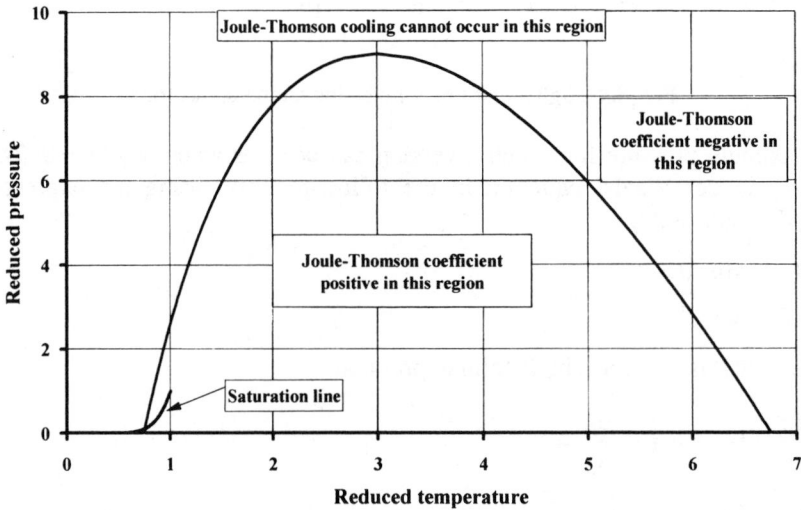

Fig. 8.10 Inversion region in terms of reduced pressure ($p_R = p/p_c$) and reduced temperature ($T_R = T/T_c$). Note the saturation line

8.4   Linde liquefaction plant

Many gases are liquefied using the Joule–Thomson effect. This approach is embodied in the Linde liquefaction process, and a schematic of the equipment is shown in Fig 8.11(a), while the thermodynamic processes are depicted in Fig 8.11(b). The Linde process is similar to a refrigerator operating on a vapour compression cycle (i.e. the typical refrigeration cycle), except that it includes a heat exchanger to transfer energy equal to Q_A and Q_B from the high temperature working fluid to that which has already been cooled through throttling processes. The two throttling processes depicted in Fig 8.11(a) both bring about cooling through the Joule–Thomson effect. The operating processes in the Linde plant will now be described; it will be assumed that the plant is already operating in steady state and that a supply of liquefied gas exists. The gas is supplied to a compressor at state 1 and make-up gas is supplied at state 11, which is the same as state 1. This is then

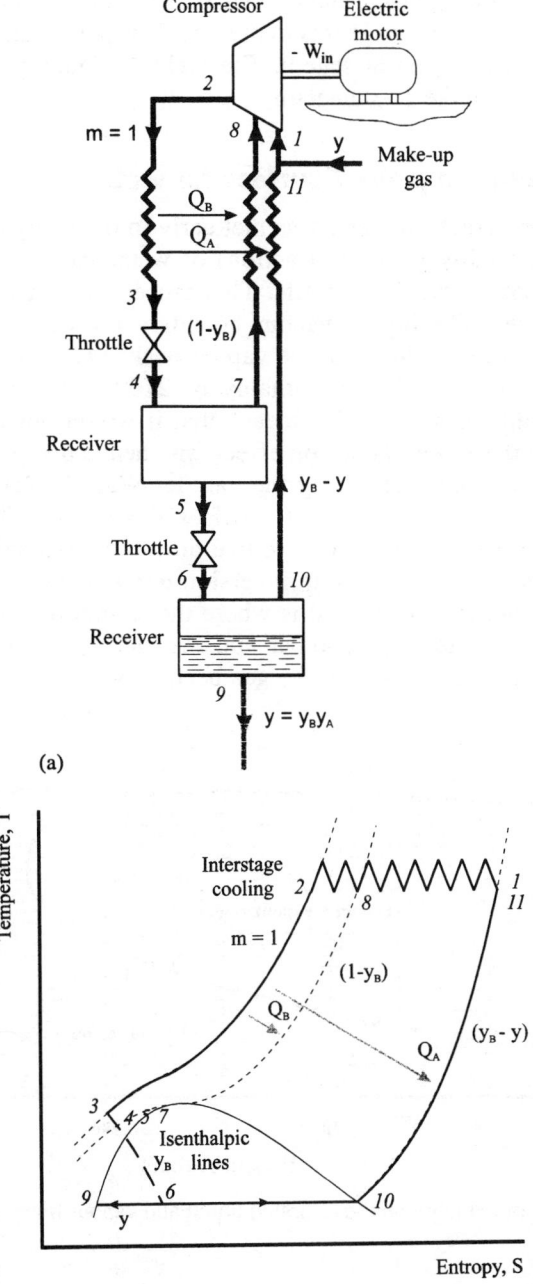

(a)

(b)

Fig. 8.11 Linde liquefaction plant: (a) schematic diagram; (b) *T–s* diagram

compressed to a high, supercritical pressure (which might be hundreds of atmospheres) by a multi-stage reciprocating compressor with inter-stage cooling. The gas finally reaches state 2, and it is then passed through a heat exchanger which cools it to state 3. At this point it is throttled for the first time, and is cooled by the Joule–Thomson effect to state 4 and passed into a receiver. Some of the gas entering the receiver is passed back to the compressor via the heat exchanger, and the remainder is passed through a second throttle until it achieves state 6. At this stage it is in the liquid–vapour region of the state diagram, and liquid gas can be removed at state 9. The yield of liquid gas is y, defined by the quality of state point 6 on the $T-s$ diagram.

8.5 Inversion point on $p-v-T$ surface for water

The Joule–Thomson effect for water was met early in the study of thermodynamics. It was shown that the quality (dryness fraction) of water vapour could be increased by passing the wet vapour through a throttle. This process is an isenthalpic one in which the pressure decreases. The liquid–vapour boundary for water is shown in Fig 8.12, and it can be seen that the enthalpy of the vapour peaks at a value of about 2800 kJ/kg. If water vapour at 20 bar, with an enthalpy of 2700 kJ/kg, is expanded through a throttle then its quality increases; this means that it would not be possible to liquefy the water vapour by the Joule–Thomson effect, and hence this point must be below the minimum inversion temperature. Using the van der Waals' relationship and substituting for the critical temperature gives $\check{T}_i = 0.75 \times T_c = 212°C$. This is equivalent to a pressure of about 20 bar, which is close to that defining the maximum enthalpy on the saturation line. This point of minimum inversion temperature can also be seen on the Mollier ($h-s$) diagram for steam, and is where the saturation curve peaks in enthalpy value. If the pressure were, say, 300 bar and the enthalpy were 2700 kJ/kg then an isenthalpic expansion would result in the gas becoming wetter, and liquefaction could occur.

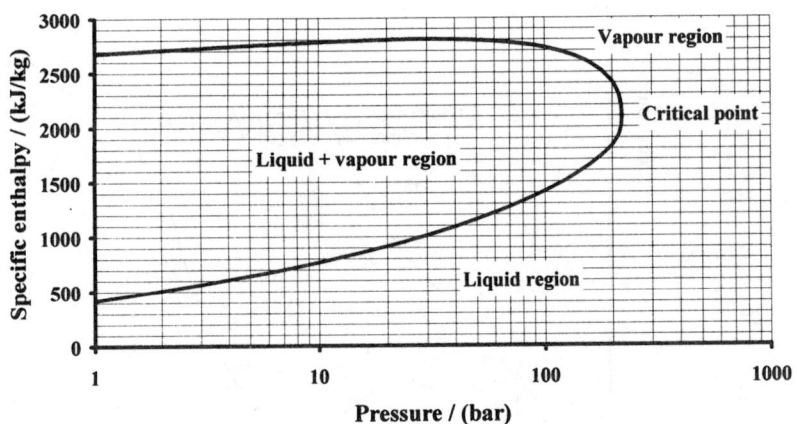

Fig. 8.12 Variation of enthalpy on the saturated liquid and vapour lines with pressure for steam

Example

A simple Linde liquefaction plant is shown in Fig 8.13. This is similar to that in Fig 8.11 except that the liquefaction takes place in a single process, and the cascade is omitted. The plant is used to liquefy air, which is fed to the compressor at 1 bar and 17°C, and compressed isothermally to 200 bar. The compressed air then transfers heat in a counterflow heat exchanger, which has no external heat losses or friction, with the stream leaving the flash chamber, and the inlet temperature of the hot stream equals the outlet

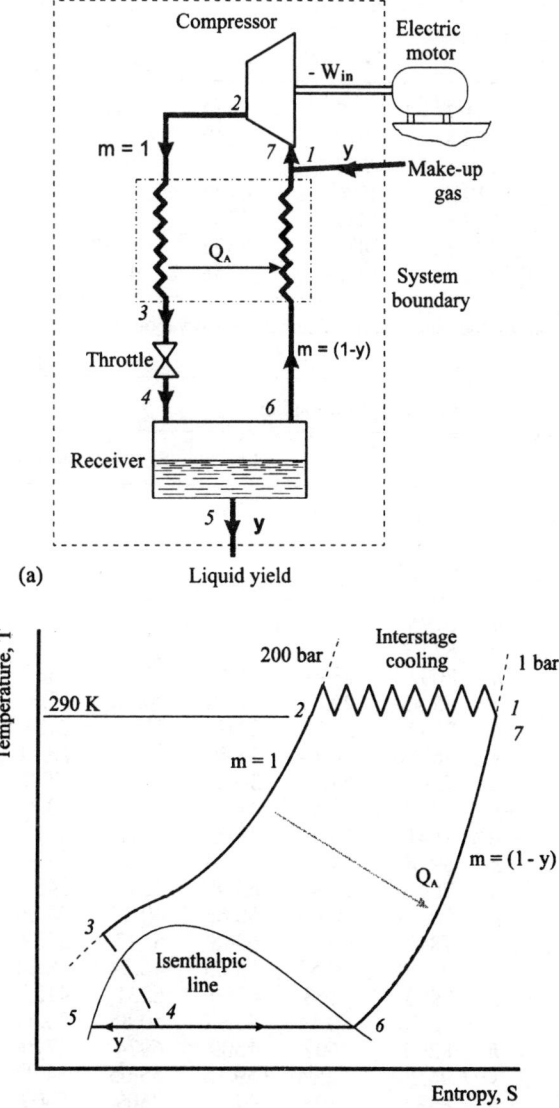

Fig. 8.13 Simplified Linde gas liquefaction plant: (a) schematic diagram of simplified plant; (b) *T−s* diagram for liquefaction process

temperature of the cold stream. Using the table of properties for air at low temperatures and high pressures (Table 8.2), and taking the dead state conditions for exergy as 1 bar and 290 K, evaluate the:

(i)   yield of liquid air per kg of compressed fluid;
(ii)   temperature of the compressed air before the Joule–Thomson expansion process;
(iii)   minimum work required per kg of liquid air;
(iv)   actual work required per kg of liquid air;
(v)   rational efficiency of the plant, and the irreversibilities introduced by the heat exchanger and the throttle.   (example based on Haywood (1980)).

Table 8.2 Properties of air (from Haywood, 1972)

Pressure (atm)	Temperature T_s (K)	Enthalpy (kJ/kmol)		Entropy (kJ/kmol K)	
		h_f	h_g	s_f	s_g
1	81.7	0	5942	0.00	74.00
5	98.5	917	6249	10.33	65.09
10	108.1	1537	6284	16.24	60.55
20	119.8	2506	6127	24.51	54.93
30	127.8	3381	5760	31.17	49.87
35	131.1	3884	5433	35.08	46.92
37.66	(T_c) 132.5	4758	4758	41.39	41.39

Properties of air on the saturated liquid and saturated vapour lines

Temperature (K)		Pressure (atmospheres)					
		1	30	40	50	200	400
90	h	6190					
	s	76.9					
100	h	6493					
	s	80.09					
110	h	6795					
	s	82.97					
120	h	7092	2343	2187	2114	2086	
	s	85.53	22.85	21.38	20.48	15.3	
125	h	7243	2994	2719	2595	2428	
	s	86.78	28.14	25.69	24.41	18.11	
130	h	7393	5976	3405	3159	2770	
	s	87.96	51.54	31.07	28.81	20.8	
135	h	7541	6355	4816	3847	3111	
	s	89.09	54.44	41.74	33.99	23.38	
140	h	7691	6664	6048	4698	3451	
	s	90.16	56.66	50.68	40.19	25.86	
145	h	7842	6928	6468	5597	3789	
	s	91.20	58.52	53.63	46.54	28.24	
150	h	7993	7168	6786	6231	4123	
	s	92.28	60.14	55.77	50.85	30.51	
160	h	8290	7602	7309	6974	4779	4944
	s	94.2	62.94	59.15	55.65	34.75	29.98
170	h	8582	7995	7761	7509	5409	5439
	s	95.95	65.31	61.90	58.89	38.57	32.98
180	h	8877	8365	8168	7966	6006	5915
	s	97.62	67.42	64.24	61.52	41.98	35.87
190	h	9172	8718	8546	8380	6583	6390
	s	99.21	69.33	66.16	63.75	45.21	38.44
200	h	9464	9056	8909	8761	7116	6858
	s	100.7	71.07	68.02	65.69	47.93	40.84
250	h	10925	10671	10581	10491	9427	9074
	s	107.3	78.27	75.61	73.44	58.28	50.73
290	h	12097	11912	11848	11784	11025	10738
	s	111.63	82.79	80.2	78.14	64.09	56.75
300	h	12364	12212	12154	12094	11405	11133
	s	112.6	83.91	81.34	79.28	65.52	58.23

Properties of superheated air at low temperatures and high pressures

h: kJ/kmol

s: kJ/kmol K

$m_w = 28.9$

Solution

(i)   *Yield of liquid per kg of compressed fluid.*

At point 1, $p_1 = 1$ bar; $T_1 = 290$ K; $h_1 = 12097$ kJ/kmol
$s_1 = 111.63$ kJ/kmol K

at point 2, $p_2 = 200$ bar; $T_2 = 290$ K; $h_2 = 11025$ kJ/kmol
$s_1 = 64.09$ kJ/kmol K

Considering the heat exchanger: since it is adiabatic

$$H_2 - H_3 = H_7 - H_6 \tag{8.30}$$

Substituting for specific enthalpies in eqn (8.30) gives

$$m_2(h_2 - h_3) = m_6(h_7 - h_6) \tag{8.31}$$

Now

$$m_6 = m_2(1 - y) \tag{8.32}$$

and hence

$$h_2 - h_3 = (1 - y)(h_7 - h_6)$$
$$\Rightarrow h_3 = h_2 - (1 - y)(h_7 - h_6) \tag{8.33}$$

The process from 3 to 4 is isenthalpic, and is a Joule–Thomson process, thus

$$h_4 = h_3 \tag{8.34}$$

But

$$h_4 = x_4 h_g + (1 - x_4)h_f = (1 - y)h_6 + yh_5 \tag{8.35}$$

Combining eqns (8.33) and (8.34) gives

$$y = \frac{h_2 - h_7}{h_7 - h_5} = \frac{12097 - 11025}{12097 - 0} = 0.08862 \tag{8.36}$$

Hence, the yield of liquid air per kg of compressed air is 0.08862 kg.

(ii)   *The temperature before the Joule–Thomson process.*
This is the temperature of the gas at point 3. From eqns (8.34) and (8.35)

$$h_3 = h_4 = x_4 h_g + (1 - x_4)h_f = (1 - y)h_6 + yh_5$$
$$= (1 - 0.08862) \times 5942 = 5415 \text{ kJ/kmol} \tag{8.37}$$

This value of enthalpy at 200 bar is equivalent to a temperature of 170 K.

(iii)   *Minimum work required per kg liquid yield.*
Consideration of the control system in Fig 8.13 shows that only three parameters cross the system boundary; these are the make-up gas, the liquid yield, and the work input to the compressor. Hence, the minimum work

required to achieve liquefaction of the gas is

$$\check{W}_{net} = y(b_1 - b_5),$$

giving the work per unit mass of liquid as

$$\check{w}_{net} = \frac{\check{W}_{net}}{y} = b_1 - b_5 \tag{8.38}$$

In this case, $b_5 = 0$ because point 5 (the liquid point at 1 bar) was chosen as the datum for properties; hence

$$\check{w}_{net} = h_1 - T_0 s_1 - 0 = 12097 - 290 \times 111.63 = -20275 \text{ kJ/kmol liquid}$$

(iv)   *Work required per liquid yield.*
The compression process is an isothermal one, and hence the work done is

$$w = p_1 v_1 \ln \frac{v_2}{v_1} = -p_1 v_1 \ln \frac{p_2}{p_1} = -\Re T_1 \ln \frac{p_2}{p_1} \tag{8.39}$$

Substituting the values gives

$$w = -\Re T_1 \ln \frac{p_2}{p_1} = -8.3143 \times 290 \times \ln \frac{200}{1} = -12775 \text{ kJ/kmol gas}$$

$$= \frac{-12775}{28.9} = -442 \text{ kJ/kg gas} = \frac{-442}{0.08863} = -4987 \text{ kJ/kg liquid}$$

(v)   *Rational efficiency.*
The rational efficiency of the plant is defined as

$$\eta_R = \frac{\check{w}_{net}}{w} = \frac{-701}{-4987} = 0.1406 \tag{8.40}$$

Hence, the plant is only 14.1% as efficient as it could be if all the energy transfers were reversible. It is instructive to examine where the irreversibilities occur.
The heat exchanger
The irreversibility of the heat exchanger is

$$I_{HE} = (1 - y)(b_7 - b_6) - (b_2 - b_3)$$
$$= 0.91137 \times \{-20\,275 - (-15\,518)\} - \{-7561 - (-5776)\}$$
$$= -2550 \text{ kJ/kmol gas}$$

This is equivalent to 20.0% of the work required to compress the gas.
The throttle to achieve liquefaction
The irreversibility of the throttle is

$$I_{throttle} = b_4 - b_3 = 0.91137 \times (-15\,518) - (-5776) = -8366 \text{ kJ/kmol gas}$$

This is equivalent to 65.5% of the work used to compress the gas.
Hence, the work required to liquefy the gas is

$$w = \check{w}_{net} + I_{HE} + I_{throttle}$$

8.6 Concluding remarks

It has been shown that gases can be liquefied in a number of ways. Gases which are liquids at temperatures close to ambient can be liquefied by cooling in a simple refrigeration system. Carbon dioxide, which cannot be maintained as a liquid at ambient pressure, is made into dry ice which is not in equilibrium at room temperature and pressure.

If it is necessary to achieve extremely low temperatures to bring about liquefaction, the Joule–Thomson effect is employed. It is possible to analyse such liquefaction plant using equilibrium thermodynamics and suitable equations of state. The efficiency of liquefaction plant has been calculated and the major influences of irreversibilities in the processes have been illustrated.

PROBLEMS

1 Show that the Joule–Thomson coefficient, μ, is given by

$$\mu = \frac{1}{c_p}\left(T\left(\frac{\partial v}{\partial T}\right)_p - v\right)$$

Hence or otherwise show that the inversion temperature (T_i) is

$$T_i = \left(\frac{\partial T}{\partial v}\right)_p v$$

The equation of state for air may be represented by

$$p = \frac{\Re T}{v_m - 0.0367} - \frac{1.368}{v_m^2}$$

where p = pressure (bar), T = temperature (K), and v_m = molar volume (m³/kmol). Determine the maximum and minimum inversion temperatures and the maximum inversion pressure for air.

[896 K; 99.6 K; 339 bar]

2 The last stage of a liquefaction process is shown in diagrammatic form in Fig P8.2. Derive the relationship between p_1 and T_1 for the maximum yield of liquid at conditions p_L, T_L, h_L for a gas obeying the state equation

$$\left(p + \frac{1.368}{v_m^2}\right)(v_m - 0.0367) = \Re T$$

where p is pressure (bar), v_m is the molar volume (m³/kmol), and T is temperature (K).

Calculate the pressure for maximum yield at a temperature of 120 K.

$$\left[p = \frac{a}{b^2}\left(1 - \sqrt{\frac{\Re b T_i}{2a}}\right)\left(3\sqrt{\frac{\Re b T_i}{2a}} - 1\right); 62.8 \text{ bar}\right]$$

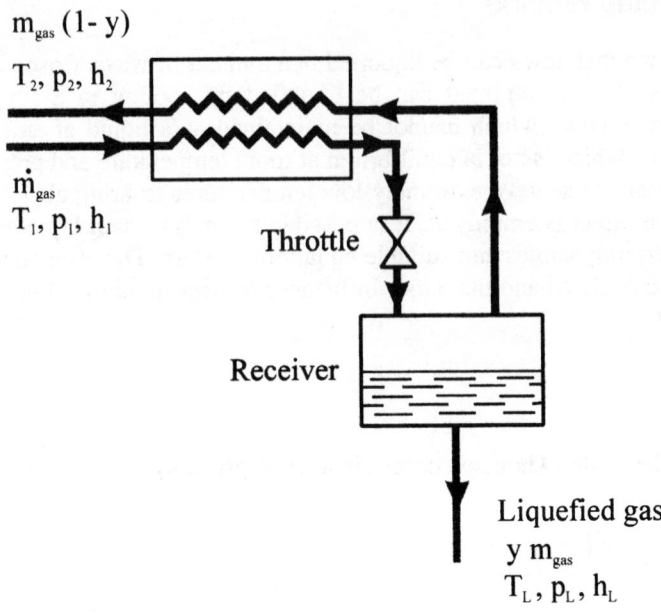

$$m_{gas}(1-y)$$
$$T_2, p_2, h_2$$

$$\dot{m}_{gas}$$
$$T_1, p_1, h_1$$

Throttle

Receiver

Liquefied gas
$$y\, m_{gas}$$
$$T_L, p_L, h_L$$

Fig. P8.2

3 The equation of state for a certain gas is

$$v_m = \frac{\Re T}{p} + \frac{k}{\Re T}$$

where k is a constant. Show that the variation of temperature with pressure for an isenthalpic process from 1 to 2 is given by

$$T_1^2 - T_2^2 = -\frac{4k}{c_p \Re}(p_1 - p_2)$$

If the initial and final pressures are 50 bar and 2 bar respectively and the initial temperature is 300 K, calculate

(a) the value of the Joule–Thomson coefficient at the initial state, and
(b) the final temperature of the gas, given that

$$k = -11.0 \text{ kJ m}^3/(\text{kmol})^2$$
$$c_{p,m} = 29.0 \text{ kJ/kmol K}.$$

$$[3.041 \times 10^{-7} \text{ m}^3 \text{K/J}; 298.5 \text{ K}]$$

4 A gas has the equation of state

$$\frac{pv_m}{\Re T} = 1 + Ap(T^3 - 9.75T_c\,T^2 + 9T_c^2\,T) + Bp^2T$$

where A and B are positive constants and T_c is the critical temperature. Determine the maximum and minimum inversion temperatures, expressed as a multiple of T_c.

$$[6T_c; 0.5T_c]$$

5 A gas has the equation of state

$$\frac{pv_m}{\Re T} = 1 + Np + Mp^2$$

where N and M are functions of temperature. Show that the equation of the inversion curve is

$$p = -\frac{dN}{dT} \bigg/ \frac{dM}{dT}$$

If the inversion curve is parabolic and of the form

$$(T - T_0)^2 = 4a(p_0 - p)$$

where T_0, p_0 and a are constants, and if the maximum inversion temperature is five times the minimum inversion temperature, show that $a = T_0^2/9p_0$ and give possible expressions for N and M.

$$\left[M = T : N = -p_0 T + \frac{(T - T_0)^3}{12a} + c \right]$$

6 In a simple Linde gas liquefaction plant (see Fig 8.13) air is taken in at the ambient conditions of 1 bar and 300 K. The water-jacketed compressor delivers the air at 200 bar and 300 K and has an *isothermal efficiency* of 70%. There is zero temperature difference at the warm end of the regenerative heat exchanger (i.e. $T_2 = T_7$). Saturated liquid air is delivered at a pressure of 1 bar. Heat leakages into the plant and pressure drops in the heat exchanger and piping can be neglected.

Calculate the yield of liquid air per unit mass of air compressed, the work input per kilogram if air is liquefied, and the *rational efficiency* of the liquefaction process.

[7.76%; 8.39 MJ/kg; 8.84%]

9

Thermodynamic Properties of Ideal Gases and Ideal Gas Mixtures of Constant Composition

It has been shown that it is necessary to use quite sophisticated equations of state to define the properties of vapours which are close the saturated vapour line. However, for gases in the superheat region, the ideal gas equation gives sufficient accuracy for most purposes. The equation of state for an ideal gas, in terms of mass, is

$$pV = mRT \tag{9.1}$$

where

p = pressure (N/m^2)
V = volume (m^3)
m = mass (kg)
R = specific gas constant (kJ/kg K)
T = absolute (or thermodynamic) temperature (K)

This can be written in more general terms using the amount of substance, when

$$pV = n\Re T \tag{9.2}$$

where

n = amount of substance, or chemical amount (kmol)
$\Re$ = universal gas constant (kJ/kmol K)

It is found that eqn (9.2) is more useful than eqn (9.1) for combustion calculations because the combustion process takes place on a molar basis. To be able to work in a molar basis it is necessary to know the molecular weights (or relative molecular masses) of the elements and compounds involved in a reaction.

9.1 Molecular weights

The molecular weight (or relative molecular mass) of a substance is the mass of its molecules relative to that of other molecules. The datum for molecular weights is carbon-12, and this is given a molecular weight of 12. All other elements and compounds have molecular weights relative to this, and their molecular weights are not integers. To be able to perform combustion calculations it is necessary to know the atomic or molecular weights of commonly encountered elements; these can be combined to give other compounds. Table 9.1 gives the data for individual elements or compounds in integral numbers (except air which is a mixture of gases); in reality only carbon-12 (used as the basis for atomic/molecular weights) has an integral value, but most values are very close to integral ones and will be quoted as such.

Table 9.1 Molecular weights of elements and compounds commonly encountered in combustion

	Air	O_2	N_2	Atmospheric N_2	H_2	CO	CO_2	H_2O	C
m_w	28.97	32	28	28.17	2	28	44	18	12

9.1.1 AIR

As stated previously, most combustion takes place between a hydrocarbon fuel and air. Air is a mixture of gases, the most abundant being oxygen and nitrogen with small proportions of other gases. In fact, in practice, air is a mixture of all elements and compounds because everything will evaporate in air until the partial pressure of its atoms, or molecules, achieves its saturated vapour pressure. In reality, this evaporation can usually be neglected, except in the case of water. Table 9.1 shows that the molecular weight of atmospheric nitrogen is higher than that of pure nitrogen. This is because 'atmospheric nitrogen' is taken to be a mixture of nitrogen and about 1.8% by mass of argon, carbon dioxide and other gases; the molecular weight of atmospheric nitrogen includes the effect of the other substances.

The composition of air is defined as 21% O_2 and 79% N_2 by volume (this can be written 21 mol% O_2 and 79 mol% N_2). This is equivalent to 23.2% O_2 and 76.8% N_2 by mass.

9.2 State equation for ideal gases

The equation of state for an ideal gas, a, is

$$pV = m_a R_a T \qquad (9.3)$$

If the mass of gas, m_a, is made equal to the molecular weight of the gas in the appropriate units, then the amount of substance a is known as a *mol* (if the mass is in kg then the amount of substance is called a *kmol*). If the volume occupied by that this amount of substance is denoted v_m then

$$v_m = \frac{m_w R_a T}{p} \qquad (9.4)$$

where v_m is the molar specific volume, and has the units of m^3/mol or $m^3/kmol$. Now Avogadro's hypothesis states that

equal volumes of all ideal gases at a particular temperature and pressure contain the same number of molecules (and hence the same amount of substance).

Hence, any other gas, b, at the same pressure and temperature will occupy the same volume as gas a, i.e.

$$\frac{pv_m}{T} = m_{w_a}R_a = m_{w_b}R_b = \cdots m_{w_i}R_i = \cdots = \mathfrak{R} \tag{9.5}$$

If a system contains an amount of substance n_a of gas a, then eqn (9.5) may be written

$$pV_m = n_a m_{w_a} R_a T = n_a \mathfrak{R} T \tag{9.6}$$

It can be seen that for ideal gases the product $m_{w_i} R_i$ is the same for all gases: it is called the *universal gas constant*, $\mathfrak{R}$. The values of the universal gas constant, $\mathfrak{R}$, together with its various units are shown in Table 9.2.

Table 9.2 Values of the universal gas constant, $\mathfrak{R}$, in various units

8314 J/kmol K
1.985 kcal/kmol K
1.985 Btu/lb-mol K
1.985 CHU/lb-mol R
2782 ft lb$_f$/lb-mol K
1545 ft lb$_f$/lb-mol R

9.2.1 IDEAL GAS EQUATION

The evaluation of the properties of an ideal gas will now be considered. It has been shown that the internal energy and enthalpy of an ideal gas are not functions of the volume or pressure, and hence these properties are simply functions of temperature alone. This means that the specific heats at constant volume and constant pressure are not partial derivatives of temperature, but can be written

$$c_v = \left(\frac{\partial u}{\partial T}\right)_v = \frac{du}{dT}$$

$$\tag{9.7}$$

$$c_p = \left(\frac{\partial h}{\partial T}\right)_p = \frac{dh}{dT}$$

Also, if c_v and c_p in molar quantities are denoted by $c_{v,m}$ and $c_{p,m}$ then, for ideal and perfect gases

$$c_{p,m} - c_{v,m} = \mathfrak{R} \tag{9.8}$$

It is possible to evaluate the properties of substances in terms of unit mass (specific properties) or unit amount of substance (molar properties). The former will be denoted by lower case letters (e.g. v, u, h, g, etc), and the latter will be denoted by lower case letters and a suffix m (e.g. v_m, u_m, h_m, g_m, etc). The molar properties are more useful for combustion calculations, and these will be considered here.

The molar internal energy and molar enthalpy can be evaluated by integrating eqns (9.7), giving

$$u_m = \int_{T_0}^{T} c_{v,m} \, dT + u_{0,m}$$
$$h_m = \int_{T_0}^{T} c_{p,m} \, dT + h_{0,m}$$

(9.9)

where $u_{0,m}$ and $h_{0,m}$ are the values of u_m and h_m at the datum temperature T_0. Now, by definition

$$h_m = u_m + pv_m$$

and for an ideal gas $pv_m = \Re T$; thus

$$h_m = u_m + \Re T$$

(9.10)

Hence, at $T = 0$, $u_m = h_m$, i.e. $u_{0,m} = h_{0,m}$. (A similar relationship exists for the specific properties, u and h.)

To be able to evaluate the internal energy or enthalpy of an ideal gas using eqn (9.10) it is necessary to know the variation of specific heat with temperature. It is possible to derive such a function from empirical data by curve-fitting techniques, if such data is available. In some regions it is necessary to use quantum mechanics to evaluate the data, but this is beyond the scope of this course. It will be assumed here that the values of c_v are known in the form

$$c_{v,m} = a + bT + cT^2 + dT^3$$

(9.11)

where a, b, c and d have been evaluated from experimental data. Since $c_{p,m} - c_{v,m} = \Re$ then an expression for $c_{p,m}$ can be easily obtained. Hence it is possible to find the values of internal energy and enthalpy at any temperature if the values at T_0 can be evaluated. This problem is not a major one if the composition of the gas remains the same, because the datum levels will cancel out. However, if the composition varies during the process it is necessary to know the individual datum values. They can be measured by calorimetric or spectrographic techniques. The 'thermal' part of the internal energy and enthalpy, i.e. that which is a function of temperature, will be denoted

$$u_m(T) = \int_{T_0}^{T} c_{v,m} \, dT$$
$$h_m(T) = \int_{T_0}^{T} c_{p,m} \, dT$$

(9.12)

and then eqns (9.9) become

$$u_m = u_m(T) + u_{0,m}$$
$$h_m = h_m(T) + h_{0,m}$$

(9.13)

assuming that the base temperature is absolute zero.

Enthalpy or internal energy data are generally presented in tabular or graphical form. The two commonest approaches are described below. First, it is useful to consider further the terms involved.

9.2.2 THE SIGNIFICANCE OF $u_{0,m}$ AND $h_{0,m}$

As previously discussed, $u_{0,m}$ and $h_{0,m}$ are the values of molar internal energy, u_m, and molar enthalpy, h_m, at the reference temperature, T_0. If $T_0 = 0$ then, for an ideal gas,

$$u_{0,m} = h_{0,m}$$

(9.14)

If $T_0 \neq 0$, then $u_{0,m}$ and $h_{0,m}$ are different. Most calculations involve changes in enthalpy or internal energy, and if the composition during a process is invariant the values of $h_{0,m}$ or $u_{0,m}$ will cancel.

However, if the composition changes during a process it is necessary to know the difference between the values of $u_{0,m}$ for the different species at the reference temperature. This is discussed below.

Obviously $u_{0,m}$ and $h_{0,m}$ are consequences of the ideal gas assumption and the equation

$$h_m = h_m(T) + h_{0,m}$$

contains the assumption that the ideal gas law applies down to T_0. If $T_0 = 0$, or a value outside the superheat region for the gas being considered, then the gas ceases to be ideal, often becoming either liquid or solid. To allow for this it is necessary to include latent heats. This will not be dealt with here, but the published data do include these corrections.

9.2.3   ENTROPY OF AN IDEAL GAS – THIRD LAW OF THERMODYNAMICS

The change in entropy during a process is defined as

$$ds_m = \frac{dh_m}{T} - \frac{\Re dp}{p} \tag{9.15}$$

If the functional relationship $h_m = h_m(T)$ is known then eqn (9.15) may be evaluated, giving

$$s_m - s_{0,m} = \int_{T_0}^{T} \frac{dh_m}{T} - \Re \ln \frac{p}{p_0} \tag{9.16}$$

where $s_{0,m}$ is the value of s_m at T_0 and p_0.

It is convenient to take the reference temperature T_0 as absolute zero. It was previously shown that

$$dh_m = c_{p,m}(T)\, dT \qquad \text{and hence} \qquad \frac{dh_m}{T} = \frac{c_{p,m}(T)\, dT}{T} \tag{9.17}$$

Consider

$$\lim_{T \to 0} \left(\frac{dh_m}{T} \right)$$

If $c_{p,m}(T_0) \geq 0$ then the expression is either infinite or indeterminate. However, before reaching absolute zero the substance will cease to be an ideal gas and will become a solid. It can be shown by the Debye theory and experiment that the specific heat of a solid is given by the law $c_p = aT^3$. Hence

$$\lim_{T \to 0} \left(\frac{dh_m}{T} \right) = \lim_{T \to 0} \left(\frac{aT^3 dT}{T} \right) = \lim_{T \to 0} (aT^2\, dT) \to 0 \tag{9.18}$$

To integrate from absolute zero to T it is necessary to include the latent heats, but still it is possible to evaluate $\int dh_m / T$.

If $s_m(T)$ is defined as

$$s_m(T) = \int_{T_0}^{T} \frac{dh_m}{T} \, dT \tag{9.19}$$

then entropy

$$s_m = s_m(T) - \Re \ln \frac{p}{p_0} + s_{0,m} \tag{9.20}$$

The term $s_{0,m}$ is the constant of integration and this can be compared with $u_{0,m}$ and $h_{0,m}$. If the composition is invariant then the value of $s_{0,m}$ will cancel out when evaluating changes in s and it is not necessary to know its value. If the composition varies then it is necessary to know, at least, the difference between s_0 values for the substances involved. It is not possible to obtain any information about s_0 from classical (macroscopic) thermodynamics but statistical thermodynamics shows that 'for an isothermal process involving *only phases in internal equilibrium the change in entropy approaches zero at absolute zero*'. This means that for substances that exist in crystalline or liquid form at low temperatures it is possible to evaluate entropy changes by assuming that differences in $s_{0,m}$ are zero.

It should also be noted that the pressure term, p, in eqn (9.20) is the *partial pressure* of the gas if it is contained in a mixture, and p_0 is a datum pressure (often chosen as 1 bar or, in the past, 1 atmosphere).

9.2.4 THE GIBBS ENERGY (FUNCTION) OF AN IDEAL GAS

The Gibbs energy will now be derived for use later when considering the equilibrium composition of mixtures (dissociation). By definition,

$$g_m = h_m - Ts_m \tag{9.21}$$

and

$$h_m = h_m(T) + h_{0,m} \tag{9.22}$$

If the pressure ratio, p/p_0, is denoted by the symbol p_r, i.e. $p_r = p/p_0$, then

$$s_m = s_m(T) - \Re \ln p_r + s_{0,m} \tag{9.23}$$

and

$$\begin{aligned} g &= h(T) + h_0 - T(s(T) - \Re \ln p_r + s_0) \\ &= (h_0 - Ts_0) + (h(T) - Ts(T)) + \Re T \ln p_r \end{aligned} \tag{9.24}$$

If the terms at T_0 are combined to give

$$g_{0,m} = h_{0,m} - Ts_{0,m} \tag{9.25}$$

and the temperature-dependent terms are combined to give

$$g_m(T) = h_m(T) - Ts_m(T) \tag{9.26}$$

then

$$g_m = g_m(T) + \Re T \ln p_r + g_{0,m} \tag{9.27}$$

Often $g_m(T)$ and $g_{0,m}$ are combined to give g_m^0 which is *the pressure-independent portion of the Gibbs energy*, i.e. $g_m^0 = g_m(T) + g_{0,m}$, and then

$$g_m = g_m^0 + \Re T \ln p_r \tag{9.28}$$

g_m^0 is the value of the molar Gibbs energy at a temperature T *and a pressure of p_0*, and it is a function of temperature alone. The datum pressure, p_0, is usually chosen as 1 bar nowadays (although much data are published based on a datum pressure of 1 atm; the difference is not usually significant in engineering problems, but it is possible to convert some of the data, e.g. equilibrium constants).

Gibbs energy presents the same difficulty when dealing with mixtures of varying composition as u_0, h_0 and s_0. If the composition is invariant, changes in Gibbs energy are easily calculated because the g_0 terms cancel. For mixtures of varying composition g_0 must be known. It can be seen from eqn (9.26) that if the reference temperature, T_0, equals zero then

$$g_{0,m} = h_{0,m} = u_{0,m} \tag{9.29}$$

9.3 Tables of $u(T)$ and $h(T)$ against T

These tables are based on polynomial equations defining the enthalpy of the gas. The number of terms can vary depending on the required accuracy and the temperature range to be covered. This section will limit the number of coefficients to six, based on Benson (1977). The equation used is of the form

$$\frac{h_m(T)}{\Re T} = \frac{h_m - h_{0,m}}{\Re T} = a_1 + a_2 T + a_3 T^2 + a_4 T^3 + a_5 T^4 \tag{9.30}$$

The values of the coefficients for various gases are listed in Table 9.3, for the range 500 to 3000 K. If used outside these ranges the accuracy of the calculation will diminish.

Table 9.3 Enthalpy coefficients for selected gases found in combustion processes (based on kJ/kmol, with the temperature in K)

Substance	a_5	a_4	a_3	a_2	a_1	a_6	$h_0 = a_0$ kJ/kmol
H_2	0.0000	-1.44392e-11	9.66990e-8	-8.18100e-6	3.43328	-3.84470	0.0000
CO	0.0000	-2.19450e-12	-3.22080e-8	3.76970e-4	3.317000	4.63284	-1.13882e5
N_2	0.0000	-6.57470e-12	1.95300e-9	2.94260e-4	3.34435	3.75863	0.0000
NO	0.0000	-4.90360e-12	-9.58800e-9	2.99380e-4	3.50174	5.11346	8.99147e4
CO_2	0.0000	8.66002e-11	-7.88542e-7	2.73114e-3	3.09590	6.58393	-3.93405e5
O_2	0.0000	1.53897e-11	-1.49524e-7	6.52350e-4	3.25304	5.71243	0.0000
H_2O	0.0000	-1.81802e-11	4.95240e-8	5.65590e-4	3.74292	9.65140e-1	-2.39082e5
CH_4	-8.58611e-15	1.62497e-10	-1.24402e-6	4.96462e-3	1.93529	8.15300	-6.6930e4
O	0.0000	-1.38670e-11	1.00187e-7	-2.51427e-4	2.76403	3.73309	2.46923e5

Hence the enthalpy, internal energy, entropy and Gibbs energy can be evaluated as follows.

Enthalpy

$$h_m(T) = \Re T(a_1 + a_2 T + a_3 T^2 + a_4 T^3 + a_5 T^4)$$
$$= \Re(a_1 T + a_2 T^2 + a_3 T^3 + a_4 T^4 + a_5 T^5)$$
$$= \Re \sum_{i=1}^{5} a_i T^i \tag{9.31}$$

Internal energy

$$u_m(T) = \Re T((a_1 - 1) + a_2 T + a_3 T^2 + a_4 T^3 + a_5 T^4)$$
$$= \Re((a_1 - 1)T + a_2 T^2 + a_3 T^3 + a_4 T^4 + a_5 T^5)$$
$$= \Re \sum_{i=1}^{5} a_i T^i - \Re T = h(T) - \Re T \tag{9.32}$$

Entropy

This is defined by eqn (9.15) as

$$ds_m = \frac{dh_m}{T} - \Re \frac{dp}{p}$$

Integrating eqn (9.15) gives

$$s_m(T) = s_m - s_{0,m} = \int_{T_0}^{T} \frac{dh_m}{T} - \Re \ln\left(\frac{p}{p_0}\right)$$

Equation (9.20) provides some problems in solution. The first is that

$$\int_{T_0}^{T} \frac{dh_m}{T} = \int_{T_0}^{T} \frac{c_{p,m} \, dT}{T}$$

and this results in $\ln(0)$ when $T = T_0$. Fortunately these problems can be overcome by use of the Van't Hoff equation, which will be derived in Chapter 12, where dissociation is introduced. This states that

$$h_m = -T^2 \left(\frac{d}{dT} (\mu_m^0/T)\right) \tag{9.33}$$

where, for $T_0 = 0$ K

$$\mu_m^0 = \text{chemical potential} = g_m(T) + g_{0,m} = g_m(T) + h_{0,m} \tag{9.34}$$

and

$$h_m = h_m(T) + h_{0,m}$$

(Note: the term μ_m^0, is similar to g_m^0, which was introduced in eqn (9.28). At this stage it is sufficient to note that the chemical potential has the same numerical value as the specific Gibbs energy. The chemical potential will be defined in Section 12.2.)

By definition, in eqn (9.31),

$$\frac{h_m(T)}{\Re T} = \frac{h_m - h_{0,m}}{\Re T} = a_1 + a_2 T + a_3 T^2 + a_4 T^3 + a_5 T^4 \tag{9.35}$$

giving

$$\frac{h_m(T)}{\Re T} = \frac{a_1}{T} + a_2 + a_3 T + a_4 T^2 + a_5 T^3 \tag{9.36}$$

Substituting eqns (9.34), (9.13) and (9.36) into eqn (9.33) and rearranging gives

$$-\left[\frac{a_1}{T} + a_2 + a_3 T + a_4 T^2 + a_5 T^3 + \frac{h_{0,m}}{\Re T^2}\right] dT = \frac{1}{\Re} d\left(\frac{g_m(T) + h_{0,m}}{T}\right) \tag{9.37}$$

which can be integrated to

$$-\left\{a_1 \ln T + a_2 T + \frac{a_3 T^2}{2} + \frac{a_4 T^3}{3} + \frac{a_5 T^4}{4} - \frac{h_{0,m}}{\Re T}\right\} + A = \frac{g_m(T) + h_{0,m}}{\Re T} \tag{9.38}$$

It is conventional to let

$$A = a_1 - a_6 \tag{9.39}$$

and then

$$\frac{g_m(T)}{\Re T} = a_1(1 - \ln T) - \left\{a_2 T + \frac{a_3 T^2}{2} + \frac{a_4 T^3}{3} + \frac{a_5 T^4}{4}\right\} - a_6 \tag{9.40}$$

The value of entropy can then be obtained from

$$s_m(T) = \frac{h_m(T) - g_m(T)}{T} \tag{9.41}$$

giving

$$s_m(T) = \Re\left(a_1 \ln T + 2a_2 T + \frac{3}{2} a_3 T^2 + \frac{4}{3} a_4 T^3 + \frac{5}{4} a_5 T^4 + a_6\right) \tag{9.42}$$

Gibbs energy

The Gibbs energy can be evaluated from eqn (9.40) or eqn (9.26):

$$g_m(T) = h_m(T) - Ts_m(T)$$

The values obtained from this approach, using eqns (9.31) to (9.42) are given in Tables 9.4 for oxygen, nitrogen, hydrogen, water, carbon monoxide, carbon dioxide, nitric oxide and methane. The tables have been evaluated up to 3500 k, slightly beyond the range stated in relation to Table 9.3, but the error is not significant. This enables most combustion problems to be solved using this data. Other commonly used tables are those of JANAF (1971).

Table 9.4

Molecular oxygen, O_2 : $h_0 = 0.0000$ kJ/kmol | | | | Molecular nitrogen, N_2 : $h_0 = 0.0000$ kJ/kmol

T	h(T)	u(T)	s(T)	g(T)	T	h(T)	u(T)	s(T)	g(T)
50	1365.74	950.03	153.84	-6326.3	50	1396.41	980.70	140.27	-5617.2
100	2757.68	1926.25	173.12	-14553.9	100	2805.07	1973.64	159.79	-13174.0
150	4174.92	2927.77	184.60	-23515.4	150	4225.96	2978.82	171.31	-21470.5
200	5616.56	3953.70	192.89	-32962.1	200	5659.09	3996.23	179.55	-30251.7
250	7081.75	5003.18	199.43	-42775.9	250	7104.43	5025.86	186.00	-39396.4
298	8509.70	6032.04	204.65	-52477.3	298	8503.43	6025.77	191.12	-48450.8
300	8569.64	6075.35	204.85	-52886.8	300	8561.97	6067.68	191.32	-48833.3
350	10079.40	7169.40	209.51	-63248.8	350	10031.66	7121.65	195.85	-58515.2
400	11610.23	8284.51	213.60	-73828.2	400	11513.46	8187.74	199.81	-68408.6
450	13161.33	9419.89	217.25	-84601.0	450	13007.34	9265.90	203.32	-78488.4
500	14731.93	10574.78	220.56	-95547.5	500	14513.22	10356.07	206.50	-88735.2
550	16321.30	11748.43	223.59	-106652.2	550	16031.05	11458.18	209.39	-99133.4
600	17928.68	12940.10	226.39	-117902.4	600	17560.74	12572.16	212.05	-109670.4
650	19553.39	14149.09	228.99	-129287.4	650	19102.23	13697.93	214.52	-120335.4
700	21194.71	15374.70	231.42	-140798.2	700	20655.41	14835.40	216.82	-131119.6
750	22851.99	16616.26	233.71	-152426.8	750	22220.19	15984.47	218.98	-142015.2
800	24524.55	17873.11	235.86	-164166.6	800	23796.47	17145.03	221.01	-153015.5
850	26211.78	19144.62	237.91	-176011.4	850	25384.12	18316.97	222.94	-164114.8
900	27913.05	20430.18	239.85	-187955.8	900	26983.03	19500.16	224.77	-175307.9
950	29627.76	21729.18	241.71	-199995.3	950	28593.06	20694.47	226.51	-186590.2
1000	31355.35	23041.05	243.48	-212125.3	1000	30214.07	21899.77	228.17	-197957.5
1050	33095.25	24365.23	245.18	-224342.1	1050	31845.92	23115.90	229.76	-209406.1
1100	34846.92	25701.19	246.81	-236642.0	1100	33488.45	24342.72	231.29	-220932.8
1150	36609.84	27048.40	248.38	-249021.9	1150	35141.49	25580.04	232.76	-232534.4
1200	38383.52	28406.36	249.89	-261478.6	1200	36804.88	26827.72	234.18	-244208.0
1250	40167.47	29774.59	251.34	-274009.5	1250	38478.43	28085.55	235.54	-255951.3
1300	41961.22	31152.63	252.75	-286612.0	1300	40161.95	29353.36	236.86	-267761.7
1350	43764.34	32540.04	254.11	-299283.6	1350	41855.26	30630.95	238.14	-279637.0
1400	45576.41	33936.39	255.43	-312022.2	1400	43558.13	31918.11	239.38	-291575.2
1450	47397.01	35341.28	256.71	-324825.7	1450	45270.36	33214.63	240.58	-303574.5
1500	49225.76	36754.31	257.95	-337692.1	1500	46991.73	34520.28	241.75	-315632.9
1550	51062.30	38175.14	259.15	-350619.6	1550	48722.01	35834.85	242.88	-327748.9
1600	52906.28	39603.40	260.32	-363606.5	1600	50460.96	37158.08	243.99	-339920.9
1650	54757.37	41038.78	261.46	-376651.1	1650	52208.33	38489.74	245.06	-352147.3
1700	56615.26	42480.95	262.57	-389752.0	1700	53963.87	39829.56	246.11	-364426.8
1750	58479.67	43929.65	263.65	-402907.6	1750	55727.32	41177.30	247.13	-376758.1
1800	60350.32	45384.58	264.70	-416116.5	1800	57498.40	42532.66	248.13	-389139.9
1850	62226.96	46845.50	265.73	-429377.5	1850	59276.85	43895.39	249.11	-401570.9
1900	64109.36	48312.19	266.74	-442689.3	1900	61062.36	45265.19	250.06	-414050.2
1950	65997.31	49784.42	267.72	-456050.7	1950	62854.64	46641.76	250.99	-426576.5
2000	67890.61	51262.01	268.68	-469460.6	2000	64653.40	48024.80	251.90	-439148.9
2050	69789.09	52744.78	269.61	-482918.0	2050	66458.32	49414.00	252.79	-451766.3
2100	71692.60	54232.57	270.53	-496421.6	2100	68269.07	50809.04	253.67	-464427.8
2150	73600.99	55725.25	271.43	-509970.7	2150	70085.34	52209.60	254.52	-477132.5
2200	75514.16	57222.70	272.31	-523564.2	2200	71906.79	53615.33	255.36	-489879.5
2250	77432.00	58724.83	273.17	-537201.2	2250	73733.06	55025.89	256.18	-502668.0
2300	79354.44	60231.55	274.02	-550880.9	2300	75563.81	56440.92	256.98	-515497.1
2350	81281.41	61742.81	274.84	-564602.5	2350	77398.68	57860.08	257.77	-528366.0
2400	83212.89	63258.57	275.66	-578365.1	2400	79237.30	59282.98	258.55	-541274.1
2450	85148.84	64778.80	276.46	-592168.0	2450	81079.29	60709.25	259.31	-554220.4
2500	87089.26	66303.51	277.24	-606010.4	2500	82924.26	62138.51	260.05	-567204.4
2550	89034.18	67832.71	278.01	-619891.8	2550	84771.83	63570.36	260.78	-580225.3
2600	90983.62	69366.44	278.77	-633811.2	2600	86621.58	65004.40	261.50	-593282.5
2650	92937.65	70904.76	279.51	-647768.3	2650	88473.12	66440.22	262.21	-606375.3
2700	94896.35	72447.74	280.24	-661762.2	2700	90326.01	67877.40	262.90	-619503.0
2750	96859.80	73995.47	280.96	-675792.5	2750	92179.84	69315.52	263.58	-632665.0
2800	98828.11	75548.07	281.67	-689858.5	2800	94034.17	70754.13	264.25	-645860.8
2850	100801.43	77105.68	282.37	-703959.7	2850	95888.56	72192.81	264.90	-659089.7
2900	102779.90	78668.43	283.06	-718095.5	2900	97742.56	73631.09	265.55	-672351.1
2950	104763.70	80236.52	283.74	-732265.5	2950	99595.70	75068.51	266.18	-685644.4
3000	106753.02	81810.12	284.41	-746469.2	3000	101447.52	76504.62	266.81	-698969.2
3050	108748.05	83389.44	285.07	-760706.1	3050	103297.54	77938.92	267.42	-712324.8
3100	110749.05	84974.72	285.72	-774975.8	3100	105145.28	79370.95	268.02	-725710.7
3150	112756.24	86566.18	286.36	-789277.8	3150	106990.25	80800.21	268.61	-739126.4
3200	114769.90	88164.14	286.99	-803611.7	3200	108831.95	82226.19	269.19	-752571.4
3250	116790.32	89768.85	287.62	-817977.1	3250	110669.87	83648.39	269.76	-766045.1
3300	118817.80	91380.61	288.24	-832373.6	3300	112503.49	85066.30	270.32	-779547.1
3350	120852.67	92999.76	288.85	-846800.9	3350	114332.28	86479.38	270.87	-793076.8
3400	122895.26	94626.64	289.46	-861258.7	3400	116155.72	87887.10	271.41	-806633.8
3450	124945.95	96261.62	290.06	-875746.5	3450	117973.26	89288.93	271.94	-820217.5
3500	127005.11	97905.06	290.65	-890264.2	3500	119784.36	90684.31	272.46	-833827.5

Table 9.4 *continued*

Carbon monoxide, CO : h_0 = -113 882.3 kJ/kmol

T	h(T)	u(T)	s(T)	g(T)
50	1386.73	971.01	146.72	-5949.2
100	2788.93	1957.50	166.15	-13825.6
150	4206.39	2959.24	177.64	-22439.0
200	5638.90	3976.04	185.88	-31536.3
250	7086.27	5007.69	192.33	-40997.3
298	8489.51	6011.84	197.47	-50356.0
300	8548.26	6053.97	197.66	-50751.1
350	10024.68	7114.67	202.22	-60750.8
400	11515.29	8189.57	206.20	-70963.2
450	13019.87	9278.44	209.74	-81363.2
500	14538.21	10381.06	212.94	-91931.5
550	16070.08	11497.21	215.86	-102652.5
600	17615.24	12626.66	218.55	-113513.6
650	19173.47	13769.17	221.04	-124504.1
700	20744.52	14924.51	223.37	-135615.0
750	22328.17	16092.44	225.56	-146838.7
800	23924.16	17272.72	227.62	-158168.5
850	25532.26	18465.11	229.57	-169598.5
900	27152.23	19669.36	231.42	-181123.5
950	28783.80	20885.22	233.18	-192738.8
1000	30426.74	22112.44	234.87	-204440.3
1050	32080.79	23350.77	236.48	-216224.3
1100	33745.68	24599.95	238.03	-228087.3
1150	35421.17	25859.72	239.52	-240026.3
1200	37106.98	27129.82	240.95	-252038.4
1250	38802.85	28409.98	242.34	-264120.9
1300	40508.52	29699.93	243.68	-276271.5
1350	42223.72	30999.41	244.97	-288487.9
1400	43948.16	32308.14	246.23	-300768.0
1450	45681.58	33625.84	247.44	-313109.8
1500	47423.69	34952.24	248.62	-325511.6
1550	49174.22	36287.06	249.77	-337971.6
1600	50932.88	37630.00	250.89	-350488.3
1650	52699.38	38980.79	251.98	-363060.0
1700	54473.44	40339.13	253.03	-375685.3
1750	56254.75	41704.73	254.07	-388363.0
1800	58043.03	43077.29	255.07	-401091.6
1850	59837.98	44456.53	256.06	-413870.1
1900	61639.29	45842.12	257.02	-426697.1
1950	63446.67	47233.79	257.96	-439571.6
2000	65259.81	48631.21	258.88	-452492.6
2050	67078.39	50034.07	259.77	-465458.9
2100	68902.11	51442.08	260.65	-478469.7
2150	70730.65	52854.90	261.51	-491523.9
2200	72563.69	54272.23	262.36	-504620.8
2250	74400.92	55693.74	263.18	-517759.3
2300	76242.01	57119.12	263.99	-530938.7
2350	78086.64	58548.03	264.79	-544158.2
2400	79934.47	59980.15	265.56	-557417.0
2450	81785.19	61415.15	266.33	-570714.3
2500	83638.45	62852.70	267.08	-584049.4
2550	85493.91	64292.45	267.81	-597421.5
2600	87351.25	65734.07	268.53	-610830.1
2650	89210.11	67177.22	269.24	-624274.4
2700	91070.16	68621.55	269.93	-637753.9
2750	92931.04	70066.71	270.62	-651267.7
2800	94792.40	71512.36	271.29	-664815.4
2850	96653.90	72958.15	271.95	-678396.4
2900	98515.18	74403.71	272.59	-692010.0
2950	100375.87	75848.68	273.23	-705655.7
3000	102235.62	77292.72	273.86	-719332.9
3050	104094.07	78735.45	274.47	-733041.1
3100	105950.84	80176.51	275.07	-746779.8
3150	107805.57	81615.53	275.67	-760548.4
3200	109657.89	83052.13	276.25	-774346.4
3250	111507.42	84485.95	276.82	-788173.3
3300	113353.79	85916.60	277.39	-802028.7
3350	115196.61	87343.71	277.94	-815912.1
3400	117035.50	88766.88	278.49	-829822.9
3450	118870.08	90185.75	279.02	-843760.7
3500	120699.96	91599.91	279.55	-857725.0

Carbon dioxide, CO_2 : h_0 = -393 404.9 kJ/kmol

T	h(T)	u(T)	s(T)	g(T)
50	1342.97	927.25	157.68	-6541.2
100	2794.62	1963.19	177.72	-14977.7
150	4350.19	3103.05	190.31	-24196.3
200	6005.05	4342.19	199.82	-33958.6
250	7754.65	5676.08	207.62	-44150.0
298	9519.29	7041.63	214.07	-54273.9
300	9594.56	7100.27	214.32	-54702.3
350	11520.46	8610.46	220.26	-65569.5
400	13528.14	10202.42	225.62	-76718.5
450	15613.47	11872.04	230.53	-88123.7
500	17772.48	13615.33	235.08	-99765.1
550	20001.26	15428.39	239.32	-111626.2
600	22296.03	17307.45	243.32	-123693.2
650	24653.12	19248.83	247.09	-135954.1
700	27068.96	21248.95	250.67	-148398.8
750	29540.09	23304.37	254.08	-161018.1
800	32063.16	25411.72	257.33	-173804.0
850	34634.93	27567.78	260.45	-186749.2
900	37252.26	29769.39	263.44	-199847.1
950	39912.12	32013.54	266.32	-213091.6
1000	42611.60	34297.30	269.09	-226477.3
1050	45347.89	36617.87	271.76	-239998.8
1100	48118.27	38972.54	274.34	-253651.6
1150	50920.17	41358.72	276.83	-267431.0
1200	53751.08	43773.92	279.24	-281332.9
1250	56608.63	46215.76	281.57	-295353.4
1300	59490.55	48681.96	283.83	-309488.7
1350	62394.68	51170.37	286.02	-323735.3
1400	65318.96	53678.94	288.15	-338089.8
1450	68261.44	56205.70	290.21	-352549.2
1500	71220.29	58748.84	292.22	-367110.3
1550	74193.77	61306.60	294.17	-381770.3
1600	77180.26	63877.38	296.07	-396526.4
1650	80178.25	66459.66	297.91	-411376.1
1700	83186.33	69052.02	299.71	-426316.8
1750	86203.20	71653.17	301.46	-441346.1
1800	89227.66	74261.92	303.16	-456461.7
1850	92258.64	76877.19	304.82	-471661.4
1900	95295.17	79498.00	306.44	-486943.2
1950	98336.36	82123.48	308.02	-502304.9
2000	101381.47	84752.87	309.56	-517744.7
2050	104429.85	87385.53	311.07	-533260.6
2100	107480.95	90020.92	312.54	-548850.9
2150	110534.33	92658.58	313.98	-564513.9
2200	113589.67	95298.21	315.38	-580248.0
2250	116646.75	97939.57	316.75	-596051.5
2300	119705.46	100582.57	318.10	-611923.0
2350	122765.79	103227.19	319.42	-627861.0
2400	125827.86	105873.54	320.70	-643864.1
2450	128891.87	108521.83	321.97	-659931.0
2500	131958.14	111172.39	323.21	-676060.5
2550	135027.11	113825.64	324.42	-692251.4
2600	138099.31	116482.13	325.62	-708502.5
2650	141175.38	119142.48	326.79	-724812.6
2700	144256.08	121807.47	327.94	-741180.9
2750	147342.28	124477.95	329.07	-757606.3
2800	150434.93	127154.89	330.19	-774087.8
2850	153535.12	129839.36	331.28	-790624.7
2900	156644.02	132532.55	332.37	-807216.0
2950	159762.95	135235.76	333.43	-823861.0
3000	162893.29	137950.39	334.48	-840558.9
3050	166036.55	140677.93	335.52	-857309.2
3100	169194.35	143420.02	336.55	-874111.0
3150	172368.42	146178.38	337.57	-890964.0
3200	175560.59	148954.83	338.57	-907867.5
3250	178772.80	151751.32	339.57	-924821.0
3300	182007.10	154569.91	340.55	-941824.0
3350	185265.64	157412.74	341.53	-958876.3
3400	188550.70	160282.08	342.51	-975977.4
3450	191864.64	163180.30	343.48	-993127.0
3500	195209.95	166109.90	344.44	-1010324.9

Table 9.4 *continued*

Molecular hydrogen, H_2 : $h_0 = 0.0000$ kJ/kmol

T	h(T)	u(T)	s(T)	g(T)
50	1427.20	1011.48	79.70	-2557.8
100	2854.64	2023.21	99.49	-7094.2
150	4282.92	3035.78	111.07	-12377.6
200	5712.58	4049.72	119.30	-18146.6
250	7144.17	5065.60	125.68	-24277.0
298	*8520.79*	*6043.13*	*130.72*	*-30434.3*
300	8578.21	6083.92	130.91	-30695.9
350	10015.20	7105.19	135.34	-37355.1
400	11455.63	8129.91	139.19	-44220.6
450	12899.96	9158.53	142.59	-51266.8
500	14348.65	10191.50	145.65	-58474.1
550	15802.13	11229.26	148.42	-65826.7
600	17260.81	12272.23	150.95	-73311.8
650	18725.08	13320.79	153.30	-80918.9
700	20195.34	14375.33	155.48	-88638.9
750	21671.92	15436.20	157.51	-96464.3
800	23155.19	16503.75	159.43	-104388.4
850	24645.46	17578.30	161.24	-112405.4
900	26143.03	18660.16	162.95	-120510.4
950	27648.20	19749.61	164.58	-128698.8
1000	29161.23	20846.93	166.13	-136966.7
1050	30682.38	21952.37	167.61	-145310.5
1100	32211.88	23066.15	169.04	-153726.9
1150	33749.95	24188.51	170.40	-162213.1
1200	35296.78	25319.62	171.72	-170766.4
1250	36852.56	26459.68	172.99	-179384.3
1300	38417.44	27608.85	174.22	-188064.6
1350	39991.57	28767.26	175.41	-196805.3
1400	41575.07	29935.05	176.56	-205604.5
1450	43168.06	31112.33	177.67	-214460.4
1500	44770.62	32299.17	178.76	-223371.5
1550	46382.83	33495.66	179.82	-232336.1
1600	48004.73	34701.85	180.85	-241352.9
1650	49636.37	35917.77	181.85	-250420.5
1700	51277.76	37143.45	182.83	-259537.7
1750	52928.94	38378.88	183.79	-268703.4
1800	54589.77	39624.03	184.73	-277916.4
1850	56260.35	40878.89	185.64	-287175.6
1900	57940.56	42143.39	186.54	-296480.1
1950	59630.34	43417.46	187.42	-305829.0
2000	61329.61	44701.01	188.28	-315221.4
2050	63038.25	45993.93	189.12	-324656.3
2100	64756.13	47296.10	189.95	-334133.0
2150	66483.11	48607.36	190.76	-343650.8
2200	68219.03	49927.57	191.56	-353208.8
2250	69963.71	51256.54	192.34	-362806.3
2300	71716.95	52594.06	193.11	-372442.8
2350	73478.53	53939.93	193.87	-382117.4
2400	75248.23	55293.91	194.62	-391829.6
2450	77025.78	56655.74	195.35	-401578.8
2500	78810.91	58025.16	196.07	-411364.3
2550	80603.35	59401.88	196.78	-421185.6
2600	82402.77	60785.59	197.48	-431042.1
2650	84208.86	62175.97	198.17	-440933.3
2700	86021.28	63572.67	198.84	-450858.6
2750	87839.67	64975.34	199.51	-460817.6
2800	89663.63	66383.59	200.17	-470809.6
2850	91492.79	67797.04	200.82	-480834.3
2900	93326.72	69215.25	201.45	-490891.1
2950	95165.00	70637.82	202.08	-500979.6
3000	97007.15	72064.27	202.70	-511099.3
3050	98852.76	73494.15	203.31	-521249.7
3100	100701.29	74926.96	203.91	-531430.3
3150	102552.25	76362.21	204.51	-541640.9
3200	104405.12	77799.36	205.09	-551880.8
3250	106259.36	79237.89	205.66	-562149.6
3300	108114.42	80677.23	206.23	-572447.1
3350	109969.70	82116.80	206.79	-582772.6
3400	111824.63	83556.01	207.34	-593125.8
3450	113678.59	84994.25	207.88	-603506.3
3500	115530.94	86430.89	208.41	-613913.6

Water, H_2O : $h_0 = -239\ 081.7$ kJ/kmol

T	h(T)	u(T)	s(T)	g(T)
50	1567.79	1152.08	130.24	-4944.1
100	3159.40	2327.97	152.28	-12068.9
150	4775.08	3527.94	165.38	-20031.6
200	6415.10	4752.24	174.81	-28547.1
250	8079.69	6001.11	182.24	-37479.7
298	*9700.99*	*7223.33*	*188.17*	*-46373.3*
300	9769.04	7274.75	188.40	-46749.9
350	11483.36	8573.35	193.68	-56304.9
400	13222.78	9897.06	198.33	-66107.3
450	14987.47	11246.03	202.48	-76129.3
500	16777.52	12620.37	206.25	-86349.1
550	18593.04	14020.18	209.71	-96749.5
600	20434.10	15445.52	212.92	-107316.2
650	22300.74	16896.45	215.91	-118037.6
700	24192.99	18372.98	218.71	-128903.7
750	26110.85	19875.13	221.36	-139905.9
800	28054.30	21402.86	223.86	-151036.9
850	30023.31	22956.15	226.25	-162290.3
900	32017.79	24534.92	228.53	-173660.3
950	34037.68	26139.09	230.72	-185141.8
1000	36082.85	27768.55	232.81	-196730.4
1050	38153.17	29423.15	234.83	-208421.8
1100	40248.48	31102.75	236.78	-220212.5
1150	42368.62	32807.17	238.67	-232099.0
1200	44513.37	34536.21	240.49	-244078.3
1250	46682.51	36289.64	242.26	-256147.4
1300	48875.80	38067.21	243.98	-268303.8
1350	51092.97	39868.66	245.66	-280545.1
1400	53333.72	41693.70	247.29	-292868.9
1450	55597.74	43542.00	248.88	-305273.2
1500	57884.69	45413.24	250.43	-317755.9
1550	60194.21	47307.04	251.94	-330315.3
1600	62525.92	49223.04	253.42	-342949.5
1650	64879.42	51160.82	254.87	-355656.9
1700	67254.27	53119.96	256.29	-368436.0
1750	69650.03	55100.00	257.68	-381285.3
1800	72066.22	57100.48	259.04	-394203.3
1850	74502.34	59120.89	260.37	-407188.7
1900	76957.88	61160.71	261.68	-420240.2
1950	79432.30	63219.41	262.97	-433356.6
2000	81925.03	65296.43	264.23	-446536.7
2050	84435.48	67391.17	265.47	-459779.4
2100	86963.05	69503.02	266.69	-473083.4
2150	89507.10	71631.36	267.89	-486447.9
2200	92066.98	73775.52	269.06	-499871.7
2250	94642.01	75934.83	270.22	-513353.9
2300	97231.49	78108.60	271.36	-526893.4
2350	99834.70	80296.09	272.48	-540489.4
2400	102450.89	82496.57	273.58	-554141.0
2450	105079.29	84709.26	274.66	-567847.1
2500	107719.12	86933.37	275.73	-581607.1
2550	110369.56	89168.10	276.78	-595419.9
2600	113029.77	91412.59	277.81	-609284.8
2650	115698.90	93666.00	278.83	-623200.9
2700	118376.06	95927.45	279.83	-637167.5
2750	121060.35	98196.02	280.82	-651183.8
2800	123750.84	100470.80	281.79	-665248.9
2850	126446.58	102750.82	282.74	-679362.1
2900	129146.60	105035.13	283.68	-693522.6
2950	131849.91	107322.73	284.60	-707729.7
3000	134555.49	109612.59	285.51	-721982.7
3050	137262.29	111903.68	286.41	-736280.8
3100	139969.26	114194.93	287.29	-750623.2
3150	142675.31	116485.26	288.15	-765009.3
3200	145379.33	118773.57	289.01	-779438.4
3250	148080.18	121058.71	289.84	-793909.6
3300	150776.73	123339.54	290.67	-808422.4
3350	153467.78	125614.88	291.48	-822976.0
3400	156152.15	127883.53	292.27	-837569.8
3450	158828.60	130144.27	293.05	-852202.9
3500	161495.91	132395.86	293.82	-866874.8

Table 9.4 *continued*

Nitric oxide, NO : $h_0 = 89\ 914.7$ kJ/kmol

T	h(T)	u(T)	s(T)	g(T)
50	1461.94	1046.22	156.66	-6371.1
100	2936.26	2104.83	177.09	-14772.6
150	4422.89	3175.75	189.14	-23948.2
200	5921.77	4258.91	197.76	-33630.9
250	7432.80	5354.22	204.51	-43693.6
298	8894.74	6417.08	209.85	-53641.9
300	8955.89	6461.60	210.06	-54061.8
350	10490.97	7580.97	214.79	-64685.9
400	12037.92	8712.20	218.92	-75530.9
450	13596.65	9855.21	222.59	-86570.5
500	15167.03	11009.88	225.90	-97784.3
550	16748.95	12176.09	228.92	-109155.9
600	18342.30	13353.72	231.69	-120672.0
650	19946.93	14542.63	234.26	-132321.5
700	21562.71	15742.70	236.65	-144095.0
750	23189.50	16953.77	238.90	-155984.4
800	24827.15	18175.71	241.01	-167982.7
850	26475.50	19408.35	243.01	-180083.8
900	28134.40	20651.53	244.91	-192282.1
950	29803.68	21905.10	246.71	-204573.0
1000	31483.16	23168.86	248.44	-216952.0
1050	33172.68	24442.66	250.08	-229415.2
1100	34872.03	25726.30	251.66	-241959.2
1150	36581.03	27019.58	253.18	-254580.7
1200	38299.48	28322.32	254.65	-267276.7
1250	40027.19	29634.31	256.06	-280044.5
1300	41763.93	30955.34	257.42	-292881.6
1350	43509.49	32285.19	258.74	-305785.7
1400	45263.66	33623.64	260.01	-318754.6
1450	47026.20	34970.47	261.25	-331786.4
1500	48796.88	36325.43	262.45	-344879.0
1550	50575.47	37688.30	263.62	-358030.9
1600	52361.70	39058.82	264.75	-371240.2
1650	54155.33	40436.74	265.86	-384505.5
1700	55956.11	41821.80	266.93	-397825.2
1750	57763.77	43213.74	267.98	-411198.0
1800	59578.03	44612.29	269.00	-424622.6
1850	61398.62	46017.17	270.00	-438097.7
1900	63225.26	47428.09	270.97	-451622.0
1950	65057.66	48844.77	271.92	-465194.5
2000	66895.51	50266.91	272.85	-478814.1
2050	68738.53	51694.22	273.76	-492479.7
2100	70586.41	53126.38	274.66	-506190.3
2150	72438.82	54563.08	275.53	-519944.9
2200	74295.46	56004.00	276.38	-533742.7
2250	76155.99	57448.81	277.22	-547582.7
2300	78020.08	58897.19	278.04	-561464.1
2350	79887.40	60348.79	278.84	-575386.1
2400	81757.59	61803.27	279.63	-589347.8
2450	83630.32	63260.29	280.40	-603348.6
2500	85505.22	64719.47	281.16	-617387.6
2550	87381.94	66180.47	281.90	-631464.0
2600	89260.09	67642.91	282.63	-645577.4
2650	91139.32	69106.42	283.35	-659726.8
2700	93019.23	70570.62	284.05	-673911.7
2750	94899.44	72035.11	284.74	-688131.4
2800	96779.56	73499.52	285.42	-702385.4
2850	98659.18	74963.43	286.08	-716672.8
2900	100537.91	76426.44	286.73	-730993.3
2950	102415.33	77888.15	287.38	-745346.1
3000	104291.02	79348.12	288.01	-759730.8
3050	106164.57	80805.95	288.63	-774146.7
3100	108035.53	82261.20	289.24	-788593.3
3150	109903.48	83713.43	289.83	-803070.0
3200	111767.97	85162.21	290.42	-817576.4
3250	113628.55	86607.08	291.00	-832111.8
3300	115484.78	88047.59	291.56	-846675.9
3350	117336.18	89483.28	292.12	-861268.1
3400	119182.30	90913.68	292.67	-875887.8
3450	121022.66	92338.32	293.21	-890534.7
3500	122856.78	93756.73	293.73	-905208.2

Methane, CH_4 : $h_0 = -66\ 930.5$ kJ/kmol

T	h(T)	u(T)	s(T)	g(T)
50	906.44	490.72	134.82	-5834.7
100	2011.62	1180.19	149.99	-12987.2
150	3308.10	2060.95	160.45	-20759.5
200	4788.60	3125.74	168.94	-29000.2
250	6446.08	4367.50	176.33	-37635.7
298	8197.36	5719.70	182.73	-46255.4
300	8273.64	5779.35	182.98	-46621.1
350	10264.62	7354.62	189.11	-55925.6
400	12412.51	9086.79	194.85	-65526.5
450	14711.00	10969.57	200.26	-75406.0
500	17153.96	12996.81	205.41	-85549.6
550	19735.44	15162.57	210.33	-95945.2
600	22449.65	17461.07	215.05	-106582.5
650	25291.00	19886.70	219.61	-117452.3
700	28254.05	22434.04	224.00	-128546.8
750	31333.53	25097.81	228.26	-139858.8
800	34524.35	27872.91	232.38	-151381.7
850	37821.57	30754.41	236.39	-163109.8
900	41220.40	33737.53	240.29	-175037.6
950	44716.23	36817.65	244.08	-187160.3
1000	48304.59	39990.29	247.78	-199473.3
1050	51981.17	43251.16	251.38	-211972.4
1100	55741.80	46596.07	254.91	-224653.9
1150	59582.46	50021.02	258.34	-237514.1
1200	63499.28	53522.12	261.71	-250549.7
1250	67488.52	57095.65	265.00	-263757.9
1300	71546.59	60738.00	268.22	-277135.7
1350	75670.02	64445.71	271.37	-290680.7
1400	79855.49	68215.47	274.46	-304390.6
1450	84099.80	72044.07	277.49	-318263.4
1500	88399.88	75928.43	280.46	-332297.0
1550	92752.80	79865.63	283.38	-346490.0
1600	97155.72	83852.84	286.25	-360840.9
1650	101605.94	87887.35	289.06	-375348.3
1700	106100.89	91966.58	291.83	-390011.4
1750	110638.09	96088.06	294.55	-404829.3
1800	115215.17	100249.43	297.23	-419801.3
1850	119829.90	104448.44	299.87	-434927.0
1900	124480.12	108682.95	302.47	-450206.2
1950	129163.80	112950.91	305.03	-465638.8
2000	133878.99	117250.39	307.55	-481225.1
2050	138623.86	121579.55	310.04	-496965.3
2100	143396.66	125936.63	312.50	-512860.1
2150	148195.74	130320.01	314.93	-528910.1
2200	153019.54	134728.08	317.33	-545116.4
2250	157866.57	139159.40	319.71	-561480.1
2300	162735.46	143612.57	322.06	-578002.6
2350	167624.89	148086.26	324.39	-594685.5
2400	172533.63	152579.31	326.69	-611530.6
2450	177460.54	157090.50	328.98	-628539.8
2500	182404.53	161618.78	331.25	-645715.4
2550	187364.61	166163.14	333.50	-663059.7
2600	192339.82	170722.64	335.74	-680575.5
2650	197329.31	175296.42	337.96	-698265.5
2700	202332.27	179883.66	340.17	-716132.8
2750	207347.95	184483.62	342.37	-734180.8
2800	212375.65	189095.61	344.57	-752412.8
2850	217414.75	193719.00	346.75	-770832.7
2900	222464.67	198353.20	348.93	-789444.4
2950	227524.88	202997.69	351.11	-808252.1
3000	232594.89	207651.99	353.29	-827260.2
3050	237674.27	212315.65	355.46	-846473.4
3100	242762.61	216988.28	357.63	-865896.5
3150	247859.57	221669.52	359.81	-885534.8
3200	252964.82	226359.06	361.99	-905393.4
3250	258078.08	231056.60	364.17	-925478.1
3300	263199.09	235761.90	366.36	-945794.7
3350	268327.62	240474.72	368.56	-966349.4
3400	273456.95	245188.33	370.77	-987145.1
3450	278606.50	249922.16	372.99	-1008198.3
3500	283756.50	254656.45	375.22	-1029506.1

9.3.1   TABLES OF MEAN SPECIFIC HEAT

Sometimes data are given in terms of mean specific heat rather than the actual specific heat at a particular temperature. This approach will now be described.

Consider the change of enthalpy, h, between temperatures T_1 and T_2. This is

$$h_2 - h_1 = \int_{T_1}^{T_2} c_p(T)\, dT = h(T_2) - h(T_1) \tag{9.43}$$

Now, this can be written

$$\bar{c}_p(T_2 - T_1) = h(T_2) - h(T_1) \tag{9.44}$$

where $\bar{c}_p$ is the mean specific heat between T_1 and T_2. Normally $\bar{c}_p$ is given at a particular temperature T, and it is then defined as the mean between the temperature T and a reference temperature T_{ref}, i.e.

$$(\bar{c}_p)_T = \frac{h(T) - h(T_{\text{ref}})}{T - T_{\text{ref}}} \tag{9.45}$$

It is also possible to write the mean specific heat as a polynomial function of temperature, in which case

$$(\bar{c}_p) = a + bT + cT^2 + \cdots \tag{9.46}$$

where a, b and c are tabulated coefficients.

If it is required to calculate the enthalpy difference between two temperatures T_1 and T_2, then

$$h_2 - h_1 = (\bar{c}_p)_{T_2}(T_2 - T_{\text{ref}}) - (\bar{c}_p)_{T_1}(T_1 - T_{\text{ref}}) \tag{9.47}$$

Having calculated the change in enthalpy, $h_2 - h_1$, then the change in internal energy, $u_2 - u_1$, may be evaluated as

$$u_2 - u_1 = h_2 - RT_2 - (h_1 - RT_1) = h_2 - h_1 - R(T_2 - T_1) \tag{9.48a}$$

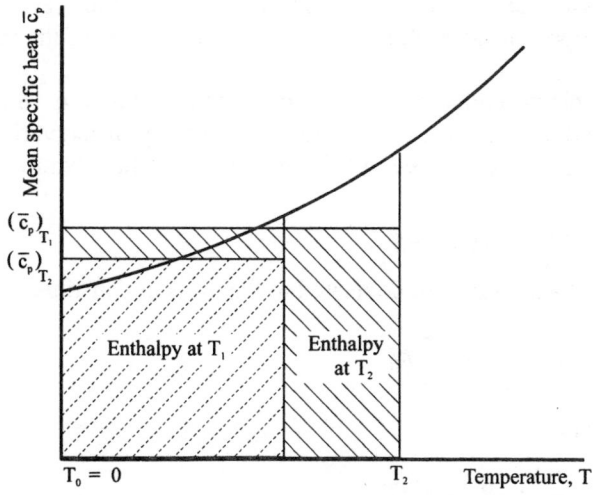

Fig. 9.1 Mean specific heat at constant pressure

or, in molar terms, by

$$u_{m,2} - u_{m,1} = h_{m,2} - \Re T_2 - (h_{m,1} - \Re T_1) = h_{m,2} - h_{m,1} - \Re(T_2 - T_1) \qquad (9.48b)$$

If the change of internal energy, $u_2 - u_1$, is known, then the change of enthalpy, $h_2 - h_1$, may be calculated in a similar way. The use of mean specific heat values enables more accurate evaluation of temperature changes because the mean specific heats enable the energy equation to be solved with an allowance for the variation of specific heats. The mean specific heats approach is depicted in Fig 9.1.

9.4   Mixtures of ideal gases

Many problems encountered in engineering involve mixtures of gases – air itself is a mixture of many gases although it can be considered to be oxygen and atmospheric nitrogen (which can be assumed to contain the nitrogen in the air and the other 'inert' gases such as argon and carbon dioxide). If the gases in a mixture are at a temperature well above their critical temperature and a pressure below the critical pressure, they act as ideal gases. Expressions relating the properties of mixtures of ideal gases will now be formulated. These expressions are a direct consequence of the Gibbs–Dalton laws.

9.4.1   DALTON PRINCIPLE

Dalton stated that

> any gas is as a vacuum to any gas mixed with it.

This statement was expanded and clarified by Gibbs to give the Gibbs–Dalton laws.

9.4.2   GIBBS–DALTON LAW

(a)   A gas mixture as a whole obeys the equation of state (eqn (9.2))

$$pV = n\Re T$$

where n is the *total* amount of substance in the mixture.

(b)   The total pressure exerted by a mixture is the sum of the pressures exerted by the individual components as each occupies the *whole volume* of the mixture at the same temperature.

(c)   The internal energy, enthalpy and entropy of the mixture are, respectively, equal to the sums of the internal energy, enthalpy and entropy of the various components as each occupies the *whole volume* at the temperature of the mixture.

9.4.3   MIXTURE RELATIONSHIPS

The total mass, m, of a mixture can be related to the mass of constituents by

$$m = m_a + m_b + m_c + \cdots = \sum_{i=1}^{n} m_i \qquad (9.49)$$

From the ideal gas law

$$pV = mRT = \frac{m}{m_w} m_w RT = n\Re T$$

and hence for the individual constituents

$$pV_a = m_a R_a T = n_a \Re T$$
$$pV_b = m_b R_b T = n_b \Re T \tag{9.50}$$
$$pV_c = m_c R_c T = n_c \Re T, \text{ etc}$$

In eqn (9.50), V_a = volume of constituent a at the pressure and temperature of the mixture, and V_b and V_c are similar volumes for constituents b and c. But the total volume, V, is given by

$$V = V_a + V_b + V_c + \cdots = \sum_{i=1}^{n} V_i \tag{9.51}$$

and therefore

$$n = n_a + n_b + n_c + \cdots = \sum_{i=1}^{n} n_i \tag{9.52}$$

Let

$$\frac{n_a}{n} = x_a = \text{mole fraction of } a \text{ in the mixture} \tag{9.53}$$

Similarly, $n_b/n = x_b$, and $n_c/n = x_c$, etc, with $n_i/n = x_i$. Therefore

$$x_a + x_b + x_c + \cdots = \sum_{i=1}^{n} x_i = 1 \tag{9.54}$$

It is possible to develop the term for the *partial pressure* of each constituent from statement (b) of the Gibbs–Dalton laws. Then, for constituent a occupying the total volume of the mixture at the pressure and temperature of the mixture,

$$p_a V = n_a \Re T$$

But, for the mixture as a whole (eqn (9.2))

$$pV = n \Re T$$

Hence, by dividing eqn (9.50) by eqn (9.2)

$$\frac{p_a}{p} = \frac{n_a}{n} = x_a \tag{9.55}$$

giving

$$p_a = x_a p, \text{ and in general } p_i = x_i p \tag{9.56}$$

and, also

$$\sum_{i=1}^{n} p_i = \sum_{i=1}^{n} x_i p = p \sum_{i=1}^{n} x_i = p \tag{9.57}$$

Often mixtures are analysed on a volumetric basis, and from the volumetric results it is possible to obtain the partial pressures and mole fractions of the components. Normally volumetric analyses are performed at constant pressure and temperature. Then, considering the *i*th component,

$$V_i = \frac{n_i \Re T}{p}$$

(9.58)

and the total volume of the mixture, V, is

$$V = \frac{n \Re T}{p}$$

(9.59)

Hence, from eqns (9.58) and (9.59)

$$\frac{V_i}{V} = \frac{n_i}{n} = x_i$$

(9.60)

If the gas composition is given in volume percentage of the mixture, the mole fraction is

$$x_i = \frac{V_i(\%)}{100}$$

(9.61)

and the partial pressure is

$$p_i = \frac{V_i(\%)}{100} p$$

(9.62)

The terms relating to the energy of a mixture can be evaluated from statement (c) of the Gibbs–Dalton laws. If e_m = molar internal energy, and h_m = molar enthalpy, then for the mixture

$$E = n e_m = \sum_{i=1}^{n} n_i e_{m,i}$$

(9.63)

and

$$H = n h_m = \sum_{i=1}^{n} n_i h_{m,i}$$

(9.64)

If the molar internal energy and molar enthalpy *of the mixture* are required then

$$e_m = \frac{1}{n} \sum_{i=1}^{n} n_i e_{m,i} = \sum_{i=1}^{n} x_i e_{m,i}$$

(9.65)

and

$$h_m = \frac{1}{n} \sum_{i=1}^{n} n_i h_{m,i} = \sum_{i=1}^{n} x_i h_{m,i}$$

(9.66)

Neglecting motion, gravity, electricity, magnetism and capillary effects, then $e_m = u_m$, and hence

$$u_m = \sum x_i u_{m,i}$$

(9.67)

The definition of enthalpy for an ideal gas is

$$h_m = u_m + \Re T \tag{9.68}$$

Thus, for the mixture

$$
\begin{aligned}
h_m &= \Sigma(x_i(u_m + \Re T)_i) \\
&= \Sigma x_i u_{m,i} + \Sigma x_i \Re T = \Sigma x_i u_{m,i} + \Re T \Sigma x_i \\
&= \Sigma x_i u_{m,i} + \Re T
\end{aligned}
\tag{9.69}
$$

9.4.4 SPECIFIC HEATS OF MIXTURES

Statement (c) of the Gibbs–Dalton law and the above expressions show that

$$u_m = \Sigma x_i u_{m,i}$$

and

$$h_m = \Sigma x_i h_{m,i}$$

By definition, the specific heats are

$$c_v = \frac{du}{dT} \quad \text{and} \quad c_p = \frac{dh}{dT}.$$

Thus

$$c_{v,m} = \left(\frac{du_m}{dT}\right) = \Sigma x_i \left(\frac{du_m}{dT}\right)_i = \Sigma x_i(c_{v,m})_i \tag{9.70}$$

and, similarly

$$c_{p,m} = \Sigma x_i(c_{p,m})_i \tag{9.71}$$

9.5 Entropy of mixtures

Consider a mixture of two ideal gases, *a* and *b*. The entropy of the mixture is equal to the sum of the entropies which each component of the mixture would have *if it alone occupied the whole volume of the mixture at the same temperature*. This concept can present some difficulty because it means that when gases are separate but all at the same temperature and pressure they have a lower entropy than when they are mixed together in a volume equal to the sums of their previous volumes; the situation can be envisaged from Fig 9.2. This can be analysed by considering the mixing process in the following way:

(i) Gases *a* and *b*, at the same pressure, are contained in a control volume but prevented from mixing by an impermeable membrane.
(ii) The membrane is then removed, and the components mix due to diffusion, i.e. due to the concentration gradient there will be a net migration of molecules from their original volumes. This means that the probability of finding a particular molecule at *any particular* point in the volume is decreased, and the system is in a less ordered state. This change in probability can be related to an increase in entropy by statistical thermodynamics. Hence, qualitatively, it can be expected that mixing will give rise to an increase in entropy.

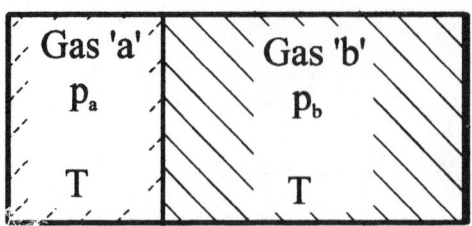

Fig. 9.2 Two gases at the same pressure, p, contained in an insulated container and separated by a membrane; $p_a = p_b = p$

Considering the mixing process from a macroscopic viewpoint, when the membrane is broken the pressure is unaffected *but the partial pressures of the individual components are decreased.*

The expression for the entropy of a gas is (eqn (9.20))

$$s_m = s_m(T) - \Re \ln \frac{p}{p_0} + s_{0,m}$$

where $s_m(T)$ is a function of T alone. Hence, considering the pressure term, which is actually the *partial* pressure of a component, a decrease in the partial pressure will cause an increase in the entropy of the gas. Equation (9.20) can be simplified by writing the pressure ratio as $p_r = p/p_0$, giving

$$s_m = s_m(T) - \Re \ln p_r + s_{0,m}$$

Now consider the change from an analytical viewpoint. Consider the two gases a and b. The total amount of substance in the gases is

$$n_T = n_a + n_b \tag{9.72}$$

The entropies of the gases before mixing are

$$S_{m_a} = n_a s_{m_a} = n_a[s_{m_a}(T) - \Re \ln p_{r_a} + s_{0,m_a}] \tag{9.73}$$

and

$$S_{m_b} = n_b s_{m_b} = n_b[s_{m_b}(T) - \Re \ln p_{r_b} + s_{0,m_b}] \tag{9.74}$$

giving the sum of the entropies before mixing (i.e. separated) as

$$(S_{m_T})_{sep} = (n_T s_{m_T})_{sep} = n_T[x_a s_{m_a}(T) + x_b s_{m_b}(T) + x_a s_{0,m_a} + x_b s_{0,m_b} - \Re(x_a \ln p_{r_a} + x_b \ln p_{r_b})] \tag{9.75}$$

Since the gases are separated and both are at the pressure, p_r, then

$$(S_{m_T})_{sep} = n_T[x_a s_{m_a}(T) + x_b s_{m_b}(T) + x_a s_{0,m_a} + x_b s_{0,m_b} - \Re \ln p] \tag{9.76}$$

Hence, the molar entropy, when the gases are separate, is

$$(s_{m_T})_{sep} = x_a s_{m_a}(T) + x_b s_{m_b}(T) + x_a s_{0,m_a} + x_b s_{0,m_b} - \Re \ln p$$
$$= \Sigma x_i s_{m_i} + \Sigma x_i s_{0,m_i} - \Re \ln p \tag{9.77}$$

The entropy of the gases after mixing is

$$(S_{m_T})_{mix} = n_T s_{m_T} = \Sigma \, n s_m = n_a s_{m_a} + n_b s_{n_b} \tag{9.78}$$

where s_{m_a} and s_{m_b} are the molar entropies if each component occupied the whole volume. Thus, substituting for s_m from eqn (9.20a) gives

$$(n_T s_{m_T})_{mix} = n_a s_{m_a}(T) + n_b s_{m_b}(T) - \Re(n_a \ln p_{r_a} + n_b \ln p_{r_b}) + n_a s_{0,m_a} + n_b s_{0,m_b} \tag{9.79}$$

$$(s_{m_T})_{mix} = x_a s_{m_a}(T) + x_b s_{m_b}(T) - \Re(x_a \ln p_{r_a} + x_b \ln p_{r_b}) + x_a s_{0,m_a} + x_b s_{0,m_b} \tag{9.80}$$

where $p_{r_i} = p_i/p_0$, and p_i = partial pressure of constituent i.

Equation (9.80) may be written

$$(s_{m_T})_{mix} = \Sigma \, x_i s_{m_i}(T) + \Sigma \, x_i s_{0,m_i} - \Re \, \Sigma \, x_i \ln p_{r_i} \tag{9.81}$$

Now, consider the final term in eqn (9.81), based on two constituents to simplify the mathematics

$$
\begin{aligned}
x_a \ln p_{r_a} + x_b \ln p_{r_b} &= x_a \ln(x_a p_r) + x_b \ln(x_b p_r) \\
&= x_a \ln x_a + x_a \ln p_r + x_b \ln x_b + x_b \ln p_r \\
&= \Sigma \, x_i \ln x_i + \ln p_r \, \Sigma \, x_i
\end{aligned}
$$

$$= \Sigma \, x_i \ln x_i + \ln p_r = \ln p_r - \Sigma \, x_i \ln \frac{1}{x_i} \tag{9.82}$$

Hence

$$(s_{m_T})_{mix} = \Sigma \, x_i s_{m_i}(T) + \Sigma \, x_i s_{0,m_i} + \Re\left(\Sigma \, x_i \ln \frac{1}{x_i} - \ln p_r\right) \tag{9.83}$$

Considering the terms on the right-hand side of eqn (9.83):

1st and 2nd terms   summation of entropies before mixing, $(S_{m_T})_{sep}$

3rd term   change of entropy due to mixing; this is due to a *change of partial pressures* when mixed

4th term   pressure term

The change of entropy due to mixing is given by

$$\Delta S = (S_{m_T})_{mix} - (S_{m_T})_{sep} \tag{9.84}$$

which is the difference between eqns (9.83) and (9.77). This gives

$$\Delta S = \Re \, \Sigma \, x_i \ln \frac{1}{x_i} \tag{9.85}$$

The numerical value of ΔS must be positive because x_i is always less than unity.

Equation (9.85) shows that there is an entropy increase due to mixing, and this is caused by the reduction in the order of the molecules. Before mixing it is possible to go into one side of the container and guarantee taking a particular molecule, because only molecules of a are in the left-hand container, and only molecules of b are in the right-hand container. After mixing it is not possible to know whether the molecule obtained will be of a or b: the order of the system has been reduced.

What happens if both a and b in Fig 9.2 are the same gas?

Superficially it might be imagined that there will still be an increase of entropy on mixing. However, since *a* and *b* are now the same there is *no change in partial pressure due to mixing*. This means that the molar fraction is unaltered and

$$\Delta S = 0 \tag{9.86}$$

9.6 Concluding remarks

A detailed study of ideal gases and ideal gas mixtures has been undertaken in preparation for later chapters. Equations have been developed for all properties, and enthalpy coefficients have been introduced for nine commonly encountered gases. Tables of gas properties have been presented for most gases occurring in combustion calculations.

Equations for gas mixtures have been developed and the effects of mixing on entropy and Gibbs energy have been shown.

PROBLEMS

Assume that air consists of 79% N_2 and 21% O_2 *by volume.*

1 A closed vessel of $0.1 \, m^3$ capacity contains a mixture of methane (CH_4) and air, the air being 20% in excess of that required for chemically correct combustion. The pressure and temperature in the vessel before combustion are, respectively, 3 bar and 100°C. Determine:
   (a) the individual partial pressures and the weights of methane, nitrogen and oxygen present before combustion;
   (b) the individual partial pressures of the burnt products, on the assumption that these are cooled to 100°C without change of volume and that all the vapour produced by combustion is condensed.

   [(a) 0.2415, 2.179, 0.579 bar; 0.01868, 0.2951, 0.08969 kg
   (b) 2.179, 0.2415, 0.0966 bar]

2 An engine runs on a rich mixture of methyl and ethyl alcohol and air. At a pressure of 1 bar and 10°C the fuel is completely vaporised. Calculate the air–fuel ratio by volume under these conditions, and the percentage of ethyl alcohol in the fuel by weight. If the total pressure of the exhaust gas is 1 bar, calculate the dew point of the water vapour in the exhaust and the percentage by volume of carbon monoxide in the dry exhaust gas, assuming all the hydrogen in the fuel forms water vapour.

   Vapour pressures at 10°C: methyl alcohol (CH_3OH), 0.0745 bar; and ethyl alcohol (C_2H_5OH), 0.310 bar.

   [63°C; 4.15%]

3 An engine working on the constant volume (Otto) cycle has a compression ratio of 6.5 to 1, and the compression follows the law $pV^{1.3} = C$, the initial pressure and temperature being 1 bar and 40°C. The specific heats at constant pressure and constant volume throughout compression and combustion are $0.96 + 0.00002T$ kJ/kg K and $0.67 + 0.00002 \, T$ kJ/kg K respectively, where T is in K.

Find
(a) the change in entropy during compression;
(b) the heat rejected per unit mass during compression;
(c) the heat rejected per unit mass during combustion if the maximum pressure is 43 bar and the energy liberated by the combustion is 2150 kJ/kg of air.
$$[(a) -0.1621 \text{ kJ/kg K}; (b) -67.8 \text{ kJ/kg}; (c) -1090.6 \text{ kJ/kg}]$$

4 A compression-ignition engine runs on a fuel of the following analysis by weight: carbon 84%, hydrogen 16%. If the pressure at the end of combustion is 55 bar, the volume ratio of expansion is 15:1, and the pressure and temperature at the end of expansion are 1.75 bar and 600°C respectively, calculate
(a) the variable specific heat at constant volume for the products of combustion; and
(b) the change in entropy during the expansion stroke per kmol.

The expansion follows the law $pV^n = C$ and there is 60% excess air. The specific heats at constant volume in kJ/kmol K between 600°C and 2400°C are:

O_2	$32 + 0.0025T$;	N_2	$29 + 0.0025T$
H_2O	$34 + 0.08T$;	CO_2	$50 + 0.0067T$

The water vapour (H_2O) may be considered to act as a perfect gas.
$$[(a) 22.1 + 0.0184T; (b) -11.42 \text{ kJ/kmol K}]$$

5 The exhaust gases of a compression–ignition engine are to be used to drive an exhaust gas turbocharger. Estimate the mean pressure ratio of expansion and the isentropic enthalpy drop per kmol of gas in the turbine if the mean exhaust temperature is 600°C and the isentropic temperature drop is 100°C. The composition of the exhaust gas by volume is CO_2, 8%; H_2O, 9.1%; O_2, 7.5%; N_2, 75.4%. The specific heats at constant volume in kJ/kmol K are:

O_2	$32 + 0.0025T$;	N_2	$29 + 0.0025T$
H_2O	$34 + 0.08T$	CO_2	$50 + 0.0067T$

The water vapour (H_2O) may be considered to act as a perfect gas.
$$[1.946; -4547.1]$$

6 (a) An amount of substance equal to 2 kmol of an ideal gas at temperature T and pressure p is contained in a compartment. In an adjacent compartment is an amount of substance equal to 1 kmol of an ideal gas at temperature $2T$ and pressure p. The gases mix adiabatically but do not react chemically when the partition separating the compartments is withdrawn. Show that, as a result of the mixing process, the entropy increases by

$$\Re\left(\ln\frac{27}{4} + \frac{\kappa}{\kappa-1}\ln\frac{32}{27}\right)$$

provided that the gases are different and that κ, the ratio of specific heats, is the same for both gases and remains a constant in the temperature range T to $2T$.
(b) What would be the entropy change if the gases being mixed were of the same species?

$$\left[\frac{\kappa\Re}{\kappa-1}\ln\frac{32}{27}\right]$$

7   The exhaust gas from a two-stroke cycle compression-ignition engine is exhausted at an elevated pressure into a large chamber. The gas from the chamber is subsequently expanded in a turbine. If the mean temperature in the chamber is 811 K and the pressure ratio of expansion in the turbine is 4 : 1, calculate the isentropic enthalpy drop in the turbine per unit mass of gas.

[256.68]

8   The following data refer to an analysis of a dual combustion cycle with a gas having specific heats varying linearly with temperature.

The pressure and temperature of the gas at the end of compression are 31 bar and 227°C respectively; the maximum pressure achieved during the cycle is 62 bar, while the maximum temperature achieved is 1700°C. The temperature at the end of the expansion stroke is 1240°C. The increases in entropy during constant volume and constant pressure combustion are 0.882 and 1.450 kJ/kg K respectively. Assuming that the fluid behaves as an ideal gas of molecular weight $m_w = 30.5$, calculate the equations for the specific heats, and also the expansion ratio if that process is isentropic.

[$0.6747 + 0.000829T$; $0.94735 + 0.000829T$; 7.814]

9   Distinguish between an ideal and a perfect gas and show that in both cases the specific entropy, s, is given by

$$s = s_0 + \int_{T_0}^{T} \frac{dh}{T} - \Re \ln\left(\frac{p}{p_0}\right)$$

Two streams of perfect gases, A and B, mix adiabatically at constant pressure and without chemical change to form a third stream. The molar specific heat at constant pressure, $c_{p,m}$, of the gas in stream A is equal to that in stream B. Stream A flows at M kmol/s and is at a temperature T_1, while stream B flows at 1 kmol/s and is at a temperature nT_1. Assuming that the gases A and B are different, show that the rate of entropy increase is

$$c_p \ln\left[\frac{1}{n}\left(\frac{M+n}{M+1}\right)^{M+1}\right] - \Re \ln\left[\frac{1}{M}\left(\frac{M}{M+1}\right)^{M+1}\right]$$

How is the above expression modified if the gases A and B are the same?
For the case $n = 1$, evaluate the rate of entropy increase

(a)   when different gases mix, and
(b)   when the gas in each stream is the same.

$$\left[c_p \ln\left[\frac{1}{n}\left(\frac{M+n}{M+1}\right)^{M+1}\right]; \text{(a)} \Re \ln\left[\frac{1}{M}\left(\frac{M}{M+1}\right)^{M+1}\right]; \text{(b)} 0\right]$$

10   A jet engine burns a weak mixture of octane (C_8H_{18}) and air, with an equivalence ratio, $\phi = 2$. The products of combustion, in which dissociation may be neglected, enter the nozzle with negligible velocity at a temperature of 1000 K. The gases, which may be considered to be ideal, leave the nozzle at the atmospheric pressure of

1.013 bar with an exit velocity of 500 m/s. The nozzle may be considered to be adiabatic and frictionless.

Determine:

(a) the specific heat at constant pressure, c_p, of the products as a function of temperature;
(b) the molecular weight, m_w, of the products;
(c) the temperature of the products at the nozzle exit;
(d) the pressure of the products at the nozzle inlet.

Specific heat at constant pressure, $c_{p,m}$, in J/kmol K, with T in K:

CO_2 $c_{p,m} = 21 \times 10^3 + 34.0T$
H_2O $c_{p,m} = 33 \times 10^3 + 8.3T$
O_2 $c_{p,m} = 28 \times 10^3 + 6.4T$
N_2 $c_{p,m} = 29 \times 10^3 + 3.4T$

$$[0.9986 \times 10^3 + 0.2105T; \ 28.71; \ 895.7; \ 1.591 \text{ bar}]$$

11 The products of combustion of a jet engine have a molecular weight, m_w, of 30 and a molar specific heat at constant pressure given by $c_{p,m} = 3.3 \times 10^4 + 15T$ J/kmol K where T is the gas temperature in Kelvin. When the jet pipe stagnation temperature is 1200 K the gases leave the nozzle at a relative speed of 600 m/s. Evaluate the static temperature of the gas at the nozzle exit and estimate the total to static pressure ratio across the nozzle. Assume that the products of combustion behave as an ideal gas and that the flow is isentropic.

In a frictional nozzle producing the mean outlet speed from the same inlet *gas* temperature, what would be the effect on
(a) the mean outlet static temperature, and
(b) the total to static pressure ratio?

$$[1092; \ 1.732; \ (a) \text{ unaffected}; \ (b) \text{ increased}]$$

10

Thermodynamics of Combustion

Combustion is an oxidation process and is usually exothermic (i.e. releases the chemical (or bond) energy contained in a fuel as thermal energy). The most common combustion processes encountered in engineering are those which convert a hydrocarbon fuel (which might range from pure hydrogen to almost pure carbon, e.g. coal) into carbon dioxide and water. This combustion is usually performed using air because it is freely available, although other oxidants can be used in special circumstances, e.g. rocket motors. The theory that will be developed here will be applicable to any mixture of fuel and oxidant and any ratio of components in the products; however, it will be described in terms of commonly available hydrocarbon fuels of the type used in combustion engines or boilers.

The simplest description of combustion is of a process that converts the *reactants* available at the beginning of combustion into *products* at the end of the process. This model presupposes that combustion is a process that can take place in only one direction and it ignores the true statistical nature of chemical change. Combustion is the combination of various atoms and molecules, and takes place when they are close enough to interact, but there is also the possibility of atoms which have previously joined together to make a product molecule separating to form reactants again. The whole mixture is really taking part in a molecular 'barn dance' and the tempo of the dance is controlled by the temperature of the mixture. The process of molecular breakdown is referred to as *dissociation*; this will be introduced in Chapter 12. In reality a true combustion process is even more complex than this because the actual rate at which the reactions can occur is finite (even if extremely fast). This rate is the basic cause of some of the pollutants produced by engines, particularly NO_x. In fact, in most combustion processes the situation is even more complex because there is an additional factor affecting combustion, which is related to the rate at which the fuel and air can mix. These ideas will be introduced in Chapter 15. Hence, the approach to combustion in this chapter is a simplified one but, in reality, it gives a reasonable assessment of what would be expected under good combustion conditions. It cannot really be used to assess emissions levels but it can be extended to this simply by the introduction of additional equations: the basic approach is still valid.

The manner in which combustion takes place is governed by the detailed design of the combustion system. The various different types of combustion process are listed in Table 10.1, and some examples are given of where the processes might be found. There is an interdependence between thermodynamics and fluid mechanics in combustion, and this interaction is the subject of current research. This book will concentrate on the thermodynamics of combustion, both in equilibrium and non-equilibrium states. The first part of the treatment of combustion will be based on equilibrium thermodynamics, and will cover combustion processes both with and without *dissociation*. It will be found that equilibrium thermodynamics enables a large number of calculations to be performed but, even with dissociation included, it does not allow the calculation of pollutants, the production of which are controlled both by mixing rates (fluid mechanics) and reaction rates (thermodynamics).

Table 10.1 Factors affecting combustion processes

Conditions of combustion	Classification	Examples
Time dependence	steady unsteady	gas turbine combustion chamber, boilers petrol engine, diesel engine
Spatial dependence	zero-dimensional	only used for modelling purposes, well-stirred reactors
	one-dimensional	approximated in pipe flows, flat flame burners
	two-dimensional three-dimensional	axisymmetric flames, e.g. Bunsen burner general combustion
Mixing of initial reactants	premixed non-premixed	petrol, or spark ignition, engine diesel engine, gas turbine combustion chamber
Flow	laminar turbulent	special cases for measuring flame speed most real engine cases, boilers
Phase of reactants	single	spark-ignited gas engines, petrol engines with fuel completely evaporated; gas-fired boilers
	multiphase	diesel engines, gas turbines, coal- and oil-fired boilers
Reaction sites	homogeneous heterogeneous	spark-ignition engines diesel engines, gas turbines, coal fired boilers
Reaction rate	equilibrium chemistry (infinite rate)	approached by some processes in which the combustion period is long compared with the reaction rate
	finite rate	all real processes: causes many pollutant emissions
Convection conditions	natural	Bunsen flame, gas cooker, central heating boiler
	forced	gas turbine combustion chamber, large boilers
Compressibility	incompressible compressible	free flames engine flames
Speed of combustion	deflagration detonation	most normal combustion processes 'knock' in spark ignition engines, explosions

10.1   Simple chemistry

Combustion is a chemical reaction and hence a knowledge of basic chemistry is required before it can be analysed. An extremely simple reaction can be written as

$$2CO + O_2 \Leftrightarrow 2CO_2 \tag{10.1}$$

This basically means that two molecules of carbon monoxide (CO) will combine with one molecule of oxygen (O_2) to create two molecules of carbon dioxide (CO_2). Both CO and O_2 are *diatomic gases*, whereas CO_2 is a *triatomic* gas. Equation (10.1) also indicates that two molecules of CO_2 will always break down into two molecules of CO and one molecule of O_2; this is signified by the symbol $\Leftrightarrow$ which indicates that the processes can go in both directions. It is conventional to refer to the mixture to the left of the arrows as the *reactants* and that to the right as the *products*; this is because exothermic combustion (i.e. in which energy is released by the process) would require CO and O_2 to combine to give CO_2. Not all reactions are exothermic and the formation of NO during dissociation occurring in an internal combustion (i.c.) engine is actually endothermic.

It should be noted from the combustion eqn (10.1) that three molecules of reactants combine to produce two molecules of products, hence there is not necessarily a balance in the number of molecules on either side of a chemical reaction. However, there is a balance in the number of atoms of each constituent in the equation and so mass is conserved.

10.1.1   FUELS

Hydrocarbon fuels are rarely single-component in nature due to the methods of formation of the raw material and its extraction from the ground. A typical barrel of crude oil contains a range of hydrocarbons, and these are separated at a refinery; the oil might produce the constituents defined in Fig 10.1. None of the products of the refinery is a single chemical compound, but each is a mixture of compounds, the constituents of which depend on the source of the fuel.

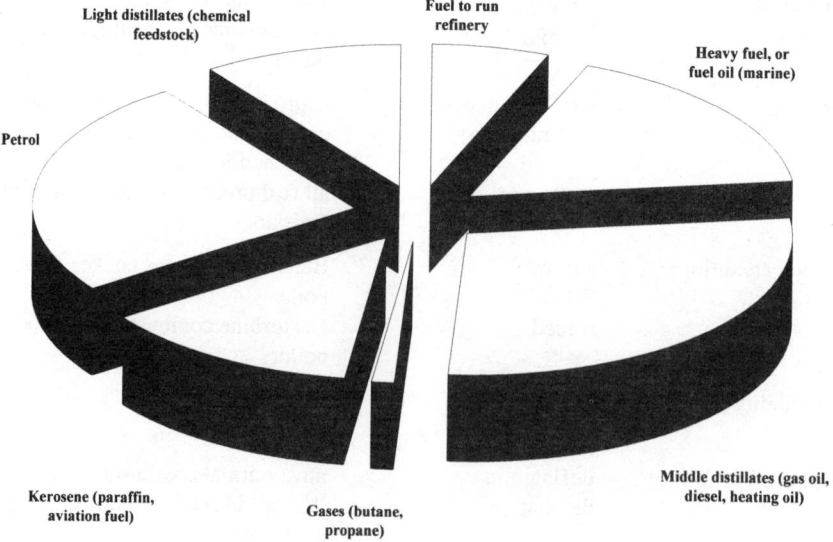

Fig. 10.1 Typical constituents of a barrel of crude oil

One fuel which approaches single-component composition is 'natural gas', which consists largely of methane (CH_4). Methane is the simplest member of a family of hydrocarbons referred to as paraffins or, more recently, *alkanes* which have a general formula C_nH_{2n+2}. The lower alkanes are methane (CH_4), ethane (C_2H_6), propane (C_3H_8) and butane (C_4H_{10}) etc. Two other alkanes that occur in discussion of liquid fuels are heptane (C_7H_{16}) and octane (C_8H_{18}). The alkanes are referred to as *saturated hydrocarbons* because it is not physically possible to add more hydrogen atoms to them. However, it is possible to find hydrocarbons with less than $2n+2$ hydrogen atoms and these are referred to as *unsaturated hydrocarbons*. A simple unsaturated hydrocarbon is acetylene (C_2H_2), which belongs to a chemical family called *alkenes*. Some fuels contain other constituents in addition to carbon and hydrogen. For example, the alcohols contain oxygen in the form of an OH radical. The chemical symbol for methanol is CH_3OH, and that for ethanol is C_2H_5OH; these are the alcohol equivalents of methane and ethane.

Often fuels are described by a mass analysis which defines the proportion by mass of the carbon and hydrogen, e.g. a typical hydrocarbon fuel might be defined as 87% C and 13% H without specifying the actual components of the liquid. Solid fuels, such as various coals, have a much higher carbon/hydrogen ratio but contain other constituents including oxygen and ash.

The molecular weights (or relative molecular masses) of fuels can be evaluated by adding together the molecular (or atomic) weights of their constituents. Three examples are given below:

Methane (CH_4)	$(m_w)_{CH_4} = 12 + 4 \times 1 = 16$
Octane (C_8H_{18})	$(m_w)_{C_8H_{18}} = 8 \times 12 + 18 \times 1 = 114$
Methanol (CH_3OH)	$(m_w)_{CH_3OH} = 12 + 3 \times 1 + 16 + 1 = 32$

10.2   Combustion of simple hydrocarbon fuels

The combustion of a hydrocarbon fuel takes place according to the constraints of chemistry. The combustion of methane with oxygen is defined by

$$CH_4 \; + \; 2O_2 \; \Rightarrow \; CO_2 \; + \; 2H_2O \tag{10.2}$$

1 kmol	2 kmol	1 kmol	2 kmol
12 + 4	2 × 32	12 × 32	2 × (2 + 16)
16 kg	64 kg	44 kg	36 kg

In this particular case there is both a molar balance and a mass balance: the latter is essential but the former is not. Usually combustion takes place between a fuel and air (a mixture of oxygen and nitrogen). It is normal to assume, at this level, that the nitrogen is an inert gas and takes no part in the process. Combustion of methane with air is given by

$$CH_4 \; + \; 2\left(O_2 + \frac{79}{21} N_2\right) \; \Rightarrow \; CO_2 \; + \; 2H_2O \; + \; 2 \times \frac{79}{21} N_2 \tag{10.3}$$

1 kmol	9.52 kmol	1 kmol	2 kmol	7.52 kmol
12 + 4	2 × (32 + 105.2)	12 + 32	2 × (2 + 16)	7.52 × 28
16 kg	274.4 kg	44 kg	36 kg	210.67 kg

10.2.1   STOICHIOMETRY

There is a clearly defined, and fixed, ratio of the masses of air and fuel that will result in complete combustion of the fuel. This mixture is known as a *stoichiometric* one and the ratio is referred to as the *stoichiometric air–fuel ratio*. The stoichiometric air–fuel ratio, ε_{stoic}, for methane can be evaluated from the chemical equation (eqn 10.3). This gives

$$\varepsilon_{stoic} = \frac{\text{mass of air}}{\text{mass of fuel}} = \frac{2 \times (32 + 105.33)}{16} = 17.17$$

This means that to obtain complete combustion of 1 kg CH_4 it is necessary to provide 17.17 kg of air. If the quantity of air is less than 17.17 kg then complete combustion will not occur and the mixture is known as *rich*. If the quantity of air is greater than that required by the stoichiometric ratio then the mixture is *weak*.

10.2.2   COMBUSTION WITH WEAK MIXTURES

A weak mixture occurs when the quantity of air available for combustion is greater than the chemically correct quantity for complete oxidation of the fuel; this means that there is excess air available. In this simple analysis, neglecting reaction rates and dissociation etc, this excess air passes through the process without taking part in it. However, even though it does not react chemically, it has an effect on the combustion process simply because it lowers the temperatures achieved due to its capacity to absorb energy. The equation for combustion of a weak mixture is

$$CH_4 + \frac{2}{\phi}(O_2 + 3.76N_2) \Rightarrow CO_2 + 2H_2O + 2\left(\frac{1-\phi}{\phi}\right)O_2 + \frac{7.52}{\phi}N_2 \qquad (10.4)$$

where ϕ is called the *equivalence ratio*, and

$$\phi = \frac{\text{actual fuel–air ratio}}{\text{stoichiometric fuel–air ratio}} = \frac{\text{stoichiometric air–fuel ratio}}{\text{actual air–fuel ratio}} \qquad (10.5)$$

For a weak mixture ϕ is less than unity. Consider a weak mixture with $\phi = 0.8$; then

$$CH_4 + 2.5(O_2 + 3.76N_2) \Rightarrow CO_2 + 2H_2O + 0.5O_2 + 9.4N_2 \qquad (10.6)$$

10.2.3   COMBUSTION WITH RICH MIXTURES

A rich mixture occurs when the quantity of air available is less than the stoichiometric quantity; this means that there is not sufficient air to burn the fuel. In this simplified approach it is assumed that the hydrogen combines preferentially with the oxygen and the carbon does not have sufficient oxygen to be completely burned to carbon dioxide; this results in partial oxidation of part of the carbon to carbon monoxide. It will be shown in Chapter 12 that the equilibrium equations, which control the way in which the hydro-carbon fuel oxidizes, govern the proportions of oxygen taken by the carbon and hydrogen of the fuel and that the approximation of preferential combination of oxygen and hydrogen is a reasonable one. In this case, to define a rich mixture, ϕ is greater than unity. Then

$$CH_4 + \frac{2}{\phi}(O_2 + 3.76N_2) \Rightarrow \left(\frac{4-3\phi}{\phi}\right)CO_2 + 2H_2O + \frac{4(\phi-1)}{\phi}CO + \frac{7.52}{\phi}N_2$$

(10.7)

If the equivalence ratio is 1.2, then eqn (10.7) is

$$CH_4 + 1.667(O_2 + 3.76N_2) \Rightarrow 0.333CO_2 + 2H_2O + 0.667CO + 6.267N_2$$ (10.8)

It is quite obvious that operating the combustion on rich mixtures results in the production of carbon monoxide (CO), an extremely toxic gas. For this reason it is now not acceptable to operate combustion systems with rich mixtures. Note that eqn (10.7) cannot be used with values of $\phi > 4/3$, otherwise the amount of CO_2 becomes negative. At this stage it must be assumed that the carbon is converted to carbon monoxide and carbon. The resulting equation is

$$CH_4 + \frac{2}{\phi}(O_2 + 3.76N_2) \Rightarrow 2H_2O + \left(\frac{4}{\phi} - 2\right)CO + \left(3 - \frac{4}{\phi}\right)C + \frac{7.52}{\phi}N_2$$ (10.9)

Equation (10.9) is a very hypothetical one because during combustion extensive dissociation occurs and this liberates oxygen by breaking down the water molecules; this oxygen is then available to create carbon monoxide and carbon dioxide rather than carbon molecules.

In reality it is also possible to produce pollutants even when the mixture is weaker than stoichiometric, simply due to poor mixing of fuel and air, quenching of flames on cold cylinder or boiler walls, trapping of the mixture in crevices (fluid mechanics effects) and also due to thermodynamic limitations in the process.

10.3 Heats of formation and heats of reaction

Combustion of fuels takes place in either a closed system or an open system. The relevant property of the fuel to be considered is the *internal energy or enthalpy, respectively, of formation or reaction*. In a naive manner it is often considered that combustion is a process of energy addition to the system. This is not true because the energy released during a combustion process is already contained in the reactants, in the form of the chemical energy of the fuel (see Chapter 11). Hence it is possible to talk of *adiabatic combustion* as a process in which no energy (heat) is transferred to, or from, the system – the temperature of the system increases because of a rearrangement of the chemical bonds in the fuel and oxidant.

Mechanical engineers are usually concerned with the combustion of hydrocarbon fuels, such as petrol, diesel oil or methane. These fuels are commonly used because of their ready availability (at present) and high energy density in terms of both mass and volume. The combustion normally takes place in the presence of air. In some other applications, e.g. space craft, rockets, etc, fuels which are not hydrocarbons are burned in the presence of other oxidants; these will not be considered here.

Hydrocarbon fuels are stable compounds of carbon and hydrogen which have been formed through the decomposition of animal and vegetable matter over many millennia. It is also possible to synthesise hydrocarbons by a number of processes in which hydrogen is added to a carbon-rich fuel. The South African Sasol plant uses the Lurgi and Fischer–Tropsch processes to convert coal from a solid fuel to a liquid one. The chemistry of fuels is considered in Chapter 11.

10.4   Application of the energy equation to the combustion process – a macroscopic approach

Equations (10.3) to (10.6) show that combustion can take place at various air–fuel ratios, and it is necessary to be able to account for the effect of mixture strength on the combustion process, especially the temperature rise that will be achieved. It is also necessary to be able to account for the different fuel composition: not all fuels will release the same quantity of energy per unit mass and hence it is required to characterise fuels by some capacity to release chemical energy in a thermal form. Both of these effects obey the First Law of Thermodynamics, i.e. the energy equation.

10.4.1   INTERNAL ENERGIES AND ENTHALPIES OF IDEAL GASES

It was shown previously that the internal energies and enthalpies of ideal gases are functions of temperature alone (c_p and c_v might still be functions of temperature). This means that the internal energy and enthalpy can be represented on U–T and H–T diagrams. It is then possible to draw a U–T or H–T line for both reactants and products (Fig 10.2). The reactants will be basically diatomic gases (neglecting the effect of the fuel) whereas the products will be a mixture of diatomic and triatomic gases – see eqn (10.3).

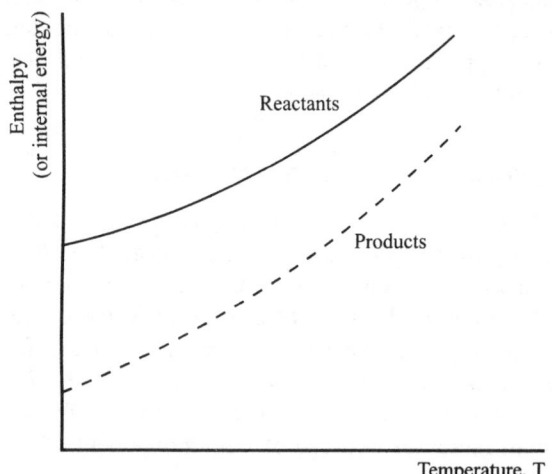

Fig. 10.2   Enthalpy (or internal energy) of reactants and products

The next question which arises is: what is the spacing between the reactants and products lines? This spacing represents the energy that can be released by the fuel.

10.4.2   HEATS OF REACTION AND FORMATION

The energy contained in the fuel can also be assessed by burning it under a specified condition; this energy is referred to as the *heat of reaction of the fuel*. The heat of reaction for a fuel is dependent on the process by which it is measured. If it is measured by a constant volume process in a combustion bomb then the *internal energy of reaction* is obtained. If it is measured in a constant pressure device then the *enthalpy of reaction* is obtained. It is more normal to measure the enthalpy of reaction because it is much easier to achieve a constant pressure process. The enthalpy of reaction of a fuel can be evaluated by burning the fuel in a stream of air, and measuring the quantity of energy that must be removed to achieve equal reactant and product temperatures (see Fig 10.3).

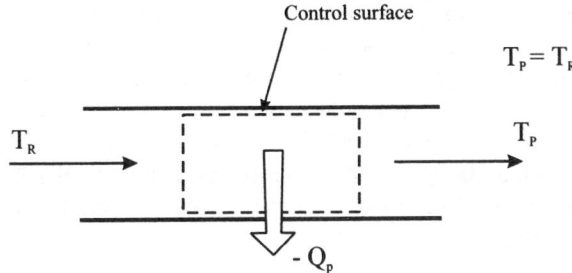

Fig. 10.3 Constant pressure measurement of enthalpy reaction

Applying the steady flow energy equation

$$\dot{Q} - \dot{W}_s = \dot{m}_e\left(h_e + \frac{V_e^2}{2} + gz_e\right) - \dot{m}_i\left(h_i + \frac{V_i^2}{2} + gz_i\right) \tag{10.10}$$

and neglecting the kinetic and potential energy terms, then

$$(Q_P)_T = (H_P)_T - (H_R)_T = n_P(h_P)_T - n_R(h_R)_T \tag{10.11}$$

where n denotes the amount of substance in either the products or reactants; this is identical to the term n which was used for the amount of substance in Chapter 9. The suffix T defines the temperature at which the enthalpy of reaction was measured. $(Q_p)_T$ is a function of this temperature and normally it is evaluated at a standard temperature of 25°C (298 K). When $(Q_p)_T$ is evaluated at a standardised temperature it will be denoted by the symbol $(Q_p)_s$. Most values of Q_p that are used in combustion calculations are the $(Q_p)_s$ ones. (In a similar way, $(Q_v)_s$ will be used for internal energy of reaction at the standard temperature.) The sign of Q_p is negative for fuels because heat must be transferred from the 'calorimeter' to achieve equal temperatures for the reactants and products (it is positive for some reactions, meaning that heat has to be transferred *to* the calorimeter to maintain constant temperatures). The value of the constant volume heat of reaction, the internal energy of reaction, $(Q_v)_s$, can be calculated from $(Q_p)_s$ as shown below, or measured using a constant volume combustion 'bomb'; again $(Q_v)_s$ has a negative value. $(Q_p)_s$ and $(Q_v)_s$ are shown in Figs 10.4(a) and (b) respectively. The term calorific value of the fuel was used in the past to define the 'heating' value of the fuel: this is actually the *negative* value of the heat of reaction, and is usually a positive number. It is usually associated with analyses in which 'heat' is added to a system during the combustion process, e.g. the air standard cycles.

Applying the first law for a closed system to constant volume combustion gives

$$(Q_v)_T = (U_P)_T - (U_R)_T = n_P(u_P)_T - n_R(u_R)_T \tag{10.12}$$

If both the products and reactants are ideal gases then $h = \int c_{p,m}\, dT$, and $u = \int c_{v,m}\, dT$, which can be evaluated from the polynomial expressions derived in Chapter 9. Thus

$$(Q_p)_s - (Q_v)_s = n_P(h_P)_T - n_R(h_R)_T - \{n_P(u_P)_T - n_R(u_R)_T\}$$
$$= n_P\{(h_P)_T - (u_P)_T\} - n_R\{(h_R)_T - (u_R)_T\}$$
$$= \Re T(n_P - n_R) \tag{10.13}$$

This result is quite logical because the definitions of $(Q_p)_s$ and $(Q_v)_s$ require that T_P and T_R are equal. Hence the constant pressure and constant volume processes are identical if the amounts of substance in the products and the reactants are equal. If the amounts of substance change during the reaction then the processes cease to be identical and, in the case of a combustion *bomb*, a piston would have to move to maintain the conditions. The movement of the piston is work equal to $\Re T(n_P - n_R)$.

It is also possible to relate the quantity of energy that is chemically bound up in the fuel to a value at absolute zero of temperature. These values are denoted as $-\Delta H_0$ and $-\Delta U_0$ and will be returned to later.

10.4.3   HEAT OF FORMATION – HESS' LAW

The heat of formation of a compound is the quantity of energy absorbed (or released) during its formation from its elements (the end pressures and temperatures being maintained equal).

For example, if CO_2 is formed from carbon and oxygen by the reaction

$$C + O_2 \rightarrow CO_2 \tag{10.14}$$

then in a constant pressure steady flow process with equal temperature end states the reaction results in heat transfer of $(Q_p)_T$ given by

$$(Q_p)_T = H_P - H_R \tag{10.15}$$

$$= -394 \text{ MJ/kmol} \tag{10.16}$$

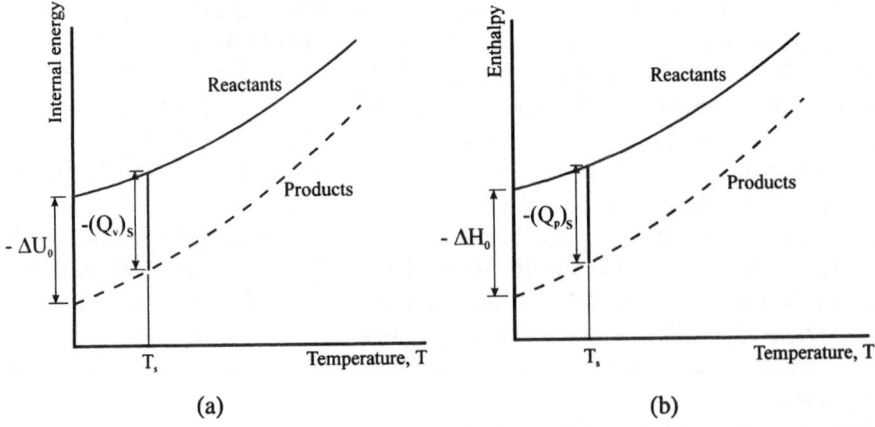

(a)   (b)

Fig. 10.4 Internal energy and enthalpy of reaction depicted on (a) internal energy–temperature and (b) enthalpy–temperature diagrams

If a slightly different reaction is performed giving the same end product, e.g.

$$CO + \frac{1}{2} O_2 \rightarrow CO_2 \tag{10.17}$$

then it is not possible to use the same simple approach because the reactants are a mixture of elements and compounds. However, Hess' law can be used to resolve this problem. This states that:

(a)   if a reaction at constant pressure or constant volume is carried out in stages, the algebraic sum of the amounts of heat evolved in the separate stages is equal to the total evolution of heat when the reaction occurs directly;

or

(b)   the heat liberated by a reaction is independent of the path of the reaction between the initial and final states.

Both of these are simply statements of the law of energy conservation and the definition of properties. However, this allows complex reactions to be built up from elemental ones. For example, the reaction

$$CO + \frac{1}{2} O_2 \rightarrow CO_2 \tag{10.18}$$

can be sub-divided into two different reactions:

$$C + \frac{1}{2} O_2 \rightarrow CO \tag{10.19a}$$

$$CO + \frac{1}{2} O_2 \rightarrow CO_2 \tag{10.19b}$$

The heat of formation of CO may be evaluated by reaction (10.19a) and then used in reaction (10.19b) to give the heat of reaction of that process. From experiment the heat of formation of carbon monoxide (CO) is -112 MJ/kmol and hence, for reaction (10.19b), the energy released is

$$(Q_p)_T = H_P - H_R = -394 - (-110.5) = -283 \text{ MJ/kmol} \tag{10.20}$$

Hence the heat of reaction for $CO + \frac{1}{2}O_2 \rightarrow CO_2$ is -283 MJ/kmol.

The heats of formation of any compounds can be evaluated by building up simple reactions, having first designated the *heats of formation of elements as zero*. This enables the enthalpy of formation of various compounds to be built up from component reactions. Enthalpies of formation are shown in Table 10.2.

These enthalpies of formation can be used to evaluate heats of reaction of more complex molecules, e.g. for methane (CH_4) the equation is

$$CH_4 + 2O_2 = CO_2 + 2H_2O(g)$$

giving

$$(Q_p)_{25} = (\Delta H_f)_{CO_2(g)} + 2(\Delta H_f)_{H_2O(g)} - (\Delta H_f)_{CH_4(g)}$$
$$= -802\ 279 \text{ kJ/kmol}$$

This gives an enthalpy of reaction per kg of CH_4 of $-50\ 142$ kJ/kg.

Table 10.2 Enthalpies of formation of some common elements and compounds

Species	Reaction	State	ΔH_f kJ/kmol
Oxygen, O_2	-	gas 25°C, 1 atm	0, element
Hydrogen, H_2	-	gas 25°C, 1 atm	0, element
Carbon, C	-	gas 25°C, 1 atm	0, element
Carbon dioxide, CO_2	$C + O_2 \rightarrow CO_2$	*gas 25°C, 1 atm*[*]	$-393\ 522$
Carbon monoxide, CO	$C + O \rightarrow CO$	*gas 25°C, 1 atm*[*]	$-110\ 529$
Water vapour, H_2O	$H_2 + 1/2O_2 \rightarrow H_2O(g)$	*gas 25°C, 1 atm*[*]	$-241\ 827$
Water (liquid), H_2O	$H_2 + 1/2O_2 \rightarrow H_2O(l)$	liquid 25°C, 1 atm	$-285\ 800$
Nitric oxide (NO)	$1/2N_2 + 1/2O_2 \rightarrow NO$	*gas 25°C, 1 atm*[*]	$+89\ 915$
Methane, CH_4	$C + 2H_2 \rightarrow CH_4(g)$	*gas 25°C, 1 atm*[*]	$-74\ 897$
Ethane, C_2H_6	$2C + 3H_2 \rightarrow C_2H_6(g)$	gas 25°C, 1 atm	$-84\ 725$
Propane, C_3H_8	$3C + 4H_2 \rightarrow C_3H_8(g)$	gas 25°C, 1 atm	$-103\ 916$
Butane, C_4H_{10}	$4C + 5H_2 \rightarrow C_4H_{10}(g)$	gas 25°C, 1 atm	$-124\ 817$
Iso-octane, C_8H_{18}	$8C + 9H_2 \rightarrow C_8H_{18}(g)$	gas 25°C, 1 atm	$-224\ 100$
Iso-octane, C_8H_{18}	$8C + 9H_2 \rightarrow C_8H_{18}(l)$	liquid 25°C, 1 atm	$-259\ 280$
Methyl alcohol, CH_3OH	$C + 2H_2 + 1/2O_2 \rightarrow CH_3OH(g)$	gas 25°C, 1 atm	$-201\ 200$
Methyl alcohol, CH_3OH	$C + 2H_2 + 1/2O_2 \rightarrow CH_3OH(l)$	liquid 25°C, 1 atm	$-238\ 600$
Ethyl alcohol, C_2H_5OH	$2C + 3H_2 + 1/2O_2 \rightarrow C_2H_5OH(g)$	gas 25°C, 1 atm	$-234\ 600$
Ethyl alcohol, C_2H_5OH	$2C + 3H_2 + 1/2O_2 \rightarrow C_2H_5OH(l)$	liquid 25°C, 1 atm	$-277\ 000$

[*] Note: the values of ΔH_f given in the tables of gas properties are based on 0 K.

Higher and lower calorific values

The enthalpy of reaction of CH_4 derived above ($-50\ 142$ kJ/kg) is the negative of the lower calorific value (LCV) because it is based on gaseous water ($H_2O(g)$) in the products. If the water in the products (exhaust) was condensed to a liquid then extra energy could be released by the process. In this case the enthalpy of reaction would be

$$(Q_p)_{25} = (\Delta H_f)_{CO_2} + 2(\Delta H_f)_{H_2O(l)} - (\Delta H_f)_{CH_4}$$
$$= -890\ 225 \text{ kJ/kmol}$$

This gives an enthalpy of reaction per kg of CH_4 of $-55\ 639$ kJ/kg. This is the negative of the higher calorific value (HCV).

Normally the lower calorific value, or lower internal energy or enthalpy of reaction, is used in engine calculations because the water in the exhaust system is usually in the vapour phase.

10.5   Combustion processes

10.5.1   ADIABATIC COMBUSTION

The heats of reaction of fuels have been described in terms of isothermal processes, i.e. the temperature of the products is made equal to the temperature of the reactants. However, it is common experience that combustion is definitely not isothermal; in fact, its major characteristic is to raise the temperature of a system. How can this be depicted on the enthalpy–temperature diagram?

Consider a constant pressure combustion process, as might occur in a gas turbine, in which there is no heat or work transfer (Fig 10.5(a)).

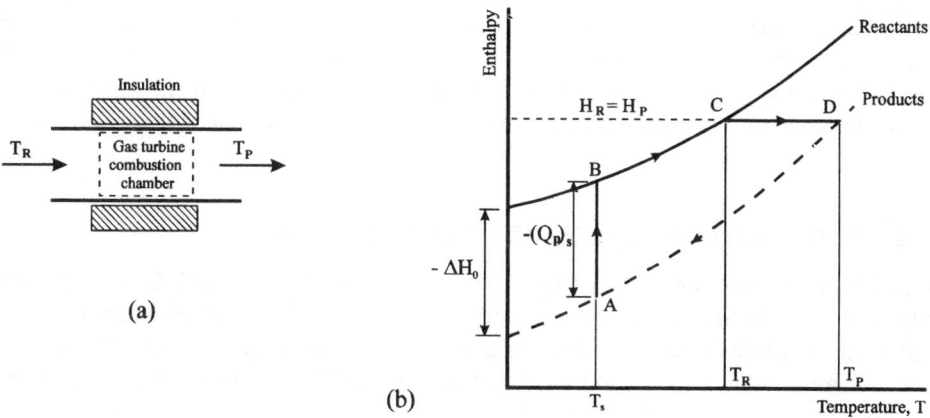

Fig. 10.5 Adiabatic combustion depicted on enthalpy–temperature diagram: (a) schematic of gas turbine combustion chamber; (b) enthalpy–temperature diagram of adiabatic combustion process

Applying the steady flow energy equation, eqn (10.11), to the combustion chamber shown in Fig 10.5(a) gives

$$H_P = H_R \tag{10.21}$$

Hence *adiabatic combustion is a process which occurs at constant enthalpy* (or internal energy, in the case of combustion at constant volume), and the criterion for equilibrium is that the enthalpies at the beginning and end of the process are equal. This process can be shown on an $H–T$ diagram (Fig 10.5(b)) and it is possible to develop from this an equation suitable for evaluating the product temperature. The sequence of reactions defining combustion are denoted by the 'cycle ABCDA'. Going clockwise around the cycle, from A, gives

$$-(Q_p)_s + [H_R(T_R) - H_R(T_s)] - [H_P(T_P) - H_P(T_s)] = 0 \tag{10.22}$$

The term $[H_R(T_R) - H_R(T_s)]$ is the difference between the enthalpy of the reactants at the temperature at the start of combustion (T_R) and the standardised temperature (T_s). The term $[H_P(T_P) - H_P(T_s)]$ is a similar one for the products. These terms can be written as

$$[H_R(T_R) - H_R(T_s)] = \sum_{i=1}^{q} n_{R_i} h_{R_i}(T_R) - \sum_{i=1}^{q} n_{R_i} h_{R_i}(T_s) = \sum_{i=1}^{q} n_{R_i} \{ h_{R_i}(T_R) - h_{R_i}(T_s) \}$$

$$\tag{10.23}$$

where i is the particular component in the reactants and q is the number of components over which the summation is made. From eqn (10.22)

$$H_P(T_P) = -(Q_P)_s + [H_R(T_R) - H_R(T_s)] + H_P(T_s) \tag{10.24}$$

If the process had been a constant volume one, as in an idealised (i.e. adiabatic) internal combustion engine, then

$$U_P(T_P) = -(Q_v)_s + [U_R(T_R) - U_R(T_s)] + U_P(T_s) \tag{10.25}$$

This possibly seems to be a complex method for evaluating a combustion process compared with the simpler heat release approach. However the advantage of this method is that it results in a true *energy balance*: the enthalpy of the products is always equal to the enthalpy of the reactants. Also, because it is written in terms of enthalpy, the variation of gas properties due to temperature and composition changes can be taken into account correctly. An approach such as this can be applied to more complex reactions which involve dissociation and rate kinetics: other simpler methods cannot give such accurate results.

10.5.2   COMBUSTION WITH HEAT AND WORK TRANSFER

The combustion process shown in Fig 10.5(b) is an adiabatic one, and the enthalpy of the products is equal before and after combustion. However, if there is heat transfer or work transfer taking place in the combustion process, as might occur during the combustion process in an internal combustion engine, then Fig 10.5(b) is modified to that shown in Fig 10.6.

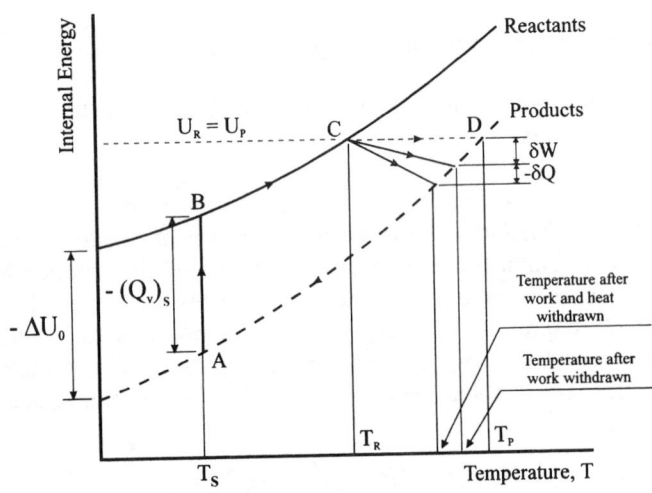

Fig. 10.6 Combustion with heat transfer and work output

If the engine provides a work output of pdV, the temperature of the products is reduced as shown in Fig 10.6, and if there is heat transfer from the cylinder, $-\Delta Q$, this reduces the temperature still further. These effects can be incorporated into the energy equation in the following way. The First Law for the process is

$$\Delta Q - \Delta W = dU = U_P(T_P) - U_R(T_R) \qquad (10.26)$$

Hence

$$U_P(T_P) = U_R(T_R) + \Delta Q - \Delta W \qquad (10.27)$$

This means that heat transfer away from the cylinder, as will normally happen in the case of an engine during the combustion phase, tends to reduce the internal energy of the products compared with the reactants. Also, if work is taken out of the cylinder due to the piston moving away from top dead centre, the internal energy is further reduced.

10.5.3 INCOMPLETE COMBUSTION

The value of enthalpy (or internal energy) of reaction for a fuel applies to complete combustion of that fuel to carbon dioxide and water. However, if the mixture is a rich one then there will be insufficient oxidant to convert all the fuel to CO_2 and H_2O, and it will be assumed here that some of the carbon is converted only to CO. The chemical equations of this case are shown in Section 10.2.3. The effect of incomplete combustion on the energy released will be considered here. Incomplete combustion can be depicted as an additional line on the enthalpy–temperature diagram – the position of this line indicates how far the combustion process has progressed (see Fig 10.7). In this case the amount of energy released can be evaluated from the energies of formation, in a similar manner to that in eqn (10.8).

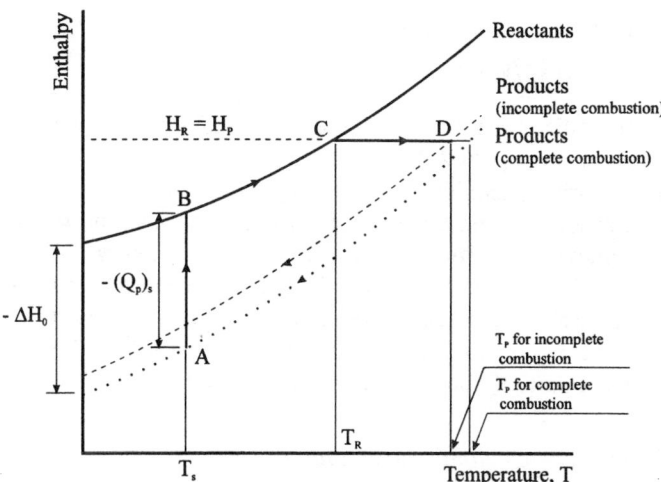

Fig. 10.7 Enthalpy–temperature diagram for incomplete combustion including rich mixtures and dissociation

10.6 Examples

These examples quote solutions to the resolution of a calculator. This is to aid readers in identifying errors in their calculations, and does not mean the data are accurate to this level.

Example 1: incomplete combustion

Evaluate the energy released from constant pressure combustion of a rich mixture of methane and air: use $\phi = 1.2$.

Solution

The chemical equation is

$$CH_4 + 1.667(O_2 + 3.76N_2) \longrightarrow 0.333CO_2 + 2H_2O + 0.667CO + 6.267N_2 \qquad (10.28)$$

The nitrogen (N_2) takes no part in this reaction and does not need to be considered. The energy liberated in the reaction (Q_p) is the difference between the enthalpies of formation of the products and reactants.

Enthalpy of formation of reactants,

$$(\Delta H_f)_R = (\Delta H_f)_{CH_4} = -74\ 897\ \text{kJ/kmol}$$

Enthalpy of formation of products,

$$(\Delta H_f)_P = 0.333(\Delta H_f)_{CO_2} + 2(\Delta H_f)_{H_2O} + 0.667(\Delta H_f)_{CO}$$
$$= 0.333(-393\ 522) + 2(-241\ 827) + 0.667(-110\ 529)$$
$$= -688\ 420.0\ \text{kJ/kmol}.$$

Hence the energy released by this combustion process is

$$Q_p = (\Delta H_f)_P - (\Delta H_f)_R$$
$$= -688\ 420.0 - (-74\ 897)$$
$$= -613\ 522.0\ \text{kJ/kmol}$$

It can be seen that this process has released only 76.5% of the energy that was available from complete combustion ($-802\ 279$ kJ/kmol). This is because the carbon has not been fully oxidised to carbon dioxide, and energy equal to 0.667×-283 MJ/kmol CH_4 (equivalent to the product of the quantity of carbon monoxide and the heat of reaction for eqn (10.18)) is unavailable in the form of thermal energy. This is equivalent to 188 761 kJ/kmol CH_4, and is the remaining 23.5% of the energy which would be available from complete combustion. While the incomplete combustion in eqn (10.34) has occurred because of the lack of oxygen in the rich mixture, a similar effect occurs with dissociation, when the products are partially broken down into reactants.

Example 2: adiabatic combustion in an engine

A stoichiometric mixture of methane (CH_4) and air is burned in an engine which has an effective compression ratio of 8 : 1. Calculate the conditions at the end of combustion at constant volume if the initial temperature and pressure are 27°C and 1 bar respectively. The lower calorific value of methane is 50 144 kJ/kg at 25°C. Assume the ratio of specific heats (κ) for the compression stroke is 1.4. See Fig 10.8.

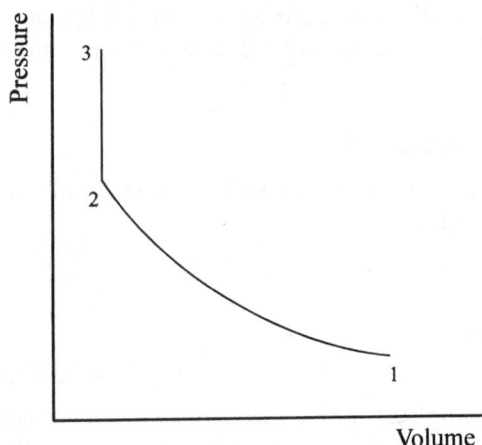

Fig. 10.8 $p - V$ diagram for constant volume combustion

Solution

$$p_1 = 1 \text{ bar}, T_1 = 300 \text{ K}$$

$$p_2 = p_1 \left(\frac{V_1}{V_2}\right)^{1.4} = 1 \times 8^{1.4} = 18.379 \text{ bar}$$

$$T_2 = T_1 \left(\frac{V_1}{V_2}\right)^{\kappa-1} = 300 \times 8^{0.4} = 689.2 \text{ K}$$

The chemical equation is

$$\underbrace{CH_4 + 2(O_2 + 3.76N_2)}_{n_R = 10.52} \longrightarrow \underbrace{CO_2 + 2H_2O + 7.52N_2}_{n_P = 10.52} \tag{10.29}$$

The combustion equation is

$$U_P(T_P) = -(Q_v)_s + [U_R(T_R) - U_R(T_s)] + U_P(T_s) \tag{10.30}$$

The reactants' energy $[U_R(T_R) - U_R(T_s)]$ is

Constituent	CH_4	O_2	N_2
$u_{689.2}$	21 883.9	15 105.9	14 587.7
u_{298}	5719.6	6032.0	6025.8
Difference	16 164.3	9073.9	8561.9
n	1	2	7.52
nu	16 164.3	18 146.4	64 385.5

Thus $U_R(T_R) - U_R(T_s) = 9.8696 \times 10^4$ kJ
The products energy at standard temperature, $U_P(T_s)$, is

Constituent	CO_2	H_2O	N_2
u_{298}	7041.6	7223.3	6025.8
n	1	2	7.52
nu_{298}	7041.6	14 446.6	45 314.0

Hence $U_P(T_s) = 66 \ 802.2$ kJ
Substituting these values into eqn (10.30), and converting the calorific value to molar terms gives

$$U_P(T_P) = 50 \ 144 \times 16 + 9.8696 \times 10^4 + 66 \ 802.2$$
$$= 9.678 \times 10^5 \text{ kJ}$$

The value of specific internal energy for the mixture of products is

$$u_P(T_P) = \frac{U_P(T_P)}{n_P} = \frac{9.678 \times 10^5}{10.52} = 92000 \text{ kJ/kmol}$$

An approximate value of the temperature can be obtained by assuming the exhaust products to be 70% N_2 and 30% CO_2. Using this approximation gives

T	$U_p(T_p)$
2500	76 848.7
3000	94 623.3

Hence T_p is around 3000 K. Assume $T_p = 3000$ K and evaluate $U_p(T_p)$:

Constituent	CO_2	H_2O	N_2
u_{3000}	137 950.4	109 612.6	76 504.6
n	1	2	7.52
nu	137 950.4	219 225.2	575 314.6

$$\sum nu = U_p(T_p) = 932\ 490.2\ \text{kJ}$$

This value is less than that obtained from eqn (10.30), and hence the temperature was underestimated.

　　Try $T_p = 3100$ K.

Constituent	CO_2	H_2O	N_2
u_{3100}	143 420.0	114 194.9	79 370.9
n	1	2	7.52
nu	143 420.0	228 389.8	596 869.2

$$\sum nu = U_p(T_p) = 968\ 678.9\ \text{kJ}$$

Hence the value T_p is between 3000 K and 3100 K.

　　Linear interpolation gives

$$T_p = 3000 + \frac{967\ 800 - 932\ 490.2}{968\ 678.9 - 932\ 490.2} \times 1000 = 3097\ \text{K}$$

Evaluating $U_p(T_p)$ at $T_p = 3097$ K gives

Constituent	CO_2	H_2O	N_2
u_{3097}	143 255.1	114 057.5	79 285.1
n	1	2	7.52
nu	143 255.1	228 114.9	596 224

$$\sum nu = U_p(T_p) = 967\ 594\ \text{kJ}$$

This is within 0.0212% of the value of $U_p(T_p)$ evaluated from eqn (10.30).

　　The final pressure, p_3, can be evaluated using the perfect gas law because

$$p_3 V_3 = n_3 \Re T_3 = n_p \Re T_3$$

and

$$p_1 V_1 = n_1 \Re T_1 = n_R \Re T_1$$

giving

$$\begin{aligned}
p_3 &= \frac{n_P \Re T_3}{n_R \Re T_1} \frac{V_1}{V_3} p_1 \\
&= \frac{3097}{300} \times 8 \times 1 = 82.59 \text{ bar}
\end{aligned}$$

Example 3: non-adiabatic combustion in an engine

In the previous calculation, Example 2, the combustion was assumed to be adiabatic. If 10% of the energy liberated by the fuel was lost by heat transfer, calculate the final temperature and pressure after combustion.

Solution

The effect of heat transfer is to reduce the amount of energy available to raise the temperature of the products. This can be introduced in eqn (10.30) to give

$$U_P(T_P) = -(Q_v)_s - Q_{HT} + [U_R(T_R) - U_R(T_s)] + U_P(T_s) \tag{10.31}$$

In this case $Q_{HT} = -0.1(Q_v)_s$ and hence

$$U_P(T_P) = -0.9(Q_v)_s + [U_R(T_R) - U_R(T_s)] + U_P(T_s) \tag{10.32}$$

Substituting values gives

$$\begin{aligned}
U_P(T_P) &= 0.9 \times 50\ 144\ \times 16 + 9.8696 \times 10^4 + 66\ 802.2 \\
&= 887\ 571.8 \text{ kJ}
\end{aligned}$$

Using a similar technique to before, the value of $u_P(T_P)$ is

$$u_P(T_P) = \frac{887\ 571.8}{10.52} = 84369.9 \text{ kJ/kmol}$$

From the previous calculation the value of T_P lies between 2500 K and 3000 K, and an estimate is

$$T_P = 2500 + \frac{84\ 369.9 - 76\ 848.7}{94\ 623.3 - 76\ 848.7} \times 500 = 2717 \text{ K}$$

Try $T_P = 2700$ K to check for energy balance:

Constituent	CO_2	H_2O	N_2
u_{2700}	121 807.5	95 927.5	67 877.4
n	1	2	7.52
nu	121 807.5	191 854.9	510 438.1

$$U_P(T_P) = 823\ 993.7 \text{ kJ}$$

This is less than the value obtained from the energy equation and hence $T_P > 2700$ K. Try $T_P = 2800$ K:

Constituent	CO_2	H_2O	N_2
u_{2800}	127 154.9	100 470.8	70 754.1
n	1	2	7.52
nu	127 154.9	200 941.6	532 071.1

$$U_P(T_P) = 860\ 167.6\ \text{kJ.}$$

Hence, by linear extrapolation, try

$$T_P = 2800 + \frac{887\ 571.8 - 860\ 167.6}{860\ 167.6 - 823\ 993.7} \times 100$$

$$= 2875.8\ \text{K}$$

Constituent	CO_2	H_2O	N_2
u_{2876}	131 227.9	103 929.1	72 935
n	1	2	7.52
nu	131 227.9	207 858.1	548 471.5

$$U_P(T_P) = 887\ 557.5\ \text{kJ}$$

This value is within 0.0016% of the value from the energy equation.

Example 4: combustion with a weak mixture of propane

A mixture of propane and excess air is burned at constant volume in an engine with a compression ratio of 10:1. The strength of the mixture (ϕ) is 0.85 and the initial conditions at the start of the compression stroke are 1.2 bar and 127°C. Assume the ratio of specific heats for compression is 1.4 and take the internal energy of reaction $(Q_v)_s$ of propane as $-46\ 440$ kJ/kg. Evaluate the final temperature and pressure after combustion neglecting dissociation. The internal energy of propane is 12 911.7 kJ/kmol at the beginning of compression and 106 690.3 kJ/kmol at the end (before combustion).

Solution

The stoichiometric chemical equation is

$$C_3H_8 + 5(O_2 + 3.76N_2) \longrightarrow 3CO + 4H_2O + 18.8N_2 \tag{10.33}$$

The weak mixture equation is

$$\underbrace{C_3H_8 + \frac{5}{0.85}(O_2 + 3.76N_2)}_{n_R = 29.0} \longrightarrow \underbrace{3O_2 + 4H_2O + 0.882O_2 + 22.12N_2}_{n_P = 30.0} \tag{10.34}$$

The combustion equation (eqn (10.30)) is

$$U_P(T_P) = -(Q_v)_s + [U_R(T_R) - U_R(T_s)] + U_P(T_s)$$

Conditions at start of combustion:

$$p_2 = p_1 \left(\frac{V_1}{V_2}\right)^\kappa = 1.2 \times 10^{1.4} = 30.142 \text{ bar}$$

$$T_2 = T_1 \left(\frac{V_1}{V_2}\right)^{\kappa-1} = 400 \times 10^{0.4} = 1004.75 \text{ K}$$

The energy of reactants $[U_R(T_R) - U_R(T_s)]$ is

Constituent	C_3H_8	O_2	N_2
$u_{1004.7}$	106 690.3	23 165.0	22 013.6
u_{298}	12 911.7	6032.0	6025.8
Difference	93 778.7	17 133.0	15 987.9
n	1	5.882	22.12
nu	93 778.7	100 776.1	353 651.4

$$[U_R(T_R) - U_R(T_s)] = 548\ 206.2 \text{ kJ}$$

The energy of products at standard temperature, T_s, is

Constituent	CO_2	H_2O	O_2	N_2
u_{298}	7041.6	7223.3	6031.7	6025.8
n	3	4	0.882	22.12
nu	21 124.8	28 893.2	5320.0	133 290.7

$$U_P(T_s) = 188\ 628.7 \text{ kJ}$$

$$U_P(T_P) = 46\ 440 \times (12 \times 3 + 8) + 548\ 206.2 + 188\ 628.7$$

$$= 2\ 043\ 360.0 + 548\ 206.2 + 188\ 628.7$$

$$= 2\ 780\ 194.9 \text{ kJ}$$

Assume $T_P = 3100$ K:

Constituent	CO_2	H_2O	O_2	N_2
u_{3100}	143 420.0	114 194.9	84 974.72	79 370.95
n	3	4	0.882	22.12
nu	430 260.1	456 779.6	74 947.7	1 755 685.0

$$U_P(T_P) = 2\ 717\ 672.4 \text{ kJ}$$

Try $T_P = 3200$ K:

Constituent	CO_2	H_2	O_2	N_2
u_{3200}	148 954.8	118 773.6	88 164.1	82 226.2
n	3	4	0.882	22.12
nu	446 864.5	475 094.3	77 760.8	1 818 843

$$U_P(T_P) = 2\ 818\ 562.6 \text{ kJ}$$

Interpolating linearly gives

$$T_P = 3100 + \frac{2\ 780\ 194.9 - 2\ 717\ 672.4}{2\ 818\ 562.6 - 2\ 717\ 672.4} \times 100 = 3162 \text{ K}$$

Check solution using $T_P = 3162$ K:

Constituent	CO_2	H_2O	O_2	N_2
u_{3162}	146 843.0	117 034.7	86 949.1	81 142.8
n	3	4	0.882	22.12
nu	440 529	468 138.7	76 689.4	1 794 878

$$U_P(T_P) = 2\ 780\ 234.8 \text{ kJ}.$$

This value is within 0.0014% of the value calculated from the energy equation. Hence $T_P = 3162$ K.

Pressure at the end of combustion is

$$p_3 = \frac{n_P \Re T_P}{n_R \Re T_1} \frac{V_1}{V_3} p_1 = \frac{30}{29} \times \frac{3162}{400} \times 10 \times 12$$

$$= 98.13 \text{ bar}$$

Example 5: combustion at constant pressure of a rich mixture of benzene

A rich mixture of benzene (C_6H_6) and air with an equivalence ratio, ϕ, of 1.2 is burned at constant pressure in an engine with a compression ratio of 15:1. If the initial conditions are 1 bar and 25°C, calculate the pressure, temperature and volume after combustion. The ratio of specific heats during compression, κ, is 1.4 and the calorific value of benzene is 40 635 kJ/kg. See Fig 10.9.

$$p_1 = 1 \text{ bar}, T_1 = 298 \text{ K}$$

$$p_2 = p_1 \left(\frac{V_1}{V_2}\right)^{1.4} = 44.31 \text{ bar}$$

$$T_2 = T_1 \left(\frac{V_1}{V_2}\right)^{0.4} = 880 \text{ K}$$

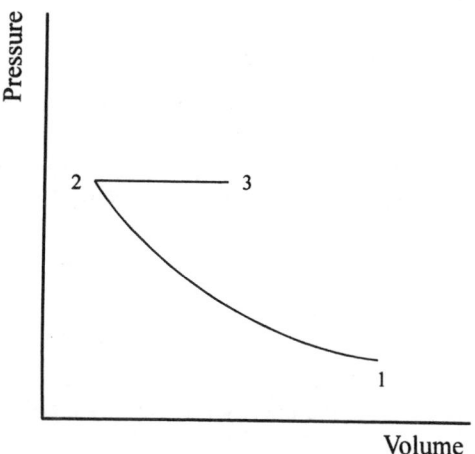

Fig. 10.9 $p-V$ diagram for constant pressure combustion

The stoichiometric chemical equation is

$$C_6H_6 + 7.5(O_2 + 3.76N_2) \rightarrow 6CO_2 + 3H_2O + 23.5N_2 \qquad (10.35)$$

The rich mixture equation is

$$C_6H_6 + \frac{7.5}{1.2}(O_2 + 3.76N_2) \rightarrow aCO_2 + bCO + 3H_2O + 23.5N_2 \qquad (10.36)$$

$$\left.\begin{array}{l} a + b = 6 \\ 2a + b + 3 = 125 \end{array}\right\} \quad a = 3.5; \quad b = 2.5$$

Hence

$$\underbrace{C_6H_6 + 6.25(O_2 + 3.76N_2)}_{n_R = 30.75} \rightarrow \underbrace{3.5CO_2 + 2.5CO + 3H_2O + 23.5N_2}_{n_P = 32.5} \qquad (10.37)$$

The energy equation for constant pressure combustion is

$$H_P(T_p) = -(Q_p)_s + [H_R(T_R) - H_R(T_s)] + H_P(T_s) \qquad (10.38)$$

The reactants energy $[H_R(T_R) - H_R(T_s)]$:

Constituent	C_6H_6	O_2	N_2
h_{880}	39 936.1	27 230.9	26 342.1
h_{298}	8201.8	8509.7	8503.4
Difference	31 734.4	18 721.2	17 838.7
n	1	6.25	23.5
nh	31 734.4	117 007.5	419 209.3

$$[H_R(T_R) - H_R(T_s)] = 567\ 951.2$$

The products enthalpy at standard temperature, T_s:

Constituent	CO_2	CO	H_2O	N_2
h_{298}	9519.3	8489.5	9701.0	8503.4
n	3.5	2.5	3	23.5
nh	33 317.5	21 223.8	29 102.9	199 830.6

$$H_P(T_s) = 283\ 474.8\ \text{kJ}$$

In this case the combustion is not complete and hence the full energy available in the fuel is not released. Applying Hess' law, the energy released by a combustion process is given by

$$(Q_p)_{25} = (\Delta H_f)_P - (\Delta H_f)_R$$

In this case

$$\begin{aligned}(Q_p)_{25} &= 3.5(\Delta h_f)_{CO_2} + 2.5(\Delta h_f)_{CO} + 3(\Delta h_f)_{H_2O} - (\Delta h_f)_{C_6H_6}\\
&= 3.5 \times -393\ 522 + 2.5 \times -110\ 529 + 3 \times -241\ 827 - (82\ 847.6)\\
&= -2\ 461\ 978.1\ \text{kJ}\end{aligned}$$

(10.39)

This value can also be calculated from the enthalpy of reaction by subtracting the energy which is not released because of the incomplete combustion of all the carbon to carbon dioxide,

$$\begin{aligned}(Q_p)_{25} &= (Q_p)_{25,\text{total}} - 2.5\{(\Delta h_f)_{CO_2} - (\Delta h_f)_{CO}\}\\
&= -40635 \times 78 - 2.5(-393\ 522 + 110\ 529)\\
&= -3\ 169\ 530 + 707\ 482.5\\
&= -2\ 462\ 047.5\ \text{kJ}\end{aligned}$$

These two values of enthalpy of combustion are the same within the accuracy of the data used (they differ by less than 0.003%). Hence, referring to Fig 10.7, the *loss of energy available due to incomplete combustion* is 707 482.5 kJ/kmol C_6H_6 because not all the carbon is converted to carbon dioxide. Thus

$$H_P(T_P) = 2\ 462\ 047.5 + 567\ 888.1 + 283\ 471.3 = 3\ 313\ 406.9\ \text{kJ}$$

Assume $T_P = 2800$ K:

Constituent	CO_2	CO	H_2O	N_2
h_{2800}	150 434.9	94 792.4	123 750.8	94 034.2
n	3.5	2.5	3	23.5
nh	526 522.2	236 981.0	371 252.5	2 209 803

$$H_P(T_P) = 3\ 344\ 558.7\ \text{kJ}$$

Hence $T_P < 2800$ K. Use $T_P = 2700$ K:

Constituent	CO_2	CO	H_2O	N_2
h_{2700}	144 256.1	91 070.2	118 376.1	90 326.0
n	3.5	2.5	3	23.5
nh	504 896.3	227 675.4	355 128.2	2 122 661

$$H_P(T_P) = 3\ 210\ 360.9 \text{ kJ}$$

Interpolate linearly:

$$T_P = 2700 + \frac{3\ 313\ 406.9 - 3\ 210\ 360.9}{3\ 344\ 558.7 - 3\ 210\ 360.9} \times 100 = 2776.8 \text{ K}$$

Constituent	CO_2	CO	H_2O	N_2
h_{2777}	149 011.4	93 936.1	122 512.5	93 181.1
n	3.5	2.5	3	23.5
nh	521 540	234 840.3	367 537.5	2 189 757

$$H_P(T_P) = 3\ 313\ 674.8 \text{ kJ}$$

Hence the equations balance within 0.010%.

The pressure at the end of combustion is the same as that at the end of compression, i.e. $p_3 = 44.31$ bar. Hence the volume at the end of combustion, V_3, is given by

$$V_3 = \frac{n_P \Re T_P}{n_R \Re T_1} \times \frac{p_1}{p_3} \times V_1 = \frac{32.5 \times 2777}{30.75 \times 298} \times \frac{1}{44.31} \times V_1$$
$$= 0.2223 V_1$$

10.7 Concluding remarks

A consistent method of analysing combustion has been introduced. This is suitable for use with all the phenomena encountered in combustion, including weak and rich mixtures, incomplete combustion, heat and work transfer, dissociation and rate kinetics. The method which is soundly based on the First Law of Thermodynamics ensures that energy is conserved.

A large number of examples of different combustion situations have been presented.

PROBLEMS

Assume that air consists of 79% N_2 and 21% O_2 *by volume*.

1 Calculate the lower heat of reaction at constant volume for benzene (C_6H_6) at 25°C. The heats of formation at 25°C are: benzene (C_6H_6), 80.3 MJ/kmol; water vapour (H_2O), −242 MJ/kmol; carbon dioxide (CO_2), −394.0 MJ/kmol.

A mixture of one part by volume of vaporised benzene to 50 parts by volume of air is ignited in a cylinder and adiabatic combustion ensues at constant volume. If the initial pressure and temperature of the mixture are 10 bar and 500 K respectively, calculate the maximum pressure and temperature after combustion neglecting dissociation. (Assume the internal energy of the fuel is 8201 kJ/kmol at 298 K and 12 998 kJ/kmol at 500 K.)

[−3170 MJ/kmol; 52.11 bar; 2580 K]

2 The heat of reaction of methane (CH_4) is determined in a constant pressure calorimeter by burning the gas as a very weak mixture. The gas flow rate is 70 litre/h, and the mean gas temperature (inlet to outlet) is 25°C. The temperature rise of the cooling water is 1.8°C, with a water flow rate of 5 kg/min. Calculate the higher and lower heats of reaction at constant volume and constant pressure in kJ/kmol if the gas pressure at inlet is 1 bar.

[−800 049.5 kJ/kmol; −887 952 kJ/kmol]

3 In an experiment to determine the calorific value of octane (C_8H_{18}) with a bomb calorimeter, the mass of octane was $5.421\ 95 \times 10^{-4}$ kg, the water equivalent of the calorimeter including water 2.677 kg and the corrected temperature rise in the water jacket 2.333 K. Calculate the lower heat of reaction of octane, in kJ/kmol at 15°C.

If the initial pressure and temperature were 25 bar and 15°C respectively, and there was 400% excess oxygen, estimate the maximum pressure and temperature reached immediately after ignition assuming no heat losses to the water jacket during this time. No air was present in the calorimeter.

[−5 099 000 kJ/kmol; 3135 K]

4 A vessel contains a mixture of ethylene (C_2H_4) and twice as much air as that required for complete combustion. If the initial pressure and temperature are 5 bar and 440 K, calculate the adiabatic temperature rise and maximum pressure when the mixture is ignited.

If the products of combustion are cooled until the water vapour is just about to condense, calculate the final temperature, pressure and heat loss per kmol of original mixture.

The enthalpy of combustion of ethylene at the absolute zero is −1 325 671 kJ/kmol, and the internal energy at 440 K is 16 529 kJ/kmol.

[1631 K; 23.53 bar; 65°C; 3.84 bar; 47 119 kJ/kmol]

5 A gas engine is operated on a stoichiometric mixture of methane (CH_4) and air. At the end of the compression stroke the pressure and temperature are 10 bar and 500 K respectively. If the combustion process is an adiabatic one at constant volume, calculate the maximum temperature and pressure reached.

[59.22 bar; 2960 K]

6 A gas injection system supplies a mixture of propane (C_3H_8) and air to a spark ignition engine, in the ratio of volumes of 1 : 30. The mixture is trapped at 1 bar and 300 K, the volumetric compression ratio is 12 : 1, and the index of compression $\kappa = 1.4$. Calculate the equivalence ratio, the maximum pressure and temperature achieved during the cycle, and also the composition (by volume) of the dry exhaust gas.

[0.79334, 119.1 bar, 2883 K; 0.8463, 0.1071, 0.0465]

7 A turbocharged, intercooled compression ignition engine is operated on octane (C_8H_{18}) and achieves constant pressure combustion. The volumetric compression ratio of the engine is $20:1$, and the pressure and temperature at the start of compression are 1.5 bar and 350 K respectively. If the air–fuel ratio is $24:1$, calculate maximum temperature and pressure achieved in the cycle, and the indicated mean effective pressure (imep, $\bar{p}_i$) of the cycle in bar. Assume that the index of compression $\kappa_C = 1.4$, while that of expansion, $\kappa_E = 1.35$

[2495 K; 99.4 bar; 15.16 bar]

8 One method of reducing the maximum temperature in an engine is to run with a rich mixture. A spark ignition engine with a compression ratio of $10:1$, operating on the Otto cycle, runs on a rich mixture of octane (C_8H_{18}) and air, with an equivalence ratio of 1.2. The trapped conditions are 1 bar and 300 K, and the index of compression is 1.4. Calculate how much lower the maximum temperature is under this condition than when the engine was operated stoichiometrically. What are the major disadvantages of operating in this mode?

[208°C]

9 A gas engine with a volumetric compression ratio of $10:1$ is run on a weak mixture of methane (CH_4) and air, with an equivalence ratio $\phi = 0.9$. If the initial temperature and pressure at the commencement of compression are 60°C and 1 bar respectively, calculate the maximum temperature and pressure reached during combustion at constant volume if compression is isentropic and 10% of the heat released during the combustion period is lost by heat transfer.

Assume the ratio of specific heats, κ, during the compression stroke is 1.4, and the heat of reaction at constant volume for methane at 25°C is -8.023×10^5 kJ/kmol CH_4.

[2817 K; 84.59 bar]

10 A jet engine burns a weak mixture ($\phi = 0.32$) of octane (C_8H_{18}) and air. The air enters the combustion chamber from the compressor at 10 bar and 500 K; assess if the temperature of the exhaust gas entering the turbine is below the limit of 1300 K. Assume that the combustion process is adiabatic and that dissociation can be neglected. The enthalpy of reaction of octane at 25°C is $-44\,880$ kJ/kg, and the enthalpy of the fuel in the reactants may be assumed to be negligible.

[Maximum temperature, $T_P = 1298$ K; value is very close to limit]

11 A gas engine is run on a chemically correct mixture of methane (CH_4) and air. The compression ratio of the engine is $10:1$, and the trapped pressure and temperature at inlet valve closure are 60°C and 1 bar respectively. Calculate the maximum temperature and pressure achieved during the cycle if:

(a) combustion occurs at constant volume;
(b) 10% of the energy added by the fuel is lost through heat transfer;
(c) the compression process is isentropic.

It can be assumed that the ratio of specific heats $\kappa = 1.4$, and that the internal energy of methane in the reactants is negligible.

[2956 K; 88.77 bar]

11

Chemistry of Combustion

The thermodynamics of combustion were considered in Chapter 10, and it was stated that adiabatic combustion could be achieved. The concept of adiabatic combustion runs counter to the experience of many engineers, who tend to relate combustion to heat addition or heat release processes. This approach is encouraged in mechanical engineers by the application of the air standard cycle to engines to enable them to be treated as heat engines. In reality combustion is not a process of energy transfer but one of energy transformation. The energy released by combustion in a spark ignition (petrol) engine is all contained in the mixture prior to combustion, and it is released by the spark. It will be shown that the energy which causes the temperature rise in a combustion process is obtained by breaking the bonds which hold the fuel atoms together.

11.1 Bond energies and heats of formation

Heats (enthalpies and internal energies) of formation can be evaluated empirically by 'burning' the fuel. They can also be evaluated by consideration of the chemical structure of the compound. Each compound consists of a number of elements held together by certain types of bond. The bond energy is the amount of energy required to separate a molecule into atoms; the energy of a particular type of bond is *similar* irrespective of the actual structure of the molecule.

This concept was introduced in Chapter 10, and heats of formation were used to evaluate heats of reaction (Hess' law). The process of breaking the chemical bonds during the combustion process can be depicted by a diagram such as Fig 11.1. It is assumed that element molecules can be atomised (in a constant pressure process) by the addition of energy equal to ΔH_a. If these atoms are then brought together they would combine, releasing dissociation energy of $\Sigma \Delta H(X-Y)_R$, to form the reactants. The sum of the dissociation and atomisation energies (taking account of the signs) results in the enthalpy of formation of the reactants. In a similar way the enthalpy of formation of the products can be evaluated. Using Hess' law, the enthalpy of reaction of the fuel can be evaluated as the difference between the enthalpies of formation of the products and the reactants These energies are essentially the bond energies of the various molecules, and some of these energies are listed in Table 11.1.

Figure 11.2 shows how the energy required to separate two atoms varies with distance: the bond energy is defined as the minimum potential energy relative to that at infinity. The point of minimum energy indicates that the molecule is in equilibrium.

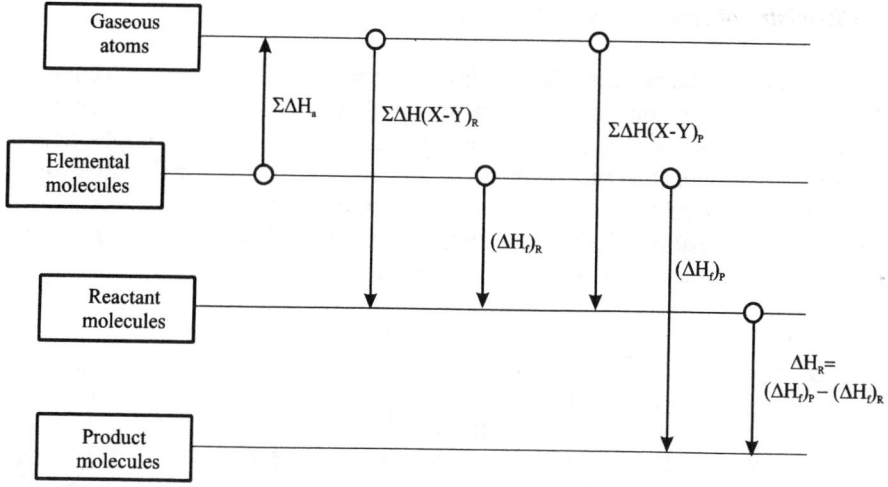

Fig. 11.1 Relationship between enthalpies of formation and reaction

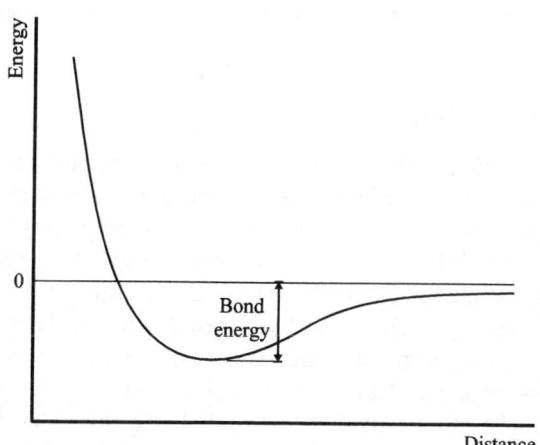

Fig. 11.2 Variation of bond energy with distance between atoms

Table 11.1 Some atomisation, dissociation and resonance energies (based on 25°C)*

Bond atomisation (ΔH_a)	Energy (MJ/kmol)	Bond dissociation ($\Delta H(X-Y)$)	Energy (MJ/kmol)	Resonance	Energy (MJ/kmol)
H—H	435.4	C—H	414.5	Benzene: C_6H_6	150.4
C (graphite)	717.2	N—H	359.5	Naphthalene: $C_{10}H_8$	255.4
$(O{=}O)_{O_2}$	498.2	O—H	428.7	Carbon dioxide: CO_2	137.9
N≡N	946.2	H—OH	497.5	-COOH group	117.0
		C—O	351.7		
		C=O	698.1		
		C—C	347.5		
		C=C	615.5		
		C≡C	812.2		

*note: these values have been taken from different sources and may not be exactly compatible.

In addition to the energy associated with particular bonds there are also energies caused by the possibility of resonance or strain in a particular molecule. For example, the ring structure of benzene, which is normally considered to be three simple double bonds between carbon atoms, together with three single ones, plus six carbon-hydrogen bonds, can reform to give bonding across the cyclic structure (see Fig 11.3).

This results in a substantially higher energy than would be estimated from the simple bond structure, and some typical values are shown in Table 11.1.

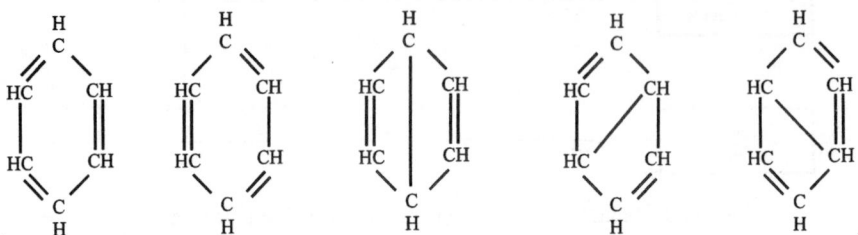

Fig. 11.3 Bonding arrangements of benzene, showing bond resonance through delocalised electrons

11.2   Energy of formation

It will be considered in this section that the processes take place at constant pressure and the property enthalpy (h or H) will be used; if the processes were constant volume ones then internal energies (u or U) would be the appropriate properties.

A hydrocarbon fuel consists of carbon and hydrogen (and possibly another element) atoms held together by chemical bonds. These atoms can be considered to have been brought together by heating carbon and hydrogen (in molecular form) under conditions that encourage the resulting atoms to bond. There are two processes involved in this: first, the carbon has to be changed from solid graphite into gaseous carbon atoms and the hydrogen also has to be atomised; then, second, the atoms must be cooled to form the hydrocarbon fuel. These two processes have two energies associated with them: the first is atomisation energy (ΔH_a) and the second is dissociation energy ($\Sigma \Delta H(X-Y)_R$) required to dissociate the chemical bond in the compound C_xH_y. Figure 11.1 shows these terms, and also indicates that the energy of formation of the fuel (ΔH_f) is the net energy required for the process, i.e.

$$\Delta H_f = \Sigma \Delta H_a - \Sigma \Delta H(X-Y)_R \qquad (11.1)$$

In reality the enthalpy of formation is more complex than given above because energy can be stored in molecules in a number of ways, including resonance energies (in the benzene ring structure) and changes of phase (latent heats). A more general representation of the *enthalpy of formation* is

$$\Delta H_f = \Sigma \Delta H_a - \Sigma \Delta H(X-Y) - \Sigma \Delta H_{res} - \Sigma \Delta H_{latent} \qquad (11.2)$$

Example

Evaluate the enthalpy of formation of CO_2 and H_2O from the atomisation and dissociation energies listed in Table 11.1.

Solution

Carbon dioxide (CO_2)

Carbon dioxide is formed from carbon and oxygen in the reaction.

$$C_{graphite} + O_2(g) \rightarrow CO_2(g) \tag{11.3}$$

where the (g) indicates that the element or compound is in the gaseous (vapour) phase, and (l) will be used to indicate that the element or compound is in the liquid phase.

The reaction in eqn (11.3) is achieved by atomisation of the individual carbon (graphite) molecules and the oxygen molecules, with subsequent recombination to form carbon dioxide. Effectively the reactant molecules, which are in a metastable state, are activated above a certain energy to produce atoms which will then combine to form the stable CO_2 molecule (see Fig 11.4).

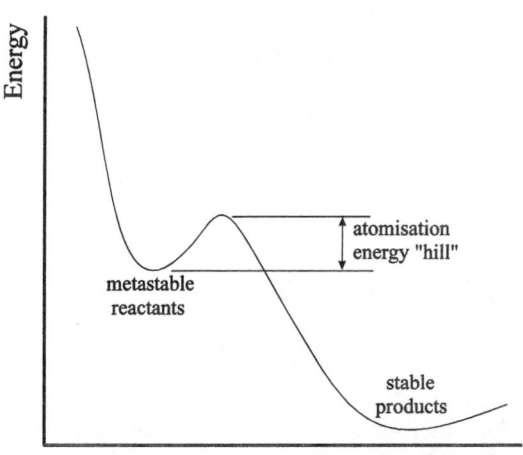

Fig. 11.4 Gibbs energy variation during a reaction

Hence, from Fig 11.1,

$$(\Delta H_f)_{CO_2} = \Sigma \Delta H_a - \Sigma \Delta H(X - Y) - \Sigma \Delta H_{res} - \Sigma \Delta H_{latent} \tag{11.4}$$

In this case there is a resonance energy but the latent energy (i.e. latent heat) is zero. Then

$$\begin{aligned}(\Delta H_f)_{CO_2} &= \Sigma \Delta H_a[C_{graphite}] + \Delta H_a[O = O] - 2(C = O) - \Delta H_{res}[CO_2] \\ &= 717.2 + 435.4 - 2 \times 698.1 - 137.9 \\ &= -381.5 \text{ MJ/kmol}\end{aligned} \tag{11.5}$$

The tabulated value is -393.5 MJ/kmol.

Water (H_2O)

This is formed from the hydrogen and oxygen molecules in the reaction

$$H_2(g) + \tfrac{1}{2}O_2(g) \rightarrow H_2O(g) \tag{11.6}$$

Thus

$$(\Delta H_f)_{H_2O} = \Sigma \Delta H_a - \Sigma \Delta H(X - Y) - \Sigma \Delta H_{res} - \Sigma \Delta H_{latent} \tag{11.7}$$

In this case $\Delta H_{res} = 0$ but ΔH_{latent} depends upon the phase of the water. First, consider the water is in vapour phase, when $\Delta H_{latent} = 0$. Then

$$
\begin{aligned}
(\Delta H_f)_{H_2O} &= \Sigma\, \Delta H_a - \Sigma\, \Delta H(X - Y) \\
&= \Delta H_a[H - H] + \tfrac{1}{2}\Sigma\, \Delta H_a[O = O] - [H - OH] - [O - H] \\
&= 435.4 + \tfrac{1}{2}\,(498.2) - 497.5 - 428.7 \\
&= -241.7 \text{ MJ/kmol}
\end{aligned}
\tag{11.8}
$$

This compares well with the tabulated values of -241.6 MJ/kmol.
   If the water was in liquid phase then eqn (11.6) becomes

$$
H_2(g) + \tfrac{1}{2}O_2(g) \longrightarrow H_2O(l)
\tag{11.9}
$$

$$
\begin{aligned}
(\Delta H_f)_{H_2O(l)} &= (\Delta H_f)_{H_2O(g)} + (m_w)_{H_2O}h_{fg} \\
&= -241.7 - 18 \times 2441.8/1000 \\
&= -285.65 \text{ MJ/kmol}
\end{aligned}
\tag{11.10}
$$

The tabulated value is -285.6 MJ/kmol.

Example

Evaluate the enthalpy of formation of methane, CH_4.

Solution

The structure of methane is tetrahedral, and of the form shown in Fig 11.5(a). Methane is often depicted in planar form as shown in Fig 11.5(b).

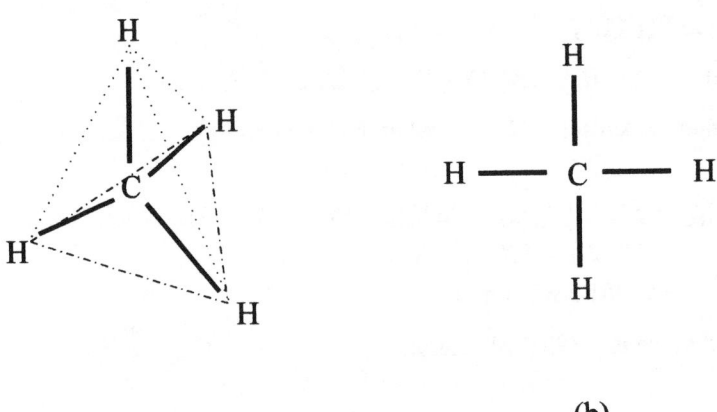

(a)   (b)

Fig. 11.5 Atomic arrangement of methane molecule

Methane can be formed by the following reaction:

$$
C_{graphite} + 2H_2(g) \longrightarrow CH_4(g)
\tag{11.11}
$$

Hence

$$(\Delta H_f)_{CH_4} = \Sigma \Delta H_a - \Sigma \Delta H(X - Y) - \Sigma \Delta H_{res} - \Sigma \Delta H_{latent}$$
$$= \Delta H_a[C_{graphite}] + 2 \Delta H_a[H - H] - 4[H - C] \qquad (11.12)$$
$$= 717.2 + 2 \times 425.4 - 4 \times 414.5$$
$$= -70 \text{ MJ/kmol}$$

The tabulated value is -74.78 MJ/kmol, and the difference occurs because the energy of the bonds in methane is affected by the structure of the methane molecule, which results in attraction forces between molecules.

Example

Evaluate the enthalpy of formation of ethanol, C_2H_5OH.

Solution

The planar representation of the structure of ethanol is shown in Fig 11.6.

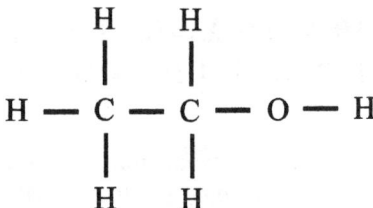

Fig. 11.6 Structure of ethanol molecule

Ethanol can be considered to be made up of the following processes:

$$2C_{graphite} + 3H_2(g) + 0.5O_2(g) \rightarrow C_2H_5OH(g)$$

giving the enthalpy of formation of ethanol as (eqn (11.12))

$$(\Delta H_f)_{C_2H_5OH} = \Sigma \Delta H_a - \Sigma \Delta H(X - Y) - \Sigma \Delta H_{res} - \Sigma \Delta H_{latent}$$

This gives the following equation:

$$2C_{graphite} + 3H_2(g) + 0.5O_2(g)$$

$$\rightarrow 2C(g) + \underbrace{2 \times 717.2}_{\substack{\text{atomisation} \\ \text{energy for C}}} + 6H + \underbrace{3 \times 435.4}_{\substack{\text{atomisation} \\ \text{energy for H}}} + O + \underbrace{0.5 \times 498.2}_{\substack{\text{atomisation} \\ \text{energy for O}}}$$

$$- 5[C-H] - [C-C] - [C-O] - [O-H]$$
$$\rightarrow C_2H_5OH(g) + 2 \times 717.2 + 3 \times 435.4 + 0.5 \times 498.2 \qquad (11.13)$$
$$- 5 \times 414.5 - 347.5 - 351.7 - 428.7$$
$$\rightarrow C_2H_5OH(g) - 210.7 \text{ MJ/kmol}$$

Hence the enthalpy of formation of gaseous ethanol is -210.7 MJ/kmol. This is equivalent to -5.853 MJ/kg of ethanol. The value obtained from tables is about -222.8 MJ/kmol, and the difference is attributable to the slight variations in bond energy that occur due to the three-dimensional nature of the chemical structure.

Example

Evaluate the *enthalpy of reaction* of methane.

Solution

This can be obtained either by using atomisation and dissociation energies, in a similar manner to that used to find the enthalpies of formation of compounds, or from the enthalpies of formation of the compounds in the reactants and products. Both methods will be used in this case. The reaction describing the combustion of methane is

$$CH_4(g) + 2O_2(g) \rightarrow CO_2(g) + 2H_2O(g) \tag{11.14}$$

The chemical structure of methane was given above. Hence the enthalpy of formation of the reactants

$$(\Delta H_f)_R = \Sigma \Delta H_a - \Sigma \Delta H(X - Y) - \Sigma \Delta H_{res} - \Sigma \Delta H_{latent}$$
$$= \Delta H_a[C_{graphite}] + 2 \Delta H_a[H - H] + 2 \Delta H_a[O = O] - 4[H - C] - 2[O = O] \tag{11.15}$$

Note that the atomisation and dissociation energies of oxygen are equal $(2 \Delta H_a[O = O] = 2[O = O])$ and cancel out, i.e. the enthalpy of formation of oxygen is zero. This value of zero is assumed as a base level for all elements. Thus

$$(\Delta H_f)_R = (\Delta H_f)_{CH_4} = 70 \text{ MJ/kmol} \tag{11.16}$$

Similarly for the products

$$(\Delta H_f)_P = \Delta H_a - \Sigma \Delta H(X - Y) - \Sigma \Delta H_{res} - \Sigma \Delta H_{latent}$$
$$= \Delta H_a[C_{graphite}] + 2 \Delta H_a[O = O] + 2 \Delta H_a[H - H]$$
$$- 2[C = O] - 2[H - OH] - 2[O - H] - \Delta H_{res}[CO_2]$$
$$= (\Delta H_f)_{CO_2} + 2(\Delta H_f)_{H_2O}$$
$$= -381.5 - 2 \times 241.7 \text{ MJ/kmol} \tag{11.17}$$

The heat of reaction is given by

$$\Delta H_R = (\Delta H_f)_P - (\Delta H_f)_R$$
$$= -381.7 - 2 \times 241.7 - (-70)$$
$$= -794.9 \text{ MJ/kmol} \tag{11.18}$$

This is close to the value of -802.9 MJ/kmol quoted as the *lower* enthalpy of reaction of methane in Table 11.2. If the higher heat of reaction of methane is required then eqn (11.14) becomes

$$CH_4(g) + 2O_2(g) \rightarrow CO_2(g) + 2H_2O(l) \tag{11.19}$$

Example

Evaluate the lower enthalpy of reaction of benzoic acid (C_6H_5COOH). The planar diagram of its structure is shown in Fig 11.7.

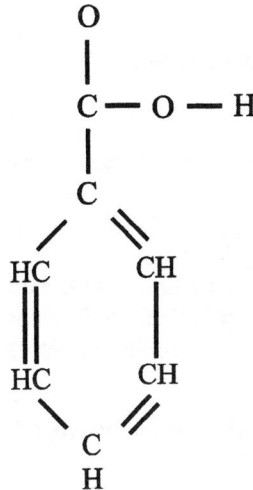

Fig. 11.7 Structure of benzoic acid molecule

Solution

The reaction of benzoic acid with oxygen is

$$C_6H_5COOH(g) + 7.5O_2(g) \longrightarrow 7CO_2(g) + 3H_2O(g) \qquad (11.20)$$

The easiest way to evaluate the enthalpy of reaction is from eqn (11.18):

$$\Delta H_R = (\Delta H_f)_P - (\Delta H_f)_R$$

The values of H_f were calculated above for CO_2 and H_2O, and hence the only unknown quantity in eqn (11.18) is the enthalpy of formation of the benzoic acid:

$$(\Delta H_f)_{C_6H_5COOH} = \Sigma \Delta H_a - \Sigma \Delta H(X-Y) - \Sigma \Delta H_{res} - \Sigma \Delta H_{latent} \qquad (11.21)$$

The acid is in gaseous form before the reaction, see eqn (11.20), and thus $\Delta H_{latent} = 0$. Substituting values into eqn (11.21) gives

$$(\Delta H_f)_{C_6H_5COOH} = \Sigma \Delta H_a - \Sigma \Delta H(X-Y) - \Sigma \Delta H_{res}$$
$$= 7\,\Delta H_a[C_{graphite}] + 3\,\Delta H_a[H-H] + \Delta H_a[O=O]$$
$$-5[H-C] - 4[C-C] - 3[C=C] - [C=O]$$
$$-[C-O] - [O-H] - [\Delta H_{res}]_{C_6H_6} - [\Delta H_{res}]_{COOH} \qquad (11.22)$$

Hence

$$(\Delta H_f)_{C_6H_5COOH} = 7 \times 717.2 + 3 \times 435.4 + 498.2 - 5 \times 414.5$$
$$- 4 \times 347.5 - 3 \times 615.5 - 698.1 - 351.7 - 428.7 - 150.4 - 117.0$$
$$= -230.1 \text{ MJ/kmol} \qquad (11.23)$$

The enthalpy of formation of the products is

$$(\Delta H_f)_P = 7(\Delta H_f)_{CO_2} + 3(\Delta H_f)_{H_2O}$$
$$= 7 \times (-381.5) + 3 \times (-241.7) \tag{11.24}$$
$$= -3395.5 \text{ MJ/kmol}$$

Thus the enthalpy of reaction of benzoic acid is

$$(\Delta H_R)_{C_6H_5COOH} = -3395.6 - (-230.1) = 3165.5 \text{ MJ/kmol} \tag{11.25}$$

The tabulated value is 3223.2 MJ/kmol, giving an inaccuracy of about 2%.

11.3 Enthalpy of reaction

The enthalpies of reaction of some commonly encountered fuels are given in Table 11.2. These have been taken from a number of sources and converted to units consistent with this text where necessary. There are a number of interesting observations that can be made from this table:

- the enthalpies of reaction of many of the hydrocarbon fuels on a basis of mass are very similar, at around 44000 kJ/kg;
- the stoichiometric air–fuel ratios of many basic hydrocarbon fuels lie in the range 13:1–17:1;
- some of the fuels have positive enthalpies of formation;
- all of the fuels have negative enthalpies of reaction;
- the enthalpies of reaction of the alcohols are less than those of the non-oxygenated fuels, simply because the oxygen cannot provide any energy of reaction;
- the commonly used hydrocarbon fuels are usually mixtures of hydrocarbon compounds.

11.4 Concluding remarks

It has been shown that the energy released by a fuel is contained in it by virtue of its structure, i.e. the bonds between the atoms. It is possible to assess the enthalpies of formation or reaction of a wide range of fuels by considering the chemical structure and the bonds in the compound.

A table of enthalpies of formation and reaction for common fuels has been given.

Table 11.2 Enthalpies of formation and reaction for commonly encountered fuels and compounds (from Goodger (1979))

Fuel	Chemical formula	Molecular weight (m_w) (integrals)	Enthalpy of formation, Δh_f (kJ/kmol)	Higher enthalpy of reaction, Δh_R (MJ/kmol)	Lower enthalpy of reaction, Δh_R (kJ/kg)	Lower enthalpy of reaction, Δh_R (MJ/kmol)	Stoichiometric air–fuel ratio, ε_{st}
Alkanes	C_nH_{2n+2}						
Methane	CH_4	16	−74898	−891	−50047	−802.9	17.16
Ethane	C_2H_6	30	−84724	−1561	−47519	−1428.8	16.02
Propane	C_3H_8	44	−103916	−2221	−46387	−2045.4	15.60
Butane	C_4H_{10}	58	−124817	−2880	−45771	−2660.2	15.38
Pentane	C_5H_{12}	72	−146538	−3538	−45384	−3274.3	15.25
Hexane	C_6H_{14}	86	−167305	−4197	−45134	−3889.3	15.16
Heptane	C_7H_{16}	100	−187946	−4856	−44955	−4504.4	15.10
Octane	C_8H_{18}	114	−208586	−5515	−44820	−5119.5	15.05
Cetane	$C_{16}H_{34}$	226	–	−10711	−44000	−9963.4	14.88
Acetylenes	C_nH_{2n-2}						
Acetylene	C_2H_2	26	+226900	−1270	−48258	−1226.4	13.20
Propyne	C_3H_4	40	+185555	−1939	−46200	−1850.9	13.73
Aromatics							
Benzene	C_6H_6	78	+82982	−3304	−40605	−3172	13.20
Toluene	C_7H_8	92	+50032	−3950	−40967	−3774	13.43
Xylene	C_8H_{10}	106	+18059	−4598	−41276	−4378	13.60
Naphthalene	$C_{10}H_8$	128	+150934	−5233	−39455	−5057	12.87
Alcohols	$C_nH_{2n+1}OH$	$C_nH_{2n+2}O$					
Methanol	CH_3OH	32	−201301	−764	−21114	−676	6.44
Ethanol	C_2H_5OH	46	−235466	−1410	−27742	−1278	8.85
Propanol	C_3H_7OH	60	−235107	−2069	−31504	−1893	10.30
Mixtures							
Gasoline (petrol)	$C_nH_{1.87n}$	≈110	–	−5203	−44000	−4840	14.6
Light diesel	$C_nH_{1.8n}$	≈170	–	−7616	−42500	−7225	14.5
Heavy diesel	$C_nH_{1.7n}$	≈200	–	−8760	−41400	−8280	14.4
Natural gas	$C_nH_{3.8n}N_{0.1n}$	≈18	–	−900	−45000	−810	14.5

12

Chemical Equilibrium and Dissociation

Up to now this book has concentrated on combustion problems which can be solved by methods based on equilibrium but which do not require an explicit statement of the fact, e.g. complete combustion of a hydrocarbon fuel in air can be analysed by assuming that the products consist only of H_2O and CO_2. These methods are not completely correct and a more rigorous analysis is necessary to obtain greater accuracy.

Consider the combustion of carbon monoxide (CO) with oxygen (O_2); up until now the reaction has been described by the equation

$$CO + 1/2 O_2 \rightarrow CO_2 \qquad (12.1)$$

It is implied in this equation that carbon monoxide combines with oxygen to form carbon dioxide, and as soon as that has happened the reaction ceases. This is not a true description of what happens in practice. The real process is one of dynamic equilibrium with some of the carbon dioxide breaking down into carbon monoxide and oxygen (or even more esoteric components) again, which might then recombine to form carbon dioxide. The breakdown of the CO_2 molecule is known as *dissociation*. To evaluate the amount of dissociation that occurs (the *degree of dissociation*) it is necessary to evolve new techniques.

12.1 Gibbs energy

The concept of Gibbs energy, G, was introduced in Chapter 1. The change in the specific Gibbs energy, g, for a system of fixed composition was defined in terms of other properties as

$$dg = v \, dp - s \, dT \qquad (12.2)$$

It was also shown that for a closed system at constant temperature and pressure, performing only mechanical work, to be in equilibrium,

$$dG)_{p,T} = 0 \qquad (12.3)$$

Equations (12.2) and (12.3) are based on the assumption that $G = mg = mg(p, T)$, and this is quite acceptable for a single component system, or one of fixed composition. If the system has more than one component, and these components can react to form other compounds, e.g. if the system contained carbon monoxide, oxygen and carbon dioxide as defined in eqn (12.1), then it is necessary to define the Gibbs energy as $G = mg = mg(p, T, m_i)$ where m_i is the mass of component i, and $m = \Sigma\, m_i$. The significance of changes of composition on the value of the Gibbs energy of a mixture will now be investigated.

If

$$G = mg = mg(p, T, m_i) \tag{12.4}$$

and if it is assumed that G is a continuous function with respect to p, and T and the masses of constituents comprising the mixture, then the change of G with changes in the independent variables is

$$dG = \left(\frac{\partial G}{\partial p}\right)_{T,\,m} dp + \left(\frac{\partial G}{\partial T}\right)_{p,\,m} dT + \left(\frac{\partial G}{\partial m_1}\right)_{p,T,m_{i\neq 1}} dm_1 + \cdots \left(\frac{\partial G}{\partial m_n}\right)_{p,T,m_{i\neq n}} dm_n \tag{12.5a}$$

where $dm_1 \ldots dm_n$ are changes in mass of the various constituents. A similar equation can be written in terms of amount of substance n, and is

$$dG = \left(\frac{\partial G}{\partial p}\right)_{T,\,n} dp + \left(\frac{\partial G}{\partial T}\right)_{p,\,n} dT + \left(\frac{\partial G}{\partial n_1}\right)_{p,T,n_{i\neq 1}} dn_1 + \cdots \left(\frac{\partial G}{\partial n_n}\right)_{p,T,n_{i\neq n}} dn_n \tag{12.5b}$$

For the initial part of the development of these equations the mass-based relationship (eqn (12.5a)) will be used because the arguments are slightly easier to understand. The term

$$\left(\frac{\partial G}{\partial m_1}\right)_{p,T,m_{i\neq 1}} dm_1$$

represents the 'quantity' of Gibbs energy introduced by the transfer of mass dm_1 of constituent 1 to the system. (This can be more readily understood by considering the change in internal energy, dU, when the term $(\partial U/\partial m_1)_{p,T,m_{i\neq 1}}\, dm_1$ has a more readily appreciated significance.)

The significance of the terms on the right of eqn (12.5a) is as follows:

1. The first term denotes the change of Gibbs energy due to a change in pressure, the temperature, total mass and composition of the system remaining constant. This is equivalent to the term derived when considering a system of constant composition, and is $V\, dp$.

2. The second term denotes the change of Gibbs energy due to a change in temperature, the pressure and total mass of the system remaining constant. This is equivalent to $-S\, dT$ derived previously.

3. The third term shows the change of Gibbs energy due to a change in the mass (or amount of substance if written in terms of n) of constituent m_1, the pressure, temperature and masses of other constituents remaining constant. It is convenient to define this as

$$\mu_1 = \left(\frac{\partial G}{\partial m_1}\right) \tag{12.6}$$

4. The fourth term is a general term of the form of term (3) in eqn (12.6).

Hence, in terms of masses

$$dG = V dp - S dT + \sum_{i=1}^{n} \mu_i dm_i \qquad (12.7a)$$

while in terms of amount of substance

$$dG = V dp - S dT + \sum_{i=1}^{n} \mu_{m_i} dn_i \qquad (12.7b)$$

12.2   Chemical potential, μ

The term μ is called the chemical potential and is defined as

$$\left(\frac{\partial G}{\partial m_i} \right)_{p,T,m_{i*1}}$$

The significance of μ will now be examined. First, it can be considered in terms of the other derived properties.
By definition

$$dG = d(H - TS) = dH - T \, dS - S \, dT \qquad (12.8)$$

Hence

$$\begin{aligned} dH &= dG + T \, dS + S \, dT \\ &= V \, dp - S \, dT + \Sigma \, \mu_i \, dm_i + T \, dS + S \, dT \\ &= V \, dp + T \, dS + \Sigma \, \mu_i \, dm_i \end{aligned} \qquad (12.9)$$

Considering each of the terms in eqn (12.9), these can be interpreted as the capacity to do work brought about by a change in a particular property. The first term is the increase in capacity to do work that is achieved by an isentropic pressure rise (cf. the work done in a feed pump of a Rankine cycle), and the second term is the increased capacity to do work that occurs as a result of reversible heat transfer. The third term is also an increase in the capacity of the system to do work, but this time it is brought about by the addition of a particular component to a mixture. For example, if oxygen is added to a mixture of carbon monoxide, carbon dioxide, water and nitrogen (the products of combustion of a hydrocarbon fuel) then the mixture could further react to convert more of the carbon monoxide to carbon dioxide, and more work output could be obtained. Thus μ_i is the increase in the capacity of a system to do work when unit mass (or, in the case of μ_{m_i}, unit amount of substance) of component i is added to the system. μ_i can be considered to be a 'chemical pressure' because it is the driving force in bringing about reactions.
Assuming that H is a continuous function,

$$dH = \left(\frac{\partial H}{\partial p} \right)_{S,m} dp + \left(\frac{\partial H}{\partial S} \right)_{p,m} dS + \left(\frac{\partial H}{\partial m_1} \right)_{S,p,m_{i*1}} dm_1 + \cdots + \left(\frac{\partial H}{\partial m_n} \right)_{S,p,m_{i*n}} dm_n$$

$$(12.10)$$

By comparison of eqns (12.9) and (12.10)

$$\mu_i = \left(\frac{\partial H}{\partial m_1} \right)_{S,p,m_{j*1}} \qquad (12.11)$$

Similarly it can be shown that

$$\mu_i = \left(\frac{\partial U}{\partial m_i}\right)_{S,v,m_{j*i}}$$

$$\mu_i = \left(\frac{\partial F}{\partial m_i}\right)_{T,v,m_{j*i}}$$

(12.12)

The following characteristics of chemical potential may be noted:

(i)   the chemical potential, μ, is a function of properties and hence is itself a thermodynamic property;

(ii) the numerical value of μ is not dependent on the property from which it is derived (all the properties have the dimensions of energy and, hence, by the conservation of energy this is reasonable);

(iii) the numerical value of μ is independent of the size of the system and is hence an *intensive* property, e.g.

$$\left(\frac{\partial U}{\partial m_i}\right)_{S,v,m_{j*i}} = \mu_i = \left(\frac{\partial(mu)}{\partial mx_i}\right)_{S,v,m_{j*i}} = \left(\frac{\partial u}{\partial x_i}\right)_{S,v,m_{j*i}}$$

(12.13)

Since μ is an intensive property it may be compared with the other intensive properties p and T etc. By the two-property rule this means that

$$p = p(\mu, T)$$

(12.14)

and similarly

$$\mu = \mu(p, T)$$

(12.15)

It can be shown that the chemical potential μ for a pure phase is equal in magnitude to the specific Gibbs energy at any given temperature and pressure, i.e.

$$\mu = g$$

(12.16)

[Note: Although $\mu = g$, it is different from g inasmuch as it is an intensive property whereas g is a specific property. Suppose there is a system of mass m, then the total Gibbs energy is $G = mg$ whereas the chemical potential of the whole system is still μ {cf. p or T}.]

12.3 Stoichiometry

Consider the reaction

$$CO + \frac{1}{2}O_2 \Rightarrow CO_2$$

(12.17)

Equation (12.17) shows the stoichiometric proportions of the reactants and products. It shows that 1 mol CO and $\frac{1}{2}$ mol O_2 could combine to form 1 mol CO_2. If the reaction proceeded to completion, no CO or O_2 would be left at the final condition.

This is the *stoichiometric* equation of the reaction and the amounts of substance in the equation give the *stoichiometric coefficients*.

The general equation for a chemical reaction is

$$v_a A + v_b B \Leftrightarrow v_c C + v_d D \qquad (12.18)$$

where v is a stoichiometric coefficient, and A, B, C and D are the substances involved in the reaction. Applying eqn (12.18) to the $CO + \frac{1}{2}O_2$ reaction gives

$$
\begin{aligned}
v_{CO} &= -1 \\
v_{O_2} &= -1/2 \\
v_{CO_2} &= 1
\end{aligned}
\qquad (12.19)
$$

It is conventional in chemistry to assign negative values to the stoichiometric coefficients on the left-hand side of the equation (nominally, the reactants) and positive signs to those on the right-hand side (nominally, the products).

12.3.1   MIXTURES

Mixtures are not necessarily stoichiometric and the following terms were introduced in Chapter 10 to describe the proportions of a mixture:

(i)   If reactants occur in proportion to the stoichiometric coefficients then the mixture is said to be *chemically correct* or *stoichiometric*.

(ii)   If the reactants have a greater proportion of fuel than the correct mixture then it is said to be *rich*.

(iii) If the reactants have a lesser proportion of fuel than the correct mixture then it is said to be *weak*.

e.g. chemically correct mixture is $CO + \frac{1}{2}O_2 \Rightarrow CO_2$

Rich mixture (excess of fuel)

$$(1 + n)CO + \frac{1}{2}O_2 \Rightarrow CO_2 + nCO \qquad (12.20)$$

Weak mixture (excess of oxidant)

$$CO + \left(\frac{1}{2} + n\right)O_2 \Rightarrow CO_2 + nO_2 \qquad (12.21)$$

NB. Equations (12.20) and (12.21) have been written neglecting dissociation.

12.4   Dissociation

The basis of dissociation is the atomic model that all mixtures of gases are in a state of dynamic equilibrium. Molecules of the compounds are being created whilst existing ones are breaking down into simpler compounds or elements (dissociating). In the equilibrium situation the rates of creation and destruction of molecules of any compounds are equal. This means that macroscopic measuring techniques do not sense the changes but give the impression that the system is in a state of 'static' equilibrium. The effect of this is that the reactions can no longer be said to be unidirectional but must be shown as

$$CO + \frac{1}{2} O_2 \Leftrightarrow CO_2 \tag{12.22}$$

On a molecular basis the above reaction can go either way. It is now necessary to consider a technique which will define the *net* direction of change for a collection of molecules. First, consider the general equation for the $CO + \frac{1}{2}O_2$ reaction, neglecting particularly esoteric and rare compounds:

$$CO + \frac{1}{2} O_2 \Leftrightarrow aCO_2 + bCO + dO_2 + eC + fO \tag{12.23}$$

where the carbon (C) and atomic oxygen (O) are formed by the breakdown of the reactants. Experience shows that in the ranges normally encountered in practice, the C and O atoms have a negligible effect. This allows the general reaction to be simplified to

$$CO + \frac{1}{2} O_2 \Leftrightarrow aCO_2 + bCO + dO_2 \tag{12.24}$$

It is possible to write this equation in a slightly different form by considering the amount of CO_2 that has dissociated. This can be defined as $\alpha = (1 - a)$, and the equation for the dissociation of CO_2 is

$$\alpha CO_2 \Leftrightarrow \alpha CO + \frac{\alpha}{2} O_2 \tag{12.25}$$

By considering the stoichiometric equation (eqn (12.22)) and the dissociation equation (eqn (12.25)) a general equation may be constructed in terms of α:

$$\left. \begin{array}{l} CO + 1/2 O_2 \Rightarrow CO_2 \\[2mm] \alpha CO_2 \Leftrightarrow \alpha CO + \dfrac{\alpha}{2} O_2 \end{array} \right\} \tag{12.26}$$

Adding eqns (12.26) gives

$$CO + \frac{1}{2} O_2 \Rightarrow (1 - \alpha)CO_2 + \alpha CO + \frac{\alpha}{2} O_2 \tag{12.27}$$

In eqn (12.27), α is known as the *degree of dissociation*. This equation shows the effect of dissociation on a chemically correct mixture. Before discussing methods of evaluating α, the effect of dissociation on non-stoichiometric mixtures will be shown, first using the carbon monoxide reaction, and then a general hydrocarbon fuel.

12.4.1 WEAK MIXTURE WITH DISSOCIATION

Equation without dissociation:

$$CO + \left(\frac{1}{2} + n \right) O_2 \Rightarrow CO_2 + nO_2 \tag{12.28}$$

Dissociation of CO_2:

$$\alpha CO_2 \Leftrightarrow \alpha CO + \frac{\alpha}{2} O_2 \tag{12.29}$$

Adding eqns (12.28) and (12.29) gives

$$CO + \left(\frac{1}{2} + n\right)O_2 \Rightarrow (1 - \alpha)CO_2 + \alpha CO + \left(\frac{\alpha}{2} + n\right)O_2 \qquad (12.30)$$

12.4.2   RICH MIXTURE WITH DISSOCIATION

Equation without dissociation:

$$(1 + n)CO + \frac{1}{2} O_2 \Rightarrow CO_2 + nCO \qquad (12.31)$$

Dissociation of CO_2:

$$\alpha CO_2 \Leftrightarrow \alpha CO + \frac{\alpha}{2} O_2$$

Total reaction:

$$(1 + n)CO + \frac{1}{2} O_2 \Rightarrow (1 + \alpha)CO_2 + (n + \alpha)CO + \frac{\alpha}{2} O_2 \qquad (12.32)$$

12.4.3   GENERAL HYDROCARBON REACTION WITH DISSOCIATION

A general hydrocarbon fuel can be defined as $C_x H_y$ and this will react with the stoichiometric quantity of air as shown in the following equation:

$$C_x H_y + \left(x + \frac{y}{4}\right)(O_2 + 3.76\,N_2) \Rightarrow xCO_2 + \frac{y}{2} H_2O + 3.76\left(x + \frac{y}{4}\right)N_2 \qquad (12.33)$$

The dissociation of the CO_2 and H_2O can be added into this equation as (eqn (12.29))

$$a_1 CO_2 \Leftrightarrow a_1 CO + \frac{a_1}{2} O_2$$

and

$$a_2 H_2O \Leftrightarrow a_2 H_2 + \frac{a_2}{2} O_2 \qquad (12.34)$$

which gives a general equation with dissociation

$$C_x H_y + \left(x + \frac{y}{4}\right)(O_2 + 3.76\,N_2)$$

$$\Rightarrow x(1 - a_1)CO_2 + \frac{y}{2}(1 - a_2)H_2O + xa_1CO + \frac{y}{2} a_2 H_2$$

$$+ \left(\frac{x}{2} a_1 + \frac{y}{4} a_2\right)O_2 + 3.76\left(x + \frac{y}{4}\right)N_2 \qquad (12.35)$$

If the fuel was benzene (C_6H_6) then eqn (12.35) would become

$$C_6H_6 + 7.5(O_2 + 3.76N_2)$$
$$\Rightarrow 6(1 - a_1)CO_2 + 3(1 - a_2)H_2O + 6a_1CO + 3a_2H_2 + (3a_1 + 1.5a_2)O_2 + 28.2N_2$$

$$(12.36)$$

If the mixture were not stoichiometric then eqn (12.33) would be modified to take account of the air–fuel ratio, and eqns (12.35) and (12.36) would also be modified. These equations are returned to in the later examples. In eqn (12.36) the combination of nitrogen and oxygen has been neglected. In many combustion processes the oxygen and nitrogen join together at high temperatures to form compounds of these elements: one of these compounds is nitric oxide (NO), and the equations can be extended to include this reaction. This will also be introduced later.

12.4.4 GENERAL OBSERVATION

As a result of dissociation there is always some oxidant in the products; hence dissociation always reduces the effect of the desired reaction, e.g. if the reaction is exothermic then dissociation reduces the energy released (see Fig 12.1).

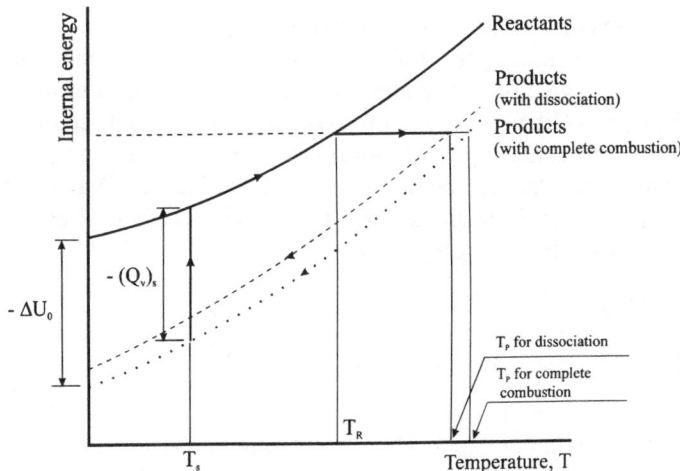

Fig. 12.1 Effect of dissociation on combustion

Having introduced the concept of dissociation it is necessary to evolve a method that allows the value of the degree of dissociation, a, to be calculated. This method will be developed in the following sections.

12.5 Calculation of chemical equilibrium and the law of mass action

General relationships will be derived, and the particular case of the $CO + \frac{1}{2}O_2$ reaction will be shown in brackets { }.

It was previously shown that for a system at constant pressure and temperature to be in an equilibrium state it must have a minimum value of Gibbs energy i.e. $dG)_{p,T} = 0$.

But, by definition, for a system at constant pressure and temperature

$$dG)_{p,T} = \mu_1 \, dm_1 + \mu_2 \, dm_2 + \cdots \mu_n \, dm_n \tag{12.37}$$

where $m_1, m_2 \ldots m_n$ are the masses (or amounts) of the *possible* constituents of the mixture.

Only four constituents will be considered during this discussion, two reactants and two products, but the theory can be extended to any number of constituents. The equilibrium equation for the *complete* reaction is

$$v_a A + v_b B \Leftrightarrow v_c C + v_d D$$

$$\left\{ CO + \frac{1}{2} \, O_2 \Leftrightarrow CO_2 \right\} \tag{12.38}$$

At some intermediate stage in the reaction the state may be represented as

$$v_a A + v_b B \longrightarrow (1 - \varepsilon) v_a A + (1 - \varepsilon) v_b B + \varepsilon v_c C + \varepsilon v_d D$$

$$\left\{ \text{e.g. } CO + \frac{1}{2} \, O_2 \longrightarrow (1 - \varepsilon) CO + (1 - \varepsilon) \frac{1}{2} \, O_2 + \varepsilon CO_2 \right\} \tag{12.39}$$

where stoichiometric coefficients equal to unity are implicit.

ε is known as the *fraction of reaction* and is an *instantaneous* value during the reaction as opposed to α, the degree of dissociation, which is a final *equilibrium* value. The use of ε allows the changes in Gibbs energy to be considered as the reaction progresses.

For the purposes of evaluating the dissociation phenomena it is possible to consider the reaction occurring at constant temperature and pressure (equal to values obtained by other means or calculated by an implicit iterative technique). The Gibbs energy of the system may be described by the following equation:

$$G = (1 - \varepsilon) \, v_a \mu_a + (1 - \varepsilon) \, v_b \mu_b + \varepsilon v_c \mu_c + \varepsilon v_d \mu_d$$

$$\left\{ G = (1 - \varepsilon) \, \mu_{CO} + \frac{1}{2} \, (1 - \varepsilon) \, \mu_{O_2} + \varepsilon \mu_{CO_2} \right\} \tag{12.40}$$

To find the equilibrium condition while maintaining pressure and temperature constant, this function has to be minimised with respect to ε, i.e. it is necessary to locate when

$$\left. \frac{\partial G}{\partial \varepsilon} \right)_{p,T} = 0$$

Now

$$dG = V \, dp - S \, dT + \Sigma \, \mu_i \, dm_i \tag{12.41}$$

and

$$dG)_{p,T} = \mu_1 \, dm_1 + \mu_2 \, dm_2 + \cdots \mu_n \, dm_n \tag{12.42}$$

From here on, the mass form of the equation (eqn 12.5a) will be replaced by the mole form of the equation (eqn 12.5b), because this is more appropriate for chemical reactions.

It is possible to relate the amount of substance of each constituent in terms of ε, namely

$$n_a = (1 - \varepsilon)v_a + A$$
$$n_b = (1 - \varepsilon)v_b + B$$
$$n_c = \varepsilon v_c + C \tag{12.43}$$
$$n_d = \varepsilon v_d + D$$

where A, B, C and D allow for the excess amount of substance in non-stoichiometric mixtures. Hence

$$dn_a = -v_a \, d\varepsilon \tag{12.44}$$

and applying similar techniques

$$-\frac{dn_a}{v_a} = -\frac{dn_b}{v_b} = \frac{dn_c}{v_c} = \frac{dn_d}{v_d} = d\varepsilon \tag{12.45}$$

This is known as the *equation of constraint* because it states that the changes of amount of substance (or mass) must be related to the stoichiometric equation (i.e. changes are *constrained* by the stoichiometry). Hence

$$dG)_{\mathrm{p,T}} = -v_a \mu_a \, d\varepsilon - v_b \mu_b \, d\varepsilon + v_c \mu_c \, d\varepsilon + v_d \mu_d \, d\varepsilon \tag{12.46}$$

giving

$$\left. \frac{\partial G}{\partial \varepsilon} \right) = -v_a \mu_a - v_b \mu_b + v_c \mu_c + v_d \mu_d$$

$$= 0 \quad \text{for equilibrium} \tag{12.47}$$

i.e. at equilibrium

$$v_a \mu_a + v_b \mu_b = v_c \mu_c + v_d \mu_d \tag{12.48}$$

Since $\mu = g$ it is possible to describe μ in a similar way to g, namely

$$\mu = \mu^0 + \Re T \ln p_r \tag{12.49}$$

where

$$\mu^0 = \mu_0 + \mu(T) \tag{12.50}$$

and is called the standard chemical potential, and is the value of μ at temperature T and the datum pressure, p_0.

The value of pressure to be used in eqn (12.49) is the ratio of the partial pressure of the individual constituent to the datum pressure.

Substituting for μ in the equilibrium equation, eqn (12.48)

$$v_a(\mu_a^0 + \Re T \ln p_{r_a}) + v_b(\mu_b^0 + \Re T \ln p_{r_b}) = v_c(\mu_c^0 + \Re T \ln p_{r_c}) + v_d(\mu_d^0 + \Re T \ln p_{r_d}) \tag{12.51}$$

which can be rearranged to give

$$(v_a\mu_a^0 + v_b\mu_b^0 - v_c\mu_c^0 - v_d\mu_d^0) + \Re T(v_a \ln p_{r_a} + v_b \ln p_{r_b} - v_c \ln p_{r_c} - v_d \ln p_{r_d}) = 0 \tag{12.52}$$

Hence

$$-(v_a\mu_a^0 + v_b\mu_b^0 - v_c\mu_c^0 - v_d\mu_d^0) = \Re T \ln\left\{\frac{p_{r_a}^{v_a}\, p_{r_b}^{v_b}}{p_{r_c}^{v_c}\, p_{r_d}^{v_d}}\right\}$$

$$= \Re T \ln\left[\left\{\frac{p_a^{v_a}\, p_b^{v_b}}{p_c^{v_c}\, p_d^{v_d}}\right\}p_0^{(v_c+v_d-v_a-v_b)}\right]$$

$$= -\Re T \ln\left\{\frac{p_{r_c}^{v_c}\, p_{r_d}^{v_d}}{p_{r_a}^{v_a}\, p_{r_b}^{v_b}}\right\}$$

$$= -\Re T \ln\left[\left\{\frac{p_c^{v_c}\, p_d^{v_d}}{p_a^{v_a}\, p_b^{v_b}}\right\}p_0^{(v_a+v_b-v_c-v_d)}\right] \tag{12.53}$$

The left-hand side of eqn (12.53) is the difference in the standard chemical potentials at the reference pressure p_0 of 1 bar or 1 atmosphere. This is defined as

$$-\Delta G_T^0 = v_a\mu_a^0 + v_b\mu_b^0 - v_c\mu_c^0 - v_d\mu_d^0 \tag{12.54}$$

Hence

$$-\frac{\Delta G_T^0}{\Re T} = \ln\left\{\frac{p_{r_c}^{v_c}\, p_{r_d}^{v_d}}{p_{r_a}^{v_a}\, p_{r_b}^{v_b}}\right\} = \ln\left[\left\{\frac{p_c^{v_c}\, p_d^{v_d}}{p_a^{v_a}\, p_b^{v_b}}\right\}p_0^{(v_a+v_b-v_c-v_d)}\right] = \ln K_{p_r} \tag{12.55}$$

giving

$$K_{p_r} = \left\{\frac{p_{r_c}^{v_c}\, p_{r_d}^{v_d}}{p_{r_a}^{v_a}\, p_{r_b}^{v_b}}\right\} = \left\{\frac{p_c^{v_c}\, p_d^{v_d}}{p_a^{v_a}\, p_b^{v_b}}\right\}p_0^{(v_a+v_b-v_c-v_d)} \tag{12.56}$$

K_{p_r} is called the *equilibrium constant*, and is a dimensionless value. Sometimes the equilibrium constant is defined as

$$K_p = \left\{\frac{p_c^{v_c}\, p_d^{v_d}}{p_a^{v_a}\, p_b^{v_b}}\right\} \tag{12.57}$$

i.e.

$$K_{p_r} = K_p p_0^{(v_a+v_b-v_c-v_d)} \quad \text{or} \quad K_p = K_{p_r}/p_0^{(v_a+v_b-v_c-v_d)} \tag{12.58}$$

K_p has the dimensions of pressure to the power of the sum of the stoichiometric coefficients, i.e. $p^{\Sigma v}$. The non-dimensional equilibrium constant is defined as

$$\ln K_{p_r} = \frac{\Delta G_T^0}{\Re T} \tag{12.59}$$

Now ΔG_T^0 is a function of T alone (having been defined at a standard pressure, p_0), therefore $K_{p_r} = f(T)$. Equation (12.56) shows that the numerical value of K_{p_r} is related to the datum pressure used to define μ^0. If the amounts of substance of reactants and products are the same then the value of K_p is not affected by the datum pressure (because $v_a + v_b - v_c - v_d = 0$). However, if the amounts of substance of products and reactants are not equal, as is the case for the $CO + \frac{1}{2}O_2 \rightarrow CO_2$ reaction, then the value of K_p will be dependent on the units of pressure. Hence, the value of equilibrium constant, K_p, is the same for the water gas reaction ($CO + H_2O \rightarrow CO_2 + H_2$) in both SI and Imperial units because K_p is dimensionless in this case.

12.5.1 K_p DEFINED IN TERMS OF MOLE FRACTION

The definition of partial pressure is

$$p_a = x_a p \tag{12.60}$$

Hence, replacing the terms for partial pressure in eqn (12.57) by the definition in eqn (12.60) gives

$$\Re T \ln \left\{ \frac{p_c^{v_c} p_d^{v_d}}{p_a^{v_a} p_b^{v_b}} \right\} = \Re T \ln \left[\left\{ \frac{x_c^{v_c} x_d^{v_d}}{x_a^{v_a} x_b^{v_b}} \right\} p^{(v_c + v_d - v_a - v_b)} \right] \tag{12.61}$$

which results in the following expression for K_p in terms of mole fraction:

$$K_p = \left\{ \frac{x_c^{v_c} x_d^{v_d}}{x_a^{v_a} x_b^{v_b}} \right\} p^{(v_c + v_d - v_a - v_b)} \tag{12.62}$$

The above expressions are known as the *law of mass action*.

12.6 Variation of Gibbs energy with composition

Equations (12.57) and (12.62) show that the equilibrium composition of a mixture is defined by the equilibrium constant, which can be defined in terms of the partial pressures or mole fractions of the constituents of the mixture; the equilibrium constant was evaluated by equating the change of Gibbs energy at constant pressure and temperature to zero, i.e. $dG)_{p,T} = 0$. It is instructive to examine how the Gibbs energy of a mixture varies with composition at constant temperature and pressure. Assume that two components of a mixture, A and B, can combine chemically to produce compound C. This is a slightly simplified form of eqns (12.38) and (12.39). If the chemical equation is

$$A + B \Rightarrow 2C \tag{12.63}$$

then at some point in the reaction, defined by the fraction of reaction, ε, the chemical composition is

$$(1 - \varepsilon)A + (1 - \varepsilon)B + 2\varepsilon C \tag{12.64}$$

and the Gibbs energy is

$$G = (1 - \varepsilon)\mu_a + (1 - \varepsilon)\mu_b + 2\varepsilon\mu_c \tag{12.65}$$

This can be written, by substituting for μ from eqns (12.49) and (12.50), as

$$G = (1 - \varepsilon)\mu_a^0 + \Re T \ln p_{r_a} + (1 - \varepsilon)\mu_b^0 + \Re T \ln p_{r_b} + 2\varepsilon(\mu_c^0 + \Re T \ln p_{r_c})$$
$$= [(1 - \varepsilon)\mu_a^0 + (1 - \varepsilon)\mu_b^0 + 2\varepsilon\mu_c^0] + \Re T[(1 - \varepsilon)\ln p_{r_a} + (1 - \varepsilon)\ln p_{r_b} + 2\varepsilon \ln p_{r_c}]$$
$$\tag{12.66}$$

The partial pressures are defined by eqn (12.60) as $p_i = x_i p$, and the mole fractions of the constituents are

$$x_a = \frac{1-\varepsilon}{2}; \; x_b = \frac{1-\varepsilon}{2}; \; x_c = \frac{2\varepsilon}{2} = \varepsilon \tag{12.67}$$

Substituting these terms in eqn (12.66) gives

$$G = [(1-\varepsilon)\mu_a^0 + (1-\varepsilon)\mu_b^0 + 2\varepsilon\mu_c^0]$$
$$+\Re T\left[(1-\varepsilon)\ln\left(\frac{1-\varepsilon}{2}\right)p_r + (1-\varepsilon)\ln\left(\frac{1-\varepsilon}{2}\right)p_r + 2\varepsilon\ln\varepsilon p_r\right]$$
$$= [(1-\varepsilon)\mu_a^0 + (1-\varepsilon)\mu_b^0 + 2\varepsilon\mu_c^0] + \Re T[(1-\varepsilon) + (1-\varepsilon) + 2\varepsilon]\ln p_r$$
$$+\Re T\left[(1-\varepsilon)\ln\left(\frac{1-\varepsilon}{2}\right) + (1-\varepsilon)\ln\left(\frac{1-\varepsilon}{2}\right) + 2\varepsilon\ln\varepsilon\right] \tag{12.68}$$

Equation (12.68) can be rearranged to show the variation of the Gibbs energy of the mixture as the reaction progresses from the reactants A and B to the products C by subtracting $2\mu_c^0$ from the left-hand side, giving

$$G - 2\mu_c^0 = (1-\varepsilon)[\mu_a^0 + \mu_b^0 - 2\mu_c^0] + 2\Re T\ln p_r + 2\Re T\left[(1-\varepsilon)\ln\left(\frac{1-\varepsilon}{2}\right) + \varepsilon\ln\varepsilon\right] \tag{12.69}$$

Equation (12.69) consists of three terms; the second one simply shows the effect of pressure and will be neglected in the following discussion. The first term is the difference between the standard chemical potentials of the separate components ($\mu_a^0 + \mu_b^0$) before any reaction has occurred, and the standard chemical potential of the mixture ($2\mu_c^0$) after the reaction is complete. Since the standard chemical potentials are constant throughout this isothermal process, this term varies linearly with the fraction of reaction, ε. The third term defines the change in chemical potential due to mixing, and is a function of the way in which the *entropy of the mixture* (not the specific entropy) varies as the reaction progresses. The manner in which the first and third terms might vary is shown in Fig 12.2, and the sum of the terms is also shown. It can be seen that for this example the equilibrium composition is at $\varepsilon = 0.78$. This figure illustrates that the Gibbs energy of the mixture initially reduces as the composition of the mixture goes from $A + B$ to C. The standard chemical potential of compound C is less than the sum of the standard chemical potentials of A and B, and hence the reaction will tend to go in the direction shown in eqn (12.63). If the standard chemical potentials were the only parameters of importance in the reaction then the reactants A and B would be completely transformed to the product, C. However, as the reaction progresses, the term based on the mole fractions varies non-monotonically, as shown by the line labelled

$$2\Re T\left[(1-\varepsilon)\ln\left(\frac{1-\varepsilon}{2}\right) + \varepsilon\ln\varepsilon\right],$$

and this affects the composition of the mixture, which obeys the law of mass action. If the two terms are added together then the variation of $G - 2\mu_c^0$ with ε is obtained. It can be seen that the larger the difference between the standard chemical potentials (i.e. the steeper the slope of the line PQ) then the larger is the fraction of reaction to achieve an equilibrium composition. This is to be expected because the driving force for the reaction has been increased. Some examples of dissociation are given later, and these can be more readily understood if this section is borne in mind.

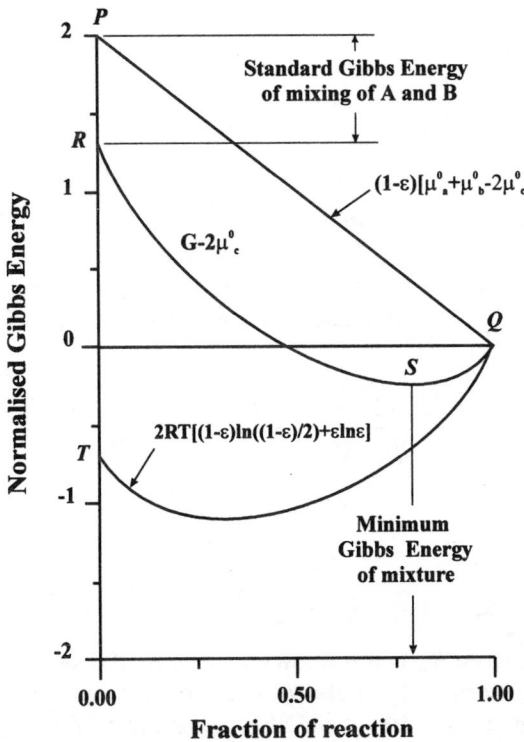

Fig. 12.2 Variation of the Gibbs energy of a mixture

12.7   Examples of significance of K_p

12.7.1   EXAMPLE 1

Consider the reaction in eqn (12.1),

$$CO + \frac{1}{2} O_2 \Rightarrow CO_2$$

$$\nu_{CO} = -1, \ \nu_{O_2} = -1/2, \ \nu_{CO_2} = 1$$

Hence

$$K_{p_r} = \frac{p_r CO_2}{p_r CO \sqrt{p_r O_2}} \tag{12.70}$$

Thus, if K_{p_r} is known, the ratio of the partial pressures in the equilibrium state is known.

It would be convenient to manipulate this expression into a more useful form. Consider the general reaction equation for the carbon monoxide and oxygen reaction (eqn (12.27)):

$$CO + \frac{1}{2}O_2 \Leftrightarrow (1 - \alpha)CO_2 + \alpha CO + \frac{\alpha}{2}O_2$$

then

$$p_{rCO_2} = \frac{n_{CO_2}}{n_P}\frac{p}{p_0} = \frac{1 - \alpha}{1 + \alpha/2}\frac{p}{p_0}$$

$$p_{rCO} = \frac{n_{CO}}{n_P}\frac{p}{p_0} = \frac{\alpha}{1 + \alpha/2}\frac{p}{p_0} \qquad (12.71)$$

$$p_{rO_2} = \frac{n_{O_2}}{n_P}\frac{p}{p_0} = \frac{\alpha/2}{1 + \alpha/2}\frac{p}{p_0}$$

Hence, the value of K_{p_r} can be related to the degree of dissociation, α, through the following equation

$$K_{p_r} = \frac{1 - \alpha}{1 + \alpha/2}\frac{1 + \alpha/2}{\alpha}\sqrt{\frac{1 + \alpha/2}{\alpha/2}}\frac{\sqrt{p_0}}{\sqrt{p}} = \frac{1 - \alpha}{\alpha}\sqrt{\frac{1 + \alpha/2}{\alpha/2}}\frac{\sqrt{p_0}}{\sqrt{p}} \qquad (12.72a)$$

and

$$K_p = \frac{1 - \alpha}{\alpha}\sqrt{\frac{1 + \alpha/2}{\alpha/2}}\frac{1}{\sqrt{p}} = \frac{K_{p_r}}{\sqrt{p_0}} \qquad (12.72b)$$

The relationship between K_{p_r} and α allows the degree of dissociation, α, to be evaluated if K_{p_r} is known. Equations (12.72) shows that, for this reaction, the value of K_{p_r} is a function of the degree of dissociation and also the pressure of the mixture. This is because the amount of substance of products is not equal to the amount of substance of reactants.

12.7.2 EXAMPLE 2

Consider the water–gas reaction

$$CO_2 + H_2 \Leftrightarrow CO + H_2O \qquad (12.73)$$

By the law of mass action

$$K_{p_r} = \frac{p_{rCO}\, p_{rH_2O}}{p_{rCO_2}\, p_{rH_2}}$$

$$= \left(\frac{x_{CO}\, x_{H_2O}}{x_{CO_2}\, x_{H_2}}\right) p^{1+1-1-1} = \left(\frac{x_{CO}\, x_{H_2O}}{x_{CO_2}\, x_{H_2}}\right) = K_p \qquad (12.74)$$

In this reaction the amount of substance in the products is equal to the amount of substance in the reactants, and there is no effect of pressure in the dissociation equation. Comparing the results for the carbon monoxide, (eqn (12.1)) and the water gas reactions (eqn (12.73)), it can be seen that if a mixture of products of the reaction in eqn (12.1) was subjected to a change in pressure, the chemical composition of the mixture would change, whereas the chemical composition of the products of the water gas reaction would be the same at any pressure.

These points are returned to later in this section.

12.7.3   *EXAMPLE 1: A ONE DEGREE OF DISSOCIATION EXAMPLE*

A spark ignition engine operates on a 10% rich mixture of carbon monoxide (CO) and air. The conditions at the end of compression are 8.5 bar and 600 K, and it can be assumed that the combustion is adiabatic and at constant volume. Calculate the maximum pressure and temperature achieved if dissociation occurs.

[There are two different approaches for solving this type of problem; both of these will be outlined below. The first approach, which develops the chemical equations from the degrees of dissociation is often the easier method for hand calculations because it is usually possible to estimate the degree of dissociation with reasonable accuracy, and it can also be assumed that the degree of dissociation of the water vapour is less than that of the carbon dioxide. The second approach is more appropriate for computer programs because it enables a set of simultaneous (usually non-linear) equations to be defined.]

General considerations

The products of combustion with dissociation have to obey all of the laws which define the conditions of the products of combustion without dissociation, described in Chapter 10, plus the ratios of constituents defined by the equilibrium constant, K_p. This means that the problem becomes one with two iterative loops: it is necessary to evaluate the degree of dissociation from the chemical equation and the equilibrium constant, and to then ensure that this obeys the energy equation (i.e. the First Law of Thermodynamics). The following examples show how this can be achieved. At this stage the value of K_p at different temperatures will simply be stated. The method of calculation K_p is described in section 12.8, and K_p values are listed in Table 12.3.

Solution

Note that there is only one reaction involved in this problem, and that the carbon monoxide converts to carbon dioxide in a single step. Most combustion processes have more than one chemical reaction.

Stoichiometric combustion equation:

$$\underbrace{CO + \frac{1}{2}(O_2 + 3.76\,N_2)}_{n_R = 3.38} \Rightarrow \underbrace{CO_2 + 1.88\,N_2}_{n_P = 2.88} \tag{12.75}$$

Rich combustion equation:

$$\underbrace{1.1CO + \frac{1}{2}(O_2 + 3.76\,N_2)}_{n_R = 3.48} \Rightarrow \underbrace{CO_2 + 0.1CO + 1.88\,N_2}_{n_P = 2.98} \tag{12.76}$$

Dissociation of CO_2 (eqn (12.29)):

$$\alpha CO_2 \Leftrightarrow \alpha CO + \frac{\alpha}{2} O_2$$

Total combustion equation with dissociation

$$1.1CO + \underbrace{\frac{1}{2}(O_2 + 3.76 N_2)}_{n_R = 3.48} \rightarrow \underbrace{(1-\alpha)CO_2 + (0.1+\alpha)CO + \frac{\alpha}{2}O_2 + 1.88 N_2}_{n_P = 2.98 + \alpha/2}$$

$$(12.77)$$

The equilibrium constant, K_{p_r}, is given by eqn (12.70):

$$K_{p_r} = \frac{(p/p_0)_{CO_2}}{(p/p_0)_{CO}\,(p/p_0)_{O_2}^{1/2}}$$

The values of partial pressures in the products are

$$\frac{p_{CO_2}}{p_0} = x_{CO_2}\frac{p_2}{p_0} = \frac{1-\alpha}{n_P}\frac{p_2}{p_0};\qquad \frac{p_{CO}}{p_0} = x_{CO}\frac{p_2}{p_0} = \frac{0.1-\alpha}{n_P}\frac{p_2}{p_0};$$

$$\frac{p_{O_2}}{p_0} = x_{O_2}\frac{p_2}{p_0} = \frac{\alpha}{2n_P}\frac{p_2}{p_0} \tag{12.78}$$

Also, from the ideal gas law,

$$p_2 V_2 = n_2 \Re T_2, \text{ and } p_1 V_1 = n_1 \Re T_1 \tag{12.79}$$

Thus

$$\frac{n_P}{p_2} = \frac{n_1 T_1}{p_1 T_2} = \frac{245.65}{T_2} \tag{12.80}$$

Hence, substituting these values in eqn (12.70) gives

$$K_{p_r} = \frac{(1-\alpha)}{(0.1+\alpha)}\left(\frac{2n_P\,p_0}{\alpha\,p_2}\right)^{1/2} \Rightarrow K_{p_r}^2 = \frac{(1-\alpha)^2}{(0.1+\alpha)^2}\left(\frac{2\times245.65}{\alpha T_2}p_0\right) \tag{12.81}$$

This is an implicit equation in the product's temperature, T_2, because $K_{p_r} = f(T_2)$. Writing $K_{p_r}^2$ as X gives

$$X(0.1+\alpha)^2 \alpha T_2 = (1-\alpha)^2 \times 2 \times 245.65 p_0 = (1-\alpha)^2 \times 2 \times 245.65 \times 1.013\,25$$

$$(12.82)$$

Expanding eqn (12.82) gives

$$0 = 497.77 - \alpha(995.54 + 0.01 X T_2) + \alpha^2(497.77 - 0.2 X T_2) - \alpha^3 X T_2 \tag{12.83}$$

The term XT_2 can be evaluated for various temperatures because $XT_2 = K_{pr}^2 T_2$:

$T_2(K)$	K_{pr}	XT_2
2800	6.582	121 303
2900	4.392	55 940
3000	3.013	27 234

Solving for α at each temperature gives

T_P	α
2800	0.090625
2900	0.1218125
3000	0.171875

These values of α all obey the chemical, i.e. dissociation, equation but they do not all obey the energy equation. It is necessary to consider the energy equation now to check which value of T_2 balances an equation of the form

$$U_P(T_P) = -(Q_v)_s + [U_R(T_R) - U_R(T_s)] + U_P(T_s)$$

This energy equation, based on the internal energy of reaction at $T = 0$, may be rewritten

$$0 = -\Delta U_0 - U_P(T_P) + U_R(T_R) \tag{12.84}$$

Now $U_R(T_R)$ is constant and is given by

Constituent	CO	O_2	N_2
u_{600}	12 626.2	12 939.6	12 571.7
n	1.1	0.5	1.88

$$U_R(T_R) = 43\ 993.4\ \text{kJ}$$

Hence,

$$U_P(T_P) = (1 - \alpha) \times 276\ 960 + 43\ 993\ \text{kJ} \tag{12.85}$$

It can be seen that the value of U_P is a function of α; the reason for this is because the combustion of the fuel (CO) is not complete when dissociation occurs. In a simple, single degree of freedom reaction like this, the reduction in energy released is directly related to the progress of the reaction:

T_P	α	$U_P(T_P)$
2800	0.090625	295 853
2900	0.1218125	287 216
3000	0.171875	273 351

Evaluating the energy which is contained in the products at 2900 K and 3000 K, allowing for the variation in α as the temperature changes, gives

$T_P = 2900$ K

Constituent	CO_2	CO	O_2	N_2
u_{2900}	131 933.3	74 388.9	78 706.6	73 597.5
n	0.871875	0.2281	0.0641	1.88

$$U_P(T_P) = 275\ 405 \text{ kJ}$$

$T_P = 3000$ K

Constituent	CO_2	CO	O_2	N_2
u_{3000}	137 320	77 277	81 863	76 468
n	0.8281	0.2719	0.085 94	1.88

$$U_P(T_P) = 285\ 521 \text{ kJ}$$

These values are plotted in Fig 12.3, and it can be seen that, if the variation of the energy terms was linear with temperature, the temperature of the products after dissociation would be 2949 K. The calculation will be repeated to show how well this result satisfies both the energy and dissociation equations.

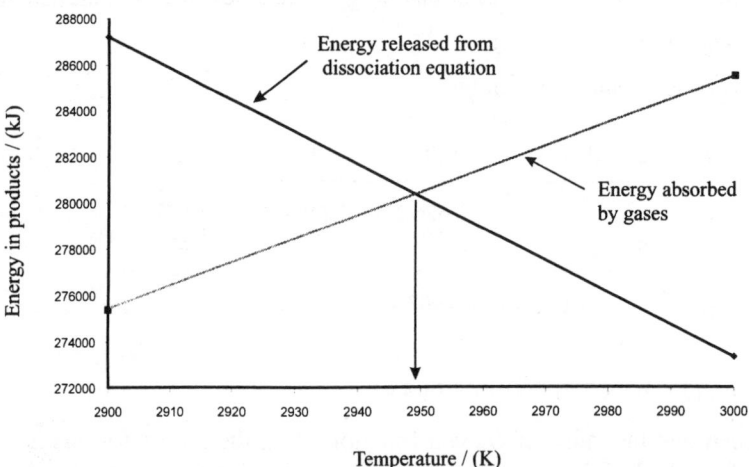

Fig. 12.3 Energy contained in products based on eqn (12.80) and the tables of energies

First, it is necessary to evaluate the degree of dissociation that will occur at this product temperature. At $T_P = 2950$ K, $K_{pr} = 3.62613$ (see Table 12.3), and this can be substituted into eqn (12.78) to give $a = 0.1494$.

Hence, the chemical equation, taking account of dissociation, is

$$1.1CO + \frac{1}{2}(O_2 + 3.76\,N_2) \Rightarrow 0.8506\,CO_2 + 0.2494\,CO + 0.0747\,O_2 + 1.88\,N_2$$

Applying the energy equation: the energy released by the combustion process gives a product energy of

$$U_P(T_P) = (1 - a) \times 276\ 960 + 43\ 993 = 279\ 575\ \text{kJ}$$

This energy is contained in the products as shown in the following table:

Constituent	CO_2	CO	O_2	N_2
u_{2949}	135 182	75 820	80 205	75 040
n	0.8506	0.2494	0.0747	1.88

$$U_P(T_P) = 280\ 961\ \text{kJ}$$

The equations are balanced to within 0.5%, and this is close enough for this example.

Alternative method

In this approach the degree of dissociation, a, will not be introduced explicitly. The chemical equation which was written in terms of a in eqn (12.77) can be written as

$$\underbrace{1.1CO + \frac{1}{2}(O_2 + 3.76\,N_2)}_{n_R = 3.48} \rightarrow \underbrace{aCO_2 + bCO + cO_2 + d\,N_2}_{n_P = a + b + c + d} \tag{12.86}$$

Considering the atomic balances,

Carbon:	$1.1 = a + b$	giving $b = 1.1 - a$	(12.87a)
Oxygen:	$2.1 = 2a + b + 2c$	giving $c = 0.5(1 - a)$	(12.87b)
Nitrogen:	$1.88 = d$		(12.87c)

Total amount of substance in the products,

$$n_P = a + b + c + d \tag{12.88}$$

The ratios of the amounts of substance in the equilibrium products are defined by the equilibrium constant

$$K_{p_r} = \frac{(p/p_0)_{CO_2}}{(p/p_0)_{CO}\,(p/p_0)_{O_2}^{1/2}} = \frac{an_P^{1/2}\,p_0^{1/2}}{bc^{1/2}p_P^{1/2}} \tag{12.89}$$

Hence

$$b^2 c\,\frac{p_P}{p_0\,n_P}\,K_{p_r}^2 = a^2 \tag{12.90}$$

and, from the atomic balances, this can be written in terms of a as

$$(1.1 - a)^2\,\frac{1}{2}(1 - a)\,\frac{p_P}{p_0\,n_P}\,K_{p_r}^2 = a^2 \tag{12.91}$$

The previous calculations showed that the temperature of the products which satisfies the governing equations is $T_P = 2950$ K, which gives a value of $K_{p_r}^2 = 13.1488$. These values will be used to demonstrate this example. Then

$$\{1.21 - 2.2\,a + a^2 - (1.21a + 2.2\,a^2 + a^3)\}\,\frac{p_P}{p_0\,n_P} \times \frac{13.1488}{2} = a^2 \tag{12.92}$$

It is possible to evaluate the ratio $p_P/p_0 n_P$ from the perfect gas relationship, giving

$$\frac{p_P}{p_0\,n_P} = \frac{p_P\,T_P}{n_R\,T_R} = \frac{8.5 \times 2950}{3.48 \times 600} = 12.009$$

This enables a cubic equation in a to be obtained:

$$95.532 - 269.23a + 251.65a^2 - 78.953a^3 = 0 \tag{12.93}$$

The solution to this equation is $a = 0.8506$, and hence the chemical equation becomes

$$1.1CO + \frac{1}{2}(O_2 + 3.76\,N_2) \rightarrow 0.8506\,CO_2 + 0.2494\,CO + 0.0747\,O_2 + 1.88\,N_2$$
$$\tag{12.94}$$

which is the same as that obtained previously. The advantage of this approach is that it is possible to derive a set of simultaneous equations which define the equilibrium state, and these can be easily solved by a computer program. The disadvantage is that it is not possible to use the intuition that most engineers can adopt to simplify the solution technique. It must be recognised that the full range of iteration was not used in this demonstration, and the solutions obtained from the original method were simply used for the first 'iteration'.

12.8 The Van't Hoff relationship between equilibrium constant and heat of reaction

It has been shown that (eqn (12.59))

$$\ln K_{p_r} = \frac{1}{\Re T}\,(\nu_a\,\mu_a^0 + \nu_b\,\mu_b^0 - \nu_c\,\mu_c^0 - \nu_d\,\mu_d^0) \tag{12.95}$$

Thus

$$\frac{d}{dT}\ln K_{p_r} = \frac{d}{dT}\left\{\frac{1}{\Re T}\,(\nu_a\,\mu_a^0 + \nu_b\,\mu_b^0 - \nu_c\,\mu_c^0 - \nu_d\,\mu_d^0)\right\}$$
$$= \frac{d}{dT}\left\{\frac{1}{\Re T}\,([\nu\mu^0]_R - [\nu\mu^0]_p)\right\} \tag{12.96}$$

Consider the definition of μ^0,

$$\mu^0 = h_0 + h(T) - T\{s_0 + s(T)\}$$
$$= h_0 + h(T) - T s_0 - T\int\frac{dh}{T} \tag{12.97}$$

Consider the terms

$$h(T) - T\int\frac{dh}{T} = T\left\{\frac{h(T)}{T} - \int\frac{dh}{T}\right\} \tag{12.98}$$

Let

$$v = h(T), \quad dv = dg$$

$$u = \frac{1}{T}, \quad du = -\frac{dT}{T^2}$$

Integrating by parts gives

$$T\left\{\frac{h(T)}{T} - \int \frac{dh(T)}{T}\right\} = -T\int \frac{h(T)dT}{T^2} \tag{12.99}$$

Thus

$$\frac{d}{dT} \ln K_{p_r} = -\sum \frac{v}{\Re} \frac{h(T)}{T^2} = -\sum \frac{vh(T)}{\Re T^2} \tag{12.100}$$

Although $\sum$ has been used as a shorthand form it does include both +ve and −ve signs; these must be taken into account when evaluating the significance of the term. Thus

$$\frac{d}{dT}(\ln K_{p_r}) = -\frac{1}{\Re T^2}(v_a h_a + v_b h_b - v_c h_c - v_d h_d)$$

$$= -\frac{1}{\Re T^2}([vh]_R - [vh]_P) \tag{12.101}$$

But, by definition

$$Q_p = (v_c h_c + v_d h_d) - (v_a h_a + v_b h_b) \tag{12.102}$$

hence

$$\frac{d}{dT}(\ln K_{p_r}) = \frac{Q_p}{\Re T^2} \tag{12.103}$$

Equation (12.103) is known as the *Van't Hoff equation*. It is useful for evaluating the heat of reaction for any particular reaction because

$$Q_p = -\frac{\Re d(\ln K_{p_r})}{d(1/T)}$$

The value of $d(\ln K_p)/d(1/T)$ may be obtained by plotting a graph of $\ln K_p$ against $1/T$.

The values of K_p have been calculated using eqn (12.95), and are listed in Table 12.3, at the end of the chapter, for four reactions. (The values have also been depicted as a graph in Fig 12.6, and it can be seen that over a small range of temperature, $\ln K_p \approx A - b/T$, where T is the temperature in K: this is to be expected from eqn (12.95)).

12.9 The effect of pressure and temperature on degree of dissociation

12.9.1 *The effect of pressure*

The effect of pressure on the degree of dissociation is defined by eqn (12.62), namely,

$$K_{p_r} = \left\{\frac{x_c^{v_c} x_d^{v_d}}{x_a^{v_a} x_b^{v_b}}\right\} p_r^{(v_c + v_d - v_a - v_b)}$$

It can be seen that the ratio of the amounts of substance is given by

$$\frac{x_a^{v_a} x_b^{v_b}}{x_c^{v_c} x_d^{v_d}} = \frac{p_r^{(v_c + v_d - v_a - v_b)}}{K_{p_r}} \tag{12.104}$$

This can be interpreted in the following way. If $v_c + v_d - v_a - v_b = 0$ then the mole fractions of the products will not be a function of pressure. However, if $v_c + v_d - v_a - v_b > 0$ then the species on the left-hand side of the chemical equation (i.e. the reactants) increase, whereas if $v_c + v_d - v_a - v_b < 0$ then the species on the right-hand side of the equation (i.e. the products) increase. The basic rule is that the effect of increasing the pressure is to shift the equilibrium to reduce the total amount of substance.

Considering the two reactions introduced previously. The equation for the carbon monoxide reaction (12.27) was

$$CO + \frac{1}{2} O_2 \Leftrightarrow (1 - \alpha)CO_2 + \alpha CO + \frac{\alpha}{2} O_2$$

and the equilibrium equation (12.72) was

$$K_{p_r} = \frac{1 - \alpha}{1 + \alpha/2} \frac{1 + \alpha/2}{\alpha} \sqrt{\frac{1 + \alpha/2}{\alpha/2}} \frac{1}{\sqrt{p_r}} = \frac{1 - \alpha}{\alpha} \sqrt{\frac{1 + \alpha/2}{\alpha/2}} \frac{1}{\sqrt{p_r}}$$

This means that $v_c + v_d - v_a - v_b < 0$, and this would result in the constituents on the 'products' side of the equation increasing. This is in agreement with the previous statement because the total amounts of reactants in eqn (12.27) is 1.5, whilst the total amounts of products is $1 + \alpha/2$, where α is less than 1.0.

A similar calculation for the combustion and dissociation of a stoichiometric methane (CH_4) and air mixture, performed using a computer program entitled EQUIL2 gave the results in Table 12.1. The chemical equation for this reaction is

$$CH_4 + 2(O_2 + 3.76N_2)$$
$$\Rightarrow (1 - \alpha_1)CO_2 + \alpha_1 CO + 2(1 - \alpha_2)H_2O + 2\alpha_2 H_2 + (\alpha_1/2 + \alpha_2)O_2 + 7.52N_2$$

$$(12.105)$$

where α_1 is the degree of dissociation of the CO_2 reaction and α_2 is the degree of dissociation of the H_2O reaction. It can be seen that dissociation tends to increase the

Table 12.1 Amount of products for constant pressure combustion of methane in air. Initial temperature = 1000 K; equivalence ratio = 1.00

Pressure (bar)	No dissociation	1	10	100
Amount of CO_2	1	0.6823	0.7829	0.8665
Amount of CO	0	0.3170	0.2170	0.1335
Amount of H_2O	2	1.8654	1.920	1.9558
Amount of H_2	0	0.1343	0.0801	0.0444
Amount of O_2	0	0.2258	0.1485	0.0889
Amount of N_2	7.52	7.52	7.52	7.52
Total amount of substance	10.52	10.7448	10.6685	10.6091

amount of substance of products, and hence the effect of an increase in pressure should be to reduce the degree of dissociation. This effect can be seen quite clearly in Table 12.1, where the amount of substance of products is compared under four sets of conditions: no dissociation, dissociation at $p = 1$ bar, dissociation at $p = 10$ bar, and dissociation at $p = 100$ bar. The minimum total amount of substance occurs when there is no dissociation, while the maximum amount of substance occurs at the lowest pressure. Figure 12.4 shows how the degrees of dissociation for the carbon dioxide and water reactions vary with pressure.

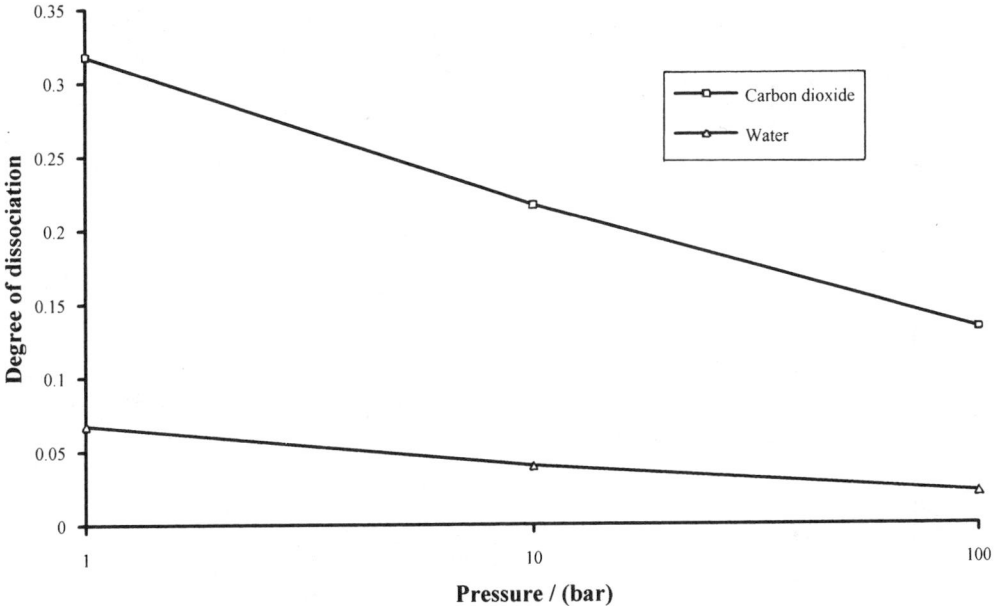

Fig. 12.4 Effect of pressure on degree of dissociation

12.9.2   THE EFFECT OF TEMPERATURE

The effect of temperature can be considered in a similar way to the effect of pressure. Basically it should be remembered that the changes in composition that take place during dissociation do so to achieve the minimum value of Gibbs energy for the mixture. The Gibbs energy of each constituent is made up of three components, the Gibbs energy at absolute zero (g_0), the Gibbs energy as a function of temperature $(g(T))$, and that related to partial pressure (see Fig 12.2). The equilibrium point is achieved when the sum of these values for all the constituents is a minimum. This means that, as the temperature rises, the constituents with the most positive heats of formation are favoured. These constituents include O_2 $(g_0 = 0)$, H_2 $(g_0 = 0)$, and CO $(g_0 = -113$ MJ/kmol$)$. Both water and carbon dioxide have larger (negative) values of g_0. This effect can be seen from the results in Table 12.2, which have been calculated for the combustion of methane in air. Another way of considering this effect is simply to study eqn (12.55), and to realise that for gases with negative heats of formation an increase in temperature leads to a *decrease* in the value of K_p. This means that the numerator of eqn (12.55) must get smaller relative to the denominator, which pushes the reaction backwards towards the reactants. This is borne out in Table 12.2: the degree of dissociation for this reaction increases with temperature, and this is shown in Fig 12.5.

Table 12.2 Amount of products for constant pressure combustion of methane in air. Pressure = 1 bar; equivalence ratio = 1

Temperature (K)	No dissociation	T = 1000 K	T = 1500 K	T = 2000 K
Amount of CO_2	1	0.6823	0.4814	0.3031
Amount of CO	0	0.3170	0.5186	0.6969
Amount of H_2O	2	1.8654	1.7320	1.5258
Amount of H_2	0	0.1343	0.2679	0.4741
Amount of O_2	0	0.2258	0.3933	0.5855
Amount of N_2	7.52	7.52	7.52	7.52
Total amount of substance	10.52	10.7448	10.9132	11.1055

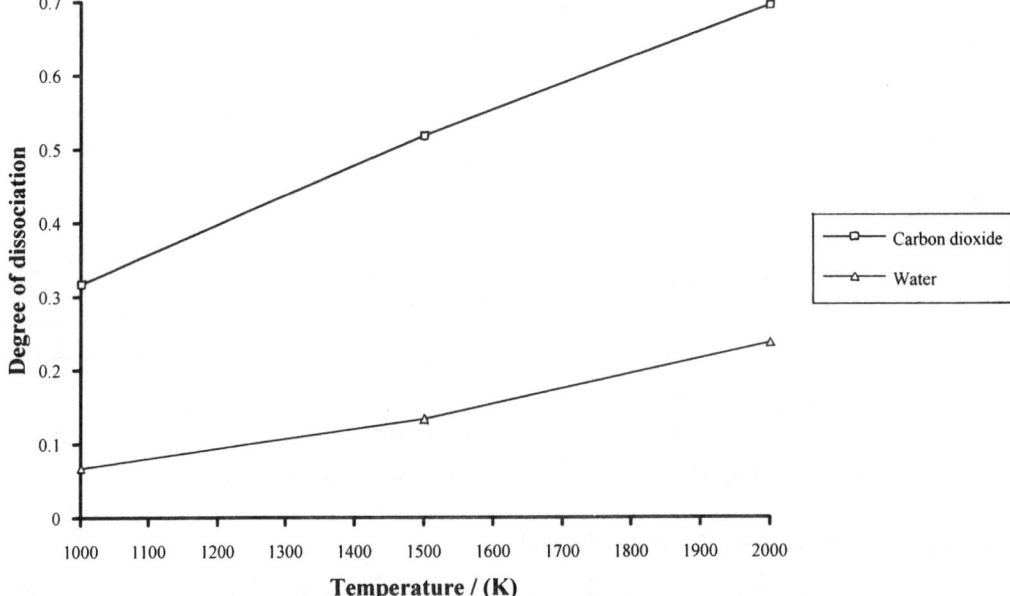

Fig. 12.5 Effect of temperature on degree of dissociation

Finally, it should be noted that in all the cases shown the degree of dissociation in the hydrogen reaction is much less than that for the carbon reaction. This supports the assumption made in previous work that the hydrogen will be favoured in the oxidation process.

12.10   Dissociation calculations for the evaluation of nitric oxide

If it is necessary to evaluate the formation of nitric oxide (NO) in a combustion chamber then the equations have to be extended to include many more species. While it is possible to add the calculation of NO to a simple dissociation problem, as is done in Example 5 below, this does not result in an accurate estimate of the quantity of NO formed. The reason for this is that NO is formed by a chain of reactions that are more complex than simply

$$N_2 + O_2 \Leftrightarrow 2NO$$

This chain of reactions has to include the formation of atomic oxygen and nitrogen, and also the OH radical. To obtain an accurate prediction of the NO concentration a total of 12 species has be considered:

[1] H_2O,	[2] H_2,	[3] OH,	[4] H,	[5] N_2,	[6] NO,
[7] N,	[8] CO_2,	[9] CO,	[10] O_2,	[11] O,	[12] Ar.

The numbers in [ ] will be used to identify the species in later equations.

The formation and breakdown of these species are defined by the following set of equations:

$$H_2 \Leftrightarrow 2H \qquad K_{p_1} = (x_4/\sqrt{x_2})\sqrt{p}$$
$$O_2 \Leftrightarrow 2O \qquad K_{p_2} = (x_{11}/\sqrt{x_{10}})\sqrt{p}$$
$$N_2 \Leftrightarrow 2N \qquad K_{p_3} = (x_7/\sqrt{x_5})\sqrt{p}$$
$$2H_2O \Leftrightarrow 2H_2 + O_2 \qquad K_{p_4} = (x_{10}/b^2)p \qquad\qquad (12.106)$$
$$2H_2O \Leftrightarrow 2OH + H_2 \qquad K_{p_5} = (x_3/(b\sqrt{x_2}))\sqrt{p}$$
$$CO_2 + H_2 \Leftrightarrow H_2O + CO \qquad K_{p_6} = bx_9/x_8$$
$$2H_2O + N_2 \Leftrightarrow 2H_2 + 2NO \qquad K_{p_7} = (x_6/(b\sqrt{x_5}))\sqrt{p}$$

where $b = x_1/x_2$.

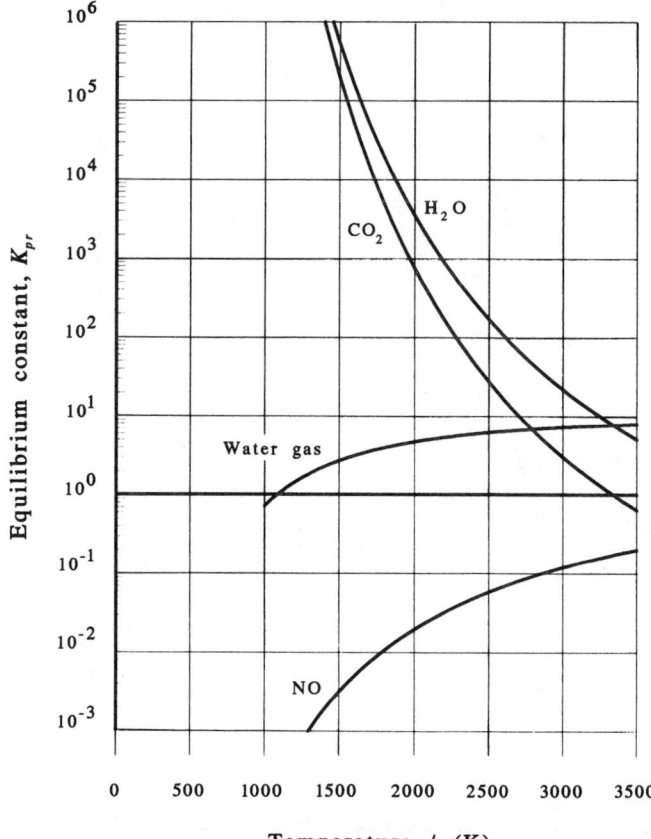

Fig. 12.6 Variation of equilibrium constant, K_{p}, at a standard pressure, $p_0 = 1$ bar, with temperature for the reactions $CO + \frac{1}{2}O_2 \Leftrightarrow CO_2$, $H_2 + \frac{1}{2}O_2 \Leftrightarrow H_2O$, $CO_2 + H_2 \Leftrightarrow CO + H_2O$, $\frac{1}{2}N_2 + \frac{1}{2}O_2 \Leftrightarrow NO$.

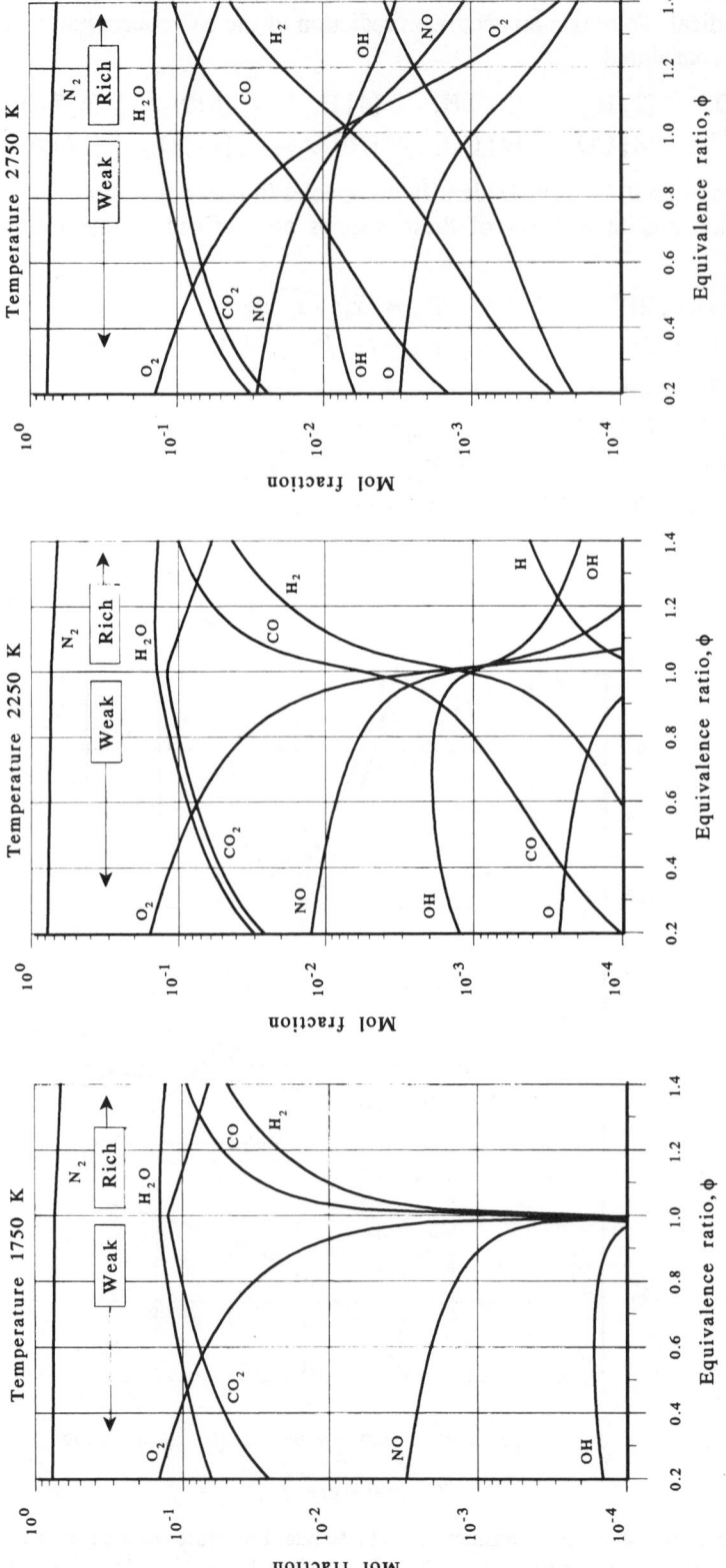

Fig. 12.7 Effect of dissociation on the products of combustion of octane and air at a pressure of 30 bar for three different temperatures

A numerical method for solving these equations is given in Horlock and Winterbone (1986), based on the original paper by Lavoie *et al.* (1970).

Figure 12.7 shows the results of performing such calculations using a simple computer program. The coefficients in the program were evaluated using the data presented in Table 9.3, but with the addition of data for OH and N. The three diagrams are based on combustion of octane in air at a constant pressure of 30 bar, and show the effect of varying equivalence ratio, ϕ, at different temperatures. The conditions are the same as quoted in Heywood (1988). Figure 12.7(a) shows the results for the lowest temperature of 1750 K, and it can be seen that the graph contains no atomic nitrogen (N) or atomic oxygen (O) because these are at very low concentrations ($<10^{-4}$). When the mixture is weak some OH is produced. It is apparent from this diagram that there is not much dissociation of the carbon dioxide or water because the concentration of oxygen drops to very low values above $\phi = 1$. Some nitric oxide is formed in the weak region ($x_{NO} = 3 \times 10^{-3}$ at the weakest mixture), but this rapidly reduces to a very low value at stoichiometric simply because there is no oxygen available to combine with the nitrogen. Figure 12.7(b) shows similar graphs for a temperature of 2250 K, and it can be seen that the NO level has increased by almost a factor of 10. There is also some NO formed by combustion with rich mixtures; this is because, by this temperature, the carbon dioxide and water are dissociating, as is indicated by the increase in the CO and the OH radical in the weak mixture region. By the time 2750 K is reached, shown in Fig 12.7(c), there is a further tripling of the NO production, and CO is prevalent throughout the weak mixture zone. There are also significant amounts of OH and oxygen over the whole range of equivalence ratio.

The diagrams shown in Fig 12.7 are relatively contrived because they do not depict real combustion situations. However, they do allow the parameters which control the production of the products of combustion to be decoupled, to show the effects of changing the parameters independently of each other. The equilibrium concentrations depicted in Fig 12.7 are the values which drive the formation of the exhaust constituents through the chemical kinetics equations, which will be discussed in Chapter 14.

12.11 Dissociation problems with two, or more, degrees of dissociation

The previous example, which considered the dissociation of carbon monoxide, shows the fundamental techniques involved in calculating dissociation but is an unrealistic example because rarely are single component fuels burned. Even if a single component fuel such as hydrogen was burned in an engine, it would be necessary to consider other 'dissociation' reactions because it is likely that nitric oxide (NO) will be formed from the combination of the oxygen and nitrogen in the combustion chamber. These more complex examples will be considered here.

12.11.1   EXAMPLE 4: COMBUSTION OF A TYPICAL HYDROCARBON FUEL

A weak mixture of octane (C_8H_{18}) and air, with an equivalence ratio of 0.9, is ignited at 10 bar and 500 K and burns at constant volume. Assuming the combustion is adiabatic, calculate the conditions at the end of combustion allowing for dissociation of the carbon dioxide and water, but neglecting any formation of NO.

Solution

Stoichiometric equation neglecting dissociation

$$\underbrace{C_8H_{18} + 12.5(O_2 + 3.76\,N_2)}_{n_R = 60.5} \Rightarrow \underbrace{8CO_2 + 9H_2O + 47\,N_2}_{n_P = 64} \tag{12.107}$$

Weak mixture neglecting dissociation

$$\underbrace{C_8H_{18} + 13.889(O_2 + 3.76\,N_2)}_{n_R = 67.111} \Rightarrow \underbrace{8CO_2 + 9H_2O + 1.3890\,O_2 + 52.222\,N_2}_{n_P = 70.611}$$

$$\tag{12.108}$$

Dissociation equations (12.25 and 12.34):

$$a_1 CO_2 \Leftrightarrow a_1 CO + \frac{a_1}{2} O_2$$

$$a_2 H_2O \Leftrightarrow a_2 H_2 + \frac{a_2}{2} O_2$$

Weak mixture including dissociation is

$$\underbrace{C_8H_{18} + 13.889(O_2 + 3.76\,N_2)}_{n_R = 67.111}$$

$$\Rightarrow 8(1 - a_1)CO_2 + 9(1 - a_2)H_2O + 8a_1 CO$$

$$\underbrace{+ 9a_2 H_2 + \left(1.389 + \frac{8}{2} a_1 + \frac{9}{2} a_2\right)O_2 + 52.222\,N_2}_{n_P = 70.611 + 4a_1 + 4.5a_2} \tag{12.109}$$

The temperature and pressure calculated for combustion without dissociation are 2874 K and 60.481 bar respectively. Dissociation will lower the final temperature: assume that $T_P = 2800$ K.

The equilibrium constants from Table 12.3 at 2800 K are

$$K_{P_{r1}} = \frac{p_{r CO_2}}{p_{r CO}\, p_{r O_2}^{1/2}} = 6.58152$$

$$\tag{12.110}$$

$$K_{P_{r2}} = \frac{p_{r CO}\, p_{r H_2O}}{p_{r CO_2}\, p_{r H_2}} = 6.8295$$

From eqn (12.109), the partial pressures of the constituents are

$$p_{r CO_2} = \frac{8(1 - a_1)}{n_P} \frac{p}{p_0}$$

$$p_{r CO} = \frac{8a_1}{n_P} \frac{p}{p_0}$$

$$p_{r O_2} = \frac{(1.389 + 4a_1 + 4.5a_2)}{n_P} \frac{p}{p_0}$$

$$p_{r H_2O} = \frac{9(1 - a_2)}{n_P} \frac{p}{p_0}$$

$$p_{r H_2} = \frac{9a_2}{n_P} \frac{p}{p_0}$$

Hence

$$K^2_{P_{r1}} = 43.3164 = \frac{[8(1-\alpha_1)]^2}{[8\alpha_1]^2} \frac{n_P}{(1.389+4\alpha_1+4.5\alpha_2)} \frac{p_0}{p} \tag{12.111}$$

and

$$K_{P_{r2}} = \frac{9(1-\alpha_2)}{9\alpha_2} \frac{8\alpha_1}{8(1-\alpha_1)} = \frac{(1-\alpha_2)}{\alpha_2} \frac{\alpha}{(1-\alpha_1)} \tag{12.112}$$

The K_{p_r} equation (eqn (12.111)) contains a pressure term, but this can be replaced by the temperature of the products by applying the perfect gas law to the mixture. Then

$$p_P V_P = n_P \Re T_P$$
$$p_R V_R = n_R \Re T_R \tag{12.113}$$

Thus

$$\frac{n_P}{p_P} = \frac{n_R \Re T_R}{p_R \Re T_P} = \frac{n_R T_R}{p_R T_P} \tag{12.114}$$

Substituting gives

$$K^2_{P_{r1}} = 43.3164 = \frac{(1-\alpha_1)^2}{\alpha_1^2} \frac{n_R T_R}{(1.389+4\alpha_1+4.5\alpha_2)p_R T_P} p_0 \tag{12.115}$$

Inserting values for these parameters gives

$$\frac{43.3164 \times 10 \times 2800}{67.111 \times 500} = \frac{(1-\alpha_1)^2}{\alpha_1^2(1.389+4\alpha_1+4.5\alpha_2)} \tag{12.116}$$

which can be expanded to give

$$144.58\alpha_1^3 + \alpha_1^2(49.2052+162.652\alpha_2) + 2\alpha_1 - 1 = 0 \tag{12.117}$$

This equation contains both α_1 and α_2. If it is assumed that $\alpha_1 \gg \alpha_2$ then eqn (12.117) is a cubic equation in α_1. Based on this assumption the value of α_1 is approximately 0.11. Substituting this value in eqn (12.112) gives $\alpha_2 = 0.017776$, which vindicates the original assumption. Recalculation around this loop gives

$$\alpha_1 = 0.108$$
$$\alpha_2 = 0.01742$$

These values are close enough solutions for the dissociation coefficients, and for the chemical equation (eqn (12.109)) which becomes

$$\underbrace{C_8H_{18} + 13.889(O_2 + 3.76N_2)}_{n_R = 67.111}$$

$$\Rightarrow \underbrace{7.136CO_2 + 8.84322H_2O + 0.846CO + 0.15678H_2 + 1.8994O_2 + 52.222N_2}_{n_P = 71.121}$$

$$\tag{12.118}$$

This equation satisfies the equilibrium constraints and the chemistry of the problem, *but it still has to be checked to see if it meets the First Law*. The First Law is defined by the equation

$$U_P(T_P) - U_P(T_s) = -(Q_v)_s + [U_R(T_R) - U_R(T_s)] \tag{12.119}$$

Note two things about this equation. First, that it is more convenient to evaluate the difference between the internal energies of the products, because now the composition of the products is also a function of temperature and hence $U_P(T_s)$ is not a constant. Second, that with dissociation, $(Q_v)_s$ is not the full value of the internal energy of reaction of octane because not all the octane has been oxidised fully to CO_2 and water. The value of $(Q_v)_s$ in this case is given by

$$(\tilde{Q}_v)_s = (Q_v)_s - 0.108 \times 8 \times (-282\ 993) - 0.01742 \times 9 \times (-241\ 827)$$
$$= -5\ 116\ 320 + 244\ 505.9 + 37\ 913.6$$
$$= -4\ 833\ 900.5\ \text{kJ/kmol octane} \tag{12.120}$$

Alternative method for calculating the energy released by combustion

The energy released from partial combustion can be calculated using Hess' law. In this case

$$\Delta \tilde{U}_0 = \sum_{\text{Products}} \Delta U_0 - \sum_{\text{Reactants}} \Delta U_0$$

$$= 7.136(\Delta U_0)_{CO_2} + 8.843\ 22(\Delta U_0)_{H_2O} + 0.846(\Delta U_0)_{CO} - (\Delta U_0)_{C_8H_{18}}$$
$$= 7.136 \times (-393\ 405) + 8.84322 \times (-239\ 082)$$
$$+ 0.846 \times (-113\ 882) - (-74\ 897)^*$$
$$= -4\ 943\ 040\ \text{kJ/kmol C}_8\text{H}_{18}\ \text{burned} \tag{12.121}$$

This value can be related to the energy released at the standard temperature of 25°C by the following equation (*note, the fuel properties have been based on methane):

$$-(\tilde{Q}_v)_s = -\Delta \tilde{U}_0 + [U_P(T_s) - U_P(T_0)] - [U_R(T_s) - U_R(T_0)]$$
$$= -4\ 943\ 040 + 7.136 \times 7041.6 + 8.84322 \times 7223.3 + 0.846 \times 6011.8$$
$$+ 0.15678 \times 6043.1 + 1.8994 \times 6032.0 + 52.222 \times 6025.8$$
$$- (5779 + 13.889 \times [6032.0 + 3.76 \times 6025.8])$$
$$= -4\ 895\ 205\ \text{kJ/kmol C}_8\text{H}_{18}\ \text{burned} \tag{12.122}$$

This is within 1.25% of that calculated from the internal energy of reaction, in eqn (12.120), and is probably because of the approximation made for the properties of octane.
Evaluating the energy terms for the reactants gives

Constituent	C_8H_{18}	O_2	N_2
u_{500}	45 783	10 574.8	10 356.1
u_{298}	0	6032.0	6025.8
Difference	45 783	4542.8	4330.3
n	1.0	13.889	52.222

$$[U_R(T_R) - U_R(T_s)] = 335\ 014\ \text{kJ}$$

Hence, the energy of the products, based on the evaluated degrees of dissociation, must be

$$U_P(T_P) - U_P(T_s) = -(\tilde{Q}_v)_s + [U_R(T_R) - U_R(T_s)]$$

$$= 4\ 833\ 716.2 + 335\ 014 \tag{12.123}$$

$$= 516\ 8731\ kJ$$

Evaluating the energy contained in the products at 2800 K gives

Constituent	CO_2	H_2O	CO	H_2	O_2	N_2
u_{2800}	127 154.9	100 470.8	71 512.4	66 383.6	75 548.1	70 754.1
u_{298}	7041.6	7223.3	6011.8	6043.1	6032.0	6025.8
Difference	120 113.3	93 247.5	65 500.5	60 340.5	69 516.1	64 728.4
n	7.1360	8.84322	0.864	0.15678	1.8994	52.222

$$U_P(T_P) - U_P(T_s) = 5\ 260\ 072.5\ kJ$$

This shows that the products at 2800 K contain more energy than was available from the energy released by the fuel and hence the energy equation has not been satisfied. It is necessary to repeat the whole calculation with a lower temperature guessed for the products. Rather than do a number of iterations, the value obtained from a computer program will be used immediately, and it will be shown that this gives good agreement in the energy equation: $T_P = 2772$ K will be used.

At $T_P = 2772$ K the values of equilibrium constant can be obtained from the tables by linear interpolation:

$$K_{P_{r1}} = 8.14828 + \frac{22}{50} \times (6.58152 - 8.14828) = 7.4589$$

$$\tag{12.124}$$

$$K_{P_{r2}} = 6.72829 + \frac{22}{50} \times (6.82925 - 6.72829) = 6.7727$$

Substituting these values gives

$$\frac{55.6352 \times 10 \times 2772}{67.111 \times 500} = \frac{(1 - \alpha_1)^2}{\alpha_1^2(1.389 + 4\alpha_1 + 4.5\alpha_2)} \tag{12.125}$$

and hence

$$183.84\alpha_1^3 + \alpha_1^2(62.838 + 206.82\alpha_2) + 2\alpha_1 - 1 = 0 \tag{12.126}$$

Assuming $\alpha_2 = 0$ for the initial iteration gives $\alpha_1 = 0.099$. Recalculation around the loop results in the values

$$\alpha_1 = 0.097$$

$$\alpha_2 = 0.0156 \tag{12.127}$$

which give a chemical equation of

$$\underbrace{C_8H_{18} + 13.889(O_2 + 3.76\,N_2)}_{n_R = 67.111}$$

$$\Rightarrow \underbrace{7.224CO_2 + 8.8596\,H_2O + 0.776\,CO + 0.1404\,H_2 + 1.8472\,O_2 + 52.222\,N_2}_{n_P = 71.061}$$

$$\tag{12.128}$$

The energy released by the fuel becomes

$$(\tilde{Q}_v)_s = (Q_v)_s - 0.097 \times 8 \times (-282\ 993) - 0.0156 \times 9 \times (-241\ 827)$$
$$= -5\ 116\ 320 + 219\ 602.7 + 33\ 952.5$$
$$= -4\ 862\ 764.8 \text{ kJ/kmol octane} \tag{12.129}$$

Hence, the energy of the products, based on the evaluated degrees of dissociation, must be

$$U_P(T_P) - U_P(T_s) = -(Q_v)_s + [U_R(T_R) - U_R(T_s)]$$
$$= 4\ 862\ 764.8 + 335\ 014$$
$$= 5\ 197\ 778.8 \text{ kJ} \tag{12.130}$$

Evaluating the energy contained in the products at 2772 K gives

Constituent	CO_2	H_2O	CO	H_2	O_2	N_2
u_{2772}	125 654.9	99 196.2	70 702.8	65 594.3	74 678.0	69 948.5
u_{298}	7041.6	7223.3	6011.8	6043.1	6032.0	6025.8
Difference	118 613.3	91 972.9	64 690.8	59 551.2	68 646.0	63 922.7
n	7.2240	8.8596	0.776	0.1404	1.8472	52.222

$$U_P(T_P) - U_P(T_s) = 5\ 195\ 240 \text{ kJ}$$

These values are within 0.05%, and hence satisfy the energy equation.

Alternative method for calculating the chemical equation

The approach used above to calculate the coefficients in the chemical equation was based on the degrees of dissociation of the two reactions occurring in this example. It was shown previously that this approach is often the best for manual solution, but that a more general approach, in which a system of simultaneous equations is developed, is better for computer solution. This second method will be outlined below, for one of the steps in the previous example.

First, eqn (12.109) can be replaced by

$$\underbrace{C_8H_{18} + 13.889(O_2 + 3.76\,N_2)}_{n_R = 67.111} \rightarrow \underbrace{aCO_2 + bH_2O + cCO + d\,H_2 + eO_2 + f\,N_2}_{n_P = a+b+c+d+e+f}$$

$$\tag{12.131}$$

If the first iteration of the previous approach is used then $T_P = 2800$ K and, as given in eqn (12.110),

$$K_{P_{r1}} = \frac{p_{rCO_2}}{p_{rCO}\ p_{rO_2}^{1/2}} = 6.58152$$

$$K_{P_{r2}} = \frac{p_{rCO}\ p_{rH_2O}}{p_{rCO_2}\ p_{rH_2}} = 6.8295$$

Now, by definition,

$$K_{P_{r1}} = \frac{p_{rCO_2}}{p_{rCO}\ p_{rO_2}^{1/2}} = \frac{a}{ce^{1/2}}\ \frac{n_P^{1/2}}{p_P^{1/2}}\ p_0^{1/2} \tag{12.132}$$

which gives

$$K_{p_{r1}}^2 = \frac{a^2}{c^2 e} \frac{n_P}{p_P} p_0 \tag{12.133}$$

For the water gas reaction

$$K_{p_{r2}} = \frac{p_{rCO} \, p_{rH_2O}}{p_{rCO_2} \, p_{rH_2}} = 6.8295 = \frac{cb}{ad} \tag{12.134}$$

and the total amount of substance in the products is

$$n_P = a + b + c + d + e + f \tag{12.135}$$

Considering the atom balances gives

Carbon: $\quad 8 = a + c \qquad\qquad\qquad \Rightarrow c = 8 - a \qquad$ (12.136a)

Hydrogen: $\quad 18 = 2b + 2d \qquad\qquad \Rightarrow b = 9 - d \qquad$ (12.136b)

Oxygen: $\quad 27.778 = 2a + b + c + 2e \Rightarrow e = 13.889 - a - (b + c)/2 \qquad$ (12.136c)

Nitrogen: $\quad f = 52.223 \qquad$ (12.136d)

From the perfect gas law

$$\frac{n_P}{p_P} = \frac{n_R T_R}{p_R T_P} \tag{12.137}$$

giving eqn (12.133) as

$$K_{p_{r1}}^2 = 43.3164 = \frac{a^2}{c^2 e} \frac{n_R T_R}{p_R T_P} p_0 \tag{12.138}$$

Equation (12.138) is a member of a non-linear set of equations in the coefficients of the chemical equation. One way to solve this is to make an assumption about the relative magnitude of d and the other parameters. If it is assumed that d is small compared with the other parameters in eqn (12.131) then eqn (12.138) becomes a cubic equation in a. This is equivalent to assuming that α_2 is small compared with α_1, as was assumed in the previous calculation. If eqns (12.136a, b and c) are substituted into eqn (12.138), with the assumption that $d = 0$, then eqn (12.138) becomes

$$(8 - a)^2 (5.389 - a/2) = \frac{a^2}{K_{p_{r1}}^2} \frac{n_R T_R}{p_R T_P} p_0 = \frac{a^2}{43.3164} \frac{67.111 \times 500}{10 \times 2800} \times 1$$

$$= 0.027667 \, a^2 \tag{12.139}$$

giving

$$344.896 - 118.224a + 13.3613a^2 - a^3/2 = 0 \tag{12.140}$$

The solution to eqn (12.140) relevant to this problem is $a = 7.125$, giving $c = 8 - a = 0.875$ and $b = 9$. It is now possible to check if this result satisfies the equilibrium condition for the water gas reaction, namely eqn (12.134), which gives

$$d = \frac{cb}{aK_{p_{r2}}} = \frac{0.875 \times 9}{7.125 \times 6.89295} = 0.16184 \tag{12.141}$$

The calculation should now be performed again based on the new value of d given in eqn (12.141). This produces a new cubic equation in a, similar to that in eqn (12.140) but with the following coefficients:

$$350.07 - 119.519a + 13.442a^2 2 - a^3/2 = 0$$

The new value of a is 7.134, which is close enough to the actual solution because the value of d is compatible with this is

$$d = \frac{cb}{aK_{p_{r2}}} = \frac{0.866 \times 8.8382}{7.134 \times 6.89295} = 0.1556 \tag{12.142}$$

and it can be seen that the change in d will have a small effect on subsequent values of the coefficients. Hence, the chemical equation is

$$\underbrace{C_8H_{18} + 13.889(O_2 + 3.76\,N_2)}_{n_R = 67.111}$$

$$\Rightarrow \underbrace{7.134CO_2 + 8.8382\,H_2O + 0.866\,CO + 0.1618\,H_2 + 1.90290\,O_2 + 52.222\,N_2}_{n_P = 71.125}$$

$$\tag{12.143}$$

Equation (12.143), which has been calculated from the general equation {eqn (12.131)}, is almost the same as that calculated using the degrees of dissociation in eqn (12.109). This is to be expected because both approaches are exactly equivalent. This solution requires further iterations to satisfy the energy equation, as done in the previous approach.

12.11.2   EXAMPLE 5: A RICH MIXTURE

A rich mixture of octane (C_8H_{18}) and air, with an equivalence ratio of 1.1, is ignited at 10 bar and 500 K and burns at constant volume. Assuming the combustion is adiabatic, calculate the conditions at the end of combustion allowing for dissociation of the carbon dioxide and water, but neglecting any formation of NO.

Solution

Stoichiometric equation neglecting dissociation:

$$\underbrace{C_8H_{18} + 12.5(O_2 + 3.76\,N_2)}_{n_R = 60.5} \Rightarrow \underbrace{8CO_2 + 9H_2O + 47\,N_2}_{n_P = 64} \tag{12.144}$$

Rich mixture neglecting dissociation:

$$\underbrace{C_8H_{18} + 11.36(O_2 + 3.76\,N_2)}_{n_R = 55.09} \Rightarrow \underbrace{5.7272CO_2 + 9H_2O + 2.2728CO + 42.71\,N_2}_{n_P = 59.73}$$

$$\tag{12.145}$$

Dissociation equations (12.25 and 12.34):

$$a_1 CO_2 \Leftrightarrow a_1 CO + \frac{a_1}{2} O_2$$

$$a_2 H_2O \Leftrightarrow a_2 H_2 + \frac{a_2}{2} O_2$$

Rich mixture including dissociation is

$$C_8 H_{18} = 11.36(O_2 + 3.76 N_2)$$

$$\Rightarrow 5.7272(1 - a_1)CO_2 + 9(1 - a_2)H_2O + (2.2728 + 5.7272 a_1)CO$$

$$+ 9a_2 H_2 + \left(\frac{5.7272}{2} a_1 + \frac{9}{2} a_2 \right) O_2 + 42.71 N_2 \quad (12.146)$$

$$\underbrace{}_{n_P = 59.73 + 2.8636 a_1 + 4.5 a_2}$$

It can be seen that this is a significantly more complex equation than for the weak mixture. Application of a computer program gives the final temperature and pressure, *without dissociation*, as 2958.6 K and 64.15 bar respectively. The equilibrium constants are

$$K_{p_{r1}} = 6.581\,52 = \frac{p_{r CO_2}}{p_{r CO}\, p_{r O_2}^{1/2}}$$

$$\quad (12.147)$$

$$K_{p_{r2}} = 6.829\,25 = \frac{p_{r CO}\, p_{r H_2O}}{p_{r CO_2}\, p_{r H_2}}$$

Substituting for the mole fractions gives

$$p_{CO_2} = \frac{5.7272(1 - a_1)}{n_P} p; \quad p_{CO} = \frac{(2.2728 + 5.7272 a_1)}{n_P} p$$

$$p_{O_2} = \frac{(2.8636 a_1 + 4.5 a_2)}{n_P} p \quad (12.148)$$

Hence

$$K_{p_{r1}}^2 = 43.3164 = \frac{5.7272^2 (1 - a_1)^2 n_P}{(2.2728 + 5.7272 a_1)^2 (2.8636 a_1 + 4.5 a_2)} \frac{p_0}{p} \quad (12.149)$$

Similarly

$$p_{H_2O} = \frac{9(1 - a_2)}{n_P} p; \quad p_{H_2} = \frac{9 a_2}{n_P} p \quad (12.150)$$

and

$$K_{p_{r2}} = \frac{(2.2728 + 5.7272 a_1) \times 9 \times (1 - a_2)}{9 a_2 \times 5.7272(1 - a_1)} = \frac{1 - a_2}{a_2} \times \frac{(2.2728 + 5.7272 a_1)}{5.7272(1 - a_1)}$$

$$\quad (12.151)$$

The $K_{p,1}$ and $K_{p,2}$ equations (eqns (12.149) and (12.151)) are again non-linear equations in α_1 and α_2, and they can be solved in the same manner as before.

The solution to this problem, allowing for dissociation is a final temperature of 2891 K and a final pressure of 62.99 bar. These result in the following degrees of dissociation: $\alpha_1 = 0.0096$, $\alpha_2 = 0.05540$. It is left to the reader to prove these values are correct.

12.11.3   EXAMPLE 6: THE FORMATION OF NITRIC OXIDE

When a mixture of oxygen and nitrogen is heated above about 2000 K these two elements will join together to form nitric oxide (NO). The elements combine in this way because the energy of formation of nitric oxide is positive, i.e. the reaction which forms it is endothermic, and hence a combination of nitrogen and oxygen tends to reduce the Gibbs energy of the mixture. Note that other compounds tend to dissociate at high temperature because their energies of formation are negative, i.e. the reactions which form them are exothermic.

Example

Methane is burned with a stoichiometric quantity of air and achieves a pressure of 99.82 bar and a temperature of 2957 K after combustion, if dissociation of the carbon dioxide and water vapour are taken into account. The mole fractions of the constituents are 7.495% CO_2; 1.890% CO; 18.183% H_2O; 0.6425% H_2; 1.267% O_2; and 70.58% N_2. Calculate the amount of nitric oxide (NO) formed at this temperature, neglecting the effect of the NO formation on the dissociation of carbon monoxide and water. Estimate the effect of the NO formation on the temperature of the products. The energy of the reactants, $U_R(T_R) = 179\ 377$ kJ.

Solution:

Chemical equation without dissociation:

$$CH_4 + 2(O_2 + 3.76N_2) \Rightarrow CO_2 + 2H_2O + 7.52N_2 \tag{12.152}$$

Chemical equation with dissociation:

$$CH_4 + 2(O_2 + 3.76N_2)$$

$$\Rightarrow n_P(0.07495\,CO_2 + 0.0189\,CO + 0.18183\,H_2O + 0.006425\,H_2$$
$$+\ 0.01267\,O_2 + 0.7058\,N_2) \tag{12.153}$$

$$\overline{\qquad\qquad n_P = 10.52 + \frac{\alpha_1}{2} + \frac{\alpha_2}{2} \qquad\qquad}$$

It is possible to evaluate the total amount of substance in the products as follows:

$$0.07495 = \frac{1 - \alpha_1}{n_P}, \quad \text{and} \quad 0.0189 = \frac{\alpha_1}{n_P} \Rightarrow n_P = 10.655$$

and hence

$$CH_4 + 2(O_2 + 3.76N_2)$$

$$\Rightarrow 0.7986\,CO_2 + 0.2014\,CO + 1.9315\,H_2O + 0.06845\,H_2$$
$$+ 0.13495\,O_2 + 7.52\,N_2 \qquad (12.154)$$

$$\underbrace{\hphantom{0.7986\,CO_2 + 0.2014\,CO + 1.9315\,H_2O}}_{n_P = 10.655}$$

Assume that combination of the nitrogen and oxygen occurs to form nitric oxide (NO):

$$\alpha_3\,NO \Leftrightarrow \frac{\alpha_3}{2}\,O_2 + \frac{\alpha_3}{2}\,N_2 \qquad (12.155)$$

The temperature of the products is $T_P = 2956$ K, giving

$$K_{P_{r3}} \approx 0.11279 = \frac{p_{rNO}}{p_{rO_2}^{1/2}\,p_{rN_2}^{1/2}} \qquad (12.156)$$

Adding in the NO equation to the chemical equation gives

$$CH_4 + 2(O_2 + 3.76N_2)$$

$$\Rightarrow 0.7986\,CO_2 + 0.2014\,CO + 1.9315\,H_2O + 0.06845\,H_2$$
$$+ \alpha_3 NO - +\left(0.13495 - \frac{\alpha_3}{2}\right)O_2 + \left(7.52 - \frac{\alpha_3}{2}\right)N_2 \qquad (12.157)$$

$$\underbrace{\hphantom{0.7986\,CO_2 + 0.2014\,CO + 1.9315\,H_2O}}_{n_P = 10.655}$$

The partial pressures of the nitrogen constituents are

$$p_{NO} = \frac{\alpha_3}{n_P}\,p; \qquad p_{N_2} = \frac{(7.52 - \alpha_3/2)}{n_P}\,p; \qquad p_{O_2} = \frac{(0.13495 - \alpha_3/2)}{n_P}\,p \qquad (12.158)$$

and hence

$$\frac{\alpha_3}{(7.52 - \alpha_3/2)^{1/2}(0.13495 - \alpha_3/2)^{1/2}} = 0.11279 \qquad (12.159)$$

Squaring gives

$$1.00318\,\alpha_3^2 + 0.048690\,\alpha_3 - 0.01291 = 0 \qquad (12.160)$$

which gives

$$\alpha_3 = 0.09174$$

Hence, the chemical equation is

$$CH_4 + 2(O_2 + 3.76N_2)$$

$$\Rightarrow 0.7986\,CO_2 + 0.2014\,CO + 1.9315\,H_2O + 0.06845\,H_2$$
$$+ 0.09174\,NO + 0.08908\,O_2 + 7.474\,N_2 \qquad (12.161)$$

$$\underbrace{\hphantom{0.7986\,CO_2 + 0.2014\,CO + 1.9315\,H_2O}}_{n_P = 10.655}$$

It can be seen that the effect of the NO formation is to reduce the amount of oxygen in the products. The effect of this is to upset the equilibrium of the carbon dioxide and water reactions, and the values of α_1 and α_2 will be changed. This will be considered later. First, the effect of the formation on the energy equation will be considered. This effect has already been shown in Fig 12.7, but a more complex set of reactions was used. It is not possible to demonstrate the more complex calculation in a hand calculation.

The energy released by the original reaction, i.e. neglecting the NO reaction, can be evaluated by considering Hess' equation:

$$\Delta \tilde{U}_0 = \sum_{\text{Products}} U_f - \sum_{\text{Reactants}} U_f$$

$$= 0.7986 \times (-393\ 405) + 0.2014 \times (-113\ 882) + 1.9315 \times (-239\ 082)$$
$$-(-66\ 930)$$
$$= -731\ 966 \text{ kJ/kmol } CH_4$$

(12.162)

The effect of the formation of NO is to reduce $\Delta \tilde{U}_0$ in the following manner:

$$\Delta \tilde{U}_0 = \sum_{\text{Products}} U_f - \sum_{\text{Reactants}} U_f$$

$$= 0.7986 \times (-393\ 405) + 0.2014 \times (-113\ 882) + 1.9315 \times (-239\ 082)$$
$$+\underbrace{0.09174 \times 89\ 915}_{\text{Formation of NO}} - (-66\ 930)$$
$$= -723\ 717 \text{ kJ/kmol } CH_4$$

(12.163)

The energy equation based on 0 K is

$$-\Delta U_0 + [U_R(T_R) - U_R(0)] - [U_P(T_P) - U_P(0)] = 0 \tag{12.164}$$

giving

$$U_P(T_P) = -\Delta U_0 + U_R(T_R) \tag{12.165}$$

Applying the energy equation to the chemical equation neglecting dissociation of NO (i.e. eqn (12.154)) gives

$$U_P(T_P) = -(-731\ 927) + 179\ 377 = 911\ 304 \text{ kJ} \tag{12.166}$$

Substituting the value of T_P given in the question namely 2957 K, gives

Constituent	CO_2	H_2O	CO	H_2	O_2	N_2
u_{2957}	135 615	107 643	76 050	70 837	80 457	75 270
n	0.7986	1.9315	0.2014	0.06845	0.13495	7.52

$$U_P(T_P) = 913\ 268 \text{ kJ}$$

This satisfies the energy equation within 0.2%.

Now considering the case with NO formation (i.e. in eqn (12.157)). First, the energy of the products is less than in the previous case because the formation of NO has reduced the energy released by combustion. In this case

$$U_P(T_P) = -(-723\ 678) + 179\ 377 = 903\ 055\ \text{kJ}$$

If it is assumed that the temperature is reduced by the ratio 903 055/911 304, then a first approximation of the products temperature is 2930 K. In addition, it is necessary to include NO in the products, and this is introduced in the table below:

Constituent	CO_2	H_2O	CO	H_2	O_2	NO	N_2
u_{2930}	134 153	106 407	75 271	70 068	79 609	77 304	74 494
n	0.7986	1.9315	0.2014	0.06845	0.08908	0.09174	7.474

$$U_P(T_P) = 903\ 567\ \text{kJ}$$

This value of the products' energy obviously satisfies the energy equation, and shows that the effect of the formation of NO has been to reduce the products temperature, in this case by about 30 K.

Table 12.3 Equilibrium constants

Equilibrium constant K_{P_r} / Temperature (K)	$\dfrac{(p/p_0)_{CO_2}}{(p/p_0)_{CO}(p/p_0)_{O_2}^{1/2}}$	$\dfrac{(p/p_0)_{H_2O}}{(p/p_0)_{H_2}(p/p_0)_{O_2}^{1/2}}$	$\dfrac{(p/p_0)_{CO}(p/p_0)_{H_2O}}{(p/p_0)_{CO_2}(p/p_0)_{H_2}}$	$\dfrac{(p/p_0)_{NO}}{(p/p_0)_{N_2}^{1/2}(p/p_0)_{O_2}^{1/2}}$
500	1.06909E + 25	7.90482E + 22	0.00739	1.5710E − 09
550	2.16076E + 22	3.78788E + 20	0.01753	1.1361E − 08
600	1.22648E + 20	4.38630E + 18	0.03576	5.9112E − 08
650	1.54176E + 18	1.00209E + 17	0.06500	2.3872E − 07
700	3.62178E + 16	3.90700E + 15	0.10788	7.8997E − 07
750	1.40361E + 15	2.33722E + 14	0.16651	2.2291E − 06
800	8.17249E + 13	1.98055E + 13	0.24234	5.5257E − 06
850	6.65502E + 12	2.23642E + 12	0.33605	1.2311E − 05
900	7.16841E + 11	3.20880E + 11	0.44763	2.5096E − 05
950	9.77343E + 10	5.63412E + 10	0.57647	4.7465E − 05
1000	1.62821E + 10	1.17472E + 10	0.72148	8.4233E − 05
1050	3.22109E + 09	2.83842E + 09	0.88120	1.4154E − 04
1100	7.39213E + 08	7.79082E + 08	1.05393	2.2687E − 04
1150	1.93024E + 08	2.38939E + 08	1.23787	3.4904E − 04
1200	5.64323E + 07	8.07620E + 07	1.43113	5.1805E − 04
1250	1.82234E + 07	2.97385E + 07	1.63189	7.4501E − 04
1300	6.42591E + 06	1.18132E + 07	1.83837	1.0419E − 03
1350	2.45016E + 06	5.02025E + 06	2.04895	1.4212E − 03
1400	1.00177E + 06	2.26609E + 06	2.26209	1.8962E − 03
1450	4.38021E + 05	1.07979E + 06	2.47646	2.4800E − 03
1500	2.00768E + 05	5.40233E + 05	2.69084	3.1860E − 03

(*continued*)

Table 12.3 (*continued*)

Equilibrium constant K_{P_r} Temperature (K)	$\dfrac{(p/p_0)_{CO_2}}{(p/p_0)_{CO}(p/p_0)_{O_2}^{1/2}}$	$\dfrac{(p/p_0)_{H_2O}}{(p/p_0)_{H_2}(p/p_0)_{O_2}^{1/2}}$	$\dfrac{(p/p_0)_{CO}(p/p_0)_{H_2O}}{(p/p_0)_{CO_2}(p/p_0)_{H_2}}$	$\dfrac{(p/p_0)_{NO}}{(p/p_0)_{N_2}^{1/2}(p/p_0)_{O_2}^{1/2}}$
1550	9.72637E + 04	2.82471E + 05	2.90418	4.0274E − 03
1600	4.93402E + 04	1.53723E + 05	3.11558	5.0168E − 03
1650	2.60986E + 04	8.67587E + 04	3.32426	6.1666E − 03
1700	1.43413E + 24	5.06191E + 24	3.52960	7.4884E − 03
1750	8.15991E + 03	3.04452E + 04	3.73107	8.9931E − 03
1800	4.79334E + 03	1.88294E + 04	3.92824	1.0691E − 02
1850	2.89945E + 03	1.19480E + 04	4.12080	1.2590E − 02
1900	1.80179E + 03	7.76297E + 03	4.30849	1.4700E − 02
1950	1.14787E + 03	5.15523E + 03	4.49113	1.7028E − 02
2000	7.48280E + 02	3.49343E + 03	4.66861	1.9579E − 02
2050	4.98289E + 02	2.41215E + 03	4.84087	2.2359E − 02
2100	3.38436E + 02	1.69484E + 03	5.00786	2.5373E − 02
2150	2.34122E + 02	1.21032E + 03	5.16961	2.8624E − 02
2200	1.64752E + 02	8.77496E + 02	5.32615	3.2115E − 02
2250	1.17799E + 02	6.45249E + 02	5.47753	3.5847E − 02
2300	8.54906E + 01	4.80785E + 02	5.62384	3.9822E − 02
2350	6.29132E + 01	3.62704E + 02	5.76515	4.4039E − 02
2400	4.69061E + 01	2.76820E + 02	5.90158	4.8498E − 02
2450	3.54023E + 01	2.13590E + 02	6.03322	5.3198E − 02
2500	2.70287E + 01	1.66502E + 02	6.16017	5.8137E − 02
2550	2.08601E + 01	1.31054E + 02	6.28255	6.3312E − 02
2600	1.62640E + 01	1.04097E + 02	6.40045	6.8721E − 02
2650	1.28029E + 01	8.33978E + 01	6.51398	7.4361E − 02
2700	1.01701E + 01	6.73591E + 01	6.62323	8.0227E − 02
2750	8.14828E + 00	5.48240E + 01	6.72829	8.6315E − 02
2800	6.58152E + 00	4.49468E + 01	6.82925	9.2620E − 02
2850	5.35700E + 00	3.71035E + 01	6.92618	9.9138E − 02
2900	4.39219E + 00	3.08294E + 01	7.01914	1.0586E − 01
2950	3.62613E + 00	2.57753E + 01	7.10821	1.1279E − 01
3000	3.01343E + 00	2.89620E + 01	7.19343	1.1992E − 01
3050	2.51997E + 00	2.47329E + 01	7.27483	1.2723E − 01
3100	2.11989E + 00	2.12369E + 01	7.35247	1.3473E − 01
3150	1.79349E + 00	1.83304E + 01	7.42636	1.4240E − 01
3200	1.52559E + 00	1.59006E + 01	7.49651	1.5025E − 01
3250	1.30444E + 00	1.38587E + 01	7.56295	1.5826E − 01
3300	1.12089E + 00	1.21342E + 01	7.62566	1.6643E − 01
3350	9.67731E − 01	1.06707E + 01	7.68464	1.7476E − 01
3400	8.39304E − 01	9.42300E + 00	7.73988	1.8322E − 01
3450	7.31095E − 01	8.35460E + 00	7.79136	1.9183E − 01
3500	6.39503E − 01	7.43587E + 00	7.83905	2.0057E − 01

12.12 Concluding remarks

It has been shown that dissociation is an equilibrium process which seeks to minimise the Gibbs energy of a mixture. In a combustion process, dissociation always reduces the temperatures and pressures achieved, and hence reduces the work output and efficiency of a device.

The thermodynamic theory of dissociation has been rigorously developed, and it will be seen later that this applies to fuel cells also (see Chapter 17).

Dissociation usually tends to break product molecules shown into intermediate reactant molecules (e.g. CO and H_2). However, the same theory shows that at high temperatures, nitrogen and oxygen will combine to form oxides of nitrogen.

A wide range of examples has been presented, covering most situations which will be encountered in respect of combustion.

PROBLEMS

Assume that air consists of 79% N_2 and 21% O_2 *by volume.*

1 A cylinder contains 1 kg carbon dioxide (CO_2), and this is compressed adiabatically. Show the pressure, temperature and specific volume are related by the equations

$$\frac{1-\alpha}{\alpha}\sqrt{\frac{(2+\alpha)}{\alpha}} = K_{Pr}\sqrt{\frac{p}{1.01325}}$$

and

$$pv = \frac{\Re(2+\alpha)T}{88}$$

where α = degree of dissociation.

Note that the equilibrium constant for the reaction $CO + \frac{1}{2}O_2 \Leftrightarrow CO_2$ is given by

$$K_{Pr} = \frac{(p_{CO_2}/p_0)}{(p_{CO}/p_0)(p_{O_2}/p_0)^{1/2}}$$

where p_0 is the datum pressure of 1 atm. CO_2, $-391\ 032$; H_2O, $-239\ 342$; CO, $-108\ 052$ kJ/kmol.

2 (a) Calculate the equilibrium constant, K_P, at 2000 K at the reference pressure of 1 atm for the reaction

$$CO + H_2O \Leftrightarrow CO_2 + H_2$$

(b) Calculate the equilibrium constant, K_P, at 2500 K at the reference pressure of 1 atm for the reaction

$$CO + \frac{1}{2}O_2 \Leftrightarrow CO_2$$

Also calculate the equilibrium constant at 2500 K at a pressure of 1 bar.

[(a) 4.6686; (b) 27.028 atm$^{-1/2}$; 27.206 bar$^{-1/2}$]

3 If it is assumed that the enthalpy of reaction, Q_p, is a constant, show that the value of K_p is given by

$$K_p = e^{-Q_p/\Re T + k}$$

where k is a constant.

For a particular reaction $Q_p = -277\,346$ kJ/kmol, and $K_p = 748.7$ at $T = 2000$ K. Evaluate K_p at the temperature of 2500 K.

[26.64]

4 A stoichiometric mixture of propane (C_3H_8) and air is burned in a constant volume bomb. The conditions just prior to combustion are 10 bar and 600 K. Evaluate the composition of the products allowing for dissociation if the peak temperature achieved is 2900 K, and show that the combustion process is adiabatic (i.e. the heat transfer is small).

The equilibrium constant for the CO_2 reaction at 2900 K is $K_{p_r} = p_{CO_2}/p_{CO}p_{O_2}^{1/2} = 4.50$, and it can be assumed that the degree of dissociation for the water reaction is one-quarter of that for the carbon dioxide one. Indicate the processes, including a heat transfer process, on the $U-T$ diagram. The internal energy of propane at 25°C is 12 380 kJ/mol, and the internal energies of formation for the various compounds are CO_2, $-391\,032$; H_2O, $-239\,342$; CO, $-108\,052$ kJ/kmol.

[0.0916; 0.0229; 0.1450; 0.0076; 0.01527; 0.7176]

5 Methane (CH_4) is burned with 50% excess air. The equilibrium products at a pressure of 10 bar and a temperature of 1600 K contain CO_2, CO, H_2O, H_2, O_2 and N_2. Calculate the partial pressures of these gases, assuming that the partial pressures of CO and H_2 are small.

[0.6545; 1.6383×10^{-5}; 1.3089; 1.0524×10^{-5}; 0.6545; 7.3822]

6 The exhaust gas of a furnace burning a hydrocarbon fuel in air is sampled and found to be 13.45% CO_2; 1.04% CO; 2.58% O_2; 7.25% H_2O; 75.68% N_2; and a negligible amount of H_2. If the temperature of the exhaust gas is 2500 K, calculate:

(a)   the carbon/hydrogen ratio of the fuel;
(b)   the equivalence ratio;
(c)   the equilibrium constant for the dissociation of CO_2 (show this as a function of pressure);
(d)   the pressure of the exhaust gas.

[1.0; 0.898; $80.52/p^{1/2}$; 8.874]

7 A weak mixture of propane (C_3H_8) and 50% excess air is ignited in a constant volume combustion chamber. The initial conditions were 1 bar and 300 K and the final composition was 7.87% CO_2, 0.081% CO, 10.58% H_2O, 0.02% H_2, 6.67% O_2 and 74.76% N_2 by volume. Evaluate the final temperature and pressure of the products. Prove that the temperature obeys the conservation of energy, if the calorific value of propane is 46 440 kJ/kg at 300 K.

The equilibrium constant for the carbon dioxide reaction is given by

$$K_p = \frac{(p_{CO})^2 p_{O_2}}{(p_{CO_2})^2} \quad \text{where } \log_{10} K_p = 8.46 - \frac{28600}{T}, \text{ and } T = \text{temperature in K.}$$

The specific heats at constant volume $(c_{v,m})$ of the constituents, in kJ/kmol K, may be taken as

CO_2	$12.7 + 22 \times 10^{-3}T$	H_2O	$22.7 + 8.3 \times 10^{-3}T$
CO	$19.7 + 2.5 \times 10^{-3}T$	N_2	$17.7 + 6.4 \times 10^{-3}T$
O_2	$20.7 + 4.4 \times 10^{-3}T$	H_2	$17.0 + 4.0 \times 10^{-3}T$

C_3H_8 neglect the internal energy in the reactants

The internal energies of formation of the constituents at 300 K are CO_2, -393×10^3; H_2O, -241×10^3; CO, -112×10^3 kJ/kmol.

[2250 K; 7.695 bar]

8 A mixture of propane and air with an equivalence ratio 0.9 (i.e. a weak mixture) is contained in a rigid vessel with a volume of 0.5 m³ at a pressure of 1 bar and 300 K. It is ignited, and after combustion the products are at a temperature of 2600 K and a pressure of 9.06 bar. Calculate:

(a) the amount of substance in the reactants;
(b) the amount of substance in the products;
(c) the amounts of individual substances in the products per unit amount of substance supplied in the fuel.

[0.02005; 0.02096; 0.0932; 0.1368; 0.01138; 0.00261; 0.02792; 0.7281]

9 A stoichiometric mixture of carbon monoxide and air reacts in a combustion chamber and forms exhaust products at 3000 K and 1 bar. If the products are in chemical equilibrium, but no reactions occur between the nitrogen and the oxygen, show that the molar fraction of carbon monoxide is approximately 0.169. The products of combustion now enter a heat exchanger where the temperature is reduced to 1000 K Calculate the heat transfer from the products per amount of substance of CO_2 in the gases leaving the heat exchanger if dissociation can be neglected at the lower temperature.

[-399.3 MJ/kmol CO_2]

10 A mixture containing hydrogen and oxygen in the ratio of 2:1 by volume is contained in a rigid vessel. This is ignited at 60°C and a pressure of 1 atm (1.013 bar), and after some time the temperature is 2227°C. Calculate the pressure and the molar composition of the mixture.

[5.13 bar, 96.5%; 2.36%; 1.18%]

11 A vessel is filled with hydrogen and carbon dioxide in equal parts by volume, and the mixture is ignited. If the initial pressure and temperature are 2 bar and 60°C respectively, and the maximum pressure is 11.8 bar, estimate

(a) the maximum temperature;
(b) the equilibrium constant, K_p; and
(c) the volumetric analysis of the products at the maximum temperature.

Use

$$\log_{10} K_p = 1.3573 - \frac{1354}{T}$$

where $K_p = \dfrac{p_{H_2O} p_{CO}}{p_{H_2} p_{CO_2}}$

and T = temperature in kelvin

[1965; 4.6573; 0.3417; 0.3417; 0.1583; 0.1583]

12 A stoichiometric mixture of hydrogen and air is compressed to 18.63 bar and 845°C. It burns adiabatically at constant volume. Show that the final equilibrium temperature is 3300 K and the degree of dissociation is 8.85%. Calculate the final pressure after combustion. Show the process, including the effect of dissociation, on a U–T diagram. The equilibrium constant is

$$K_{p_r} = \frac{(p_{H_2O}/p_0)}{(p_{H_2}/p_0)(p_{O_2}/p_0)^{1/2}} = 12.1389$$

at $T = 3300\,\text{K}$

Use the following internal energies:

	Internal energy (kJ/kmol)			
T (K)	Oxygen, O_2	Hydrogen, H_2	Nitrogen, N_2	Water, H_2O
0	0	0	0	−239 081.7
1100	25 700.3	23 065.3	24 341.8	
1150	27 047.4	24 187.6	25 579.1	
3300	91 379.8	80 633.5	85 109.6	142 527.4

[47.58 bar]

13 A mixture containing equal volumes of carbon dioxide (CO_2) and hydrogen (H_2) is contained in a rigid vessel. It is ignited at 60°C and a pressure of 1 atm, and after some time the temperature is 2227°C. Calculate the pressure and the molar composition of the mixture.

[7.508 atm; 35.6%; 35.6%; 14.4%; 14.4%]

14 A gas turbine combustion chamber receives air at 6 bar and 500 K. It is fuelled using octane (C_8H_{18}) at an equivalence ratio of 0.8 (i.e. weak), which burns at constant pressure. The amount of carbon dioxide and water in the products is 9.993% and 11.37% by volume respectively. Calculate the maximum temperature achieved after combustion and evaluate the degrees of dissociation for each of the following reactions:

$$CO + \frac{1}{2} O_2 \rightarrow CO_2$$

$$H_2 + \frac{1}{2} O_2 \rightarrow H_2O$$

Assume the octane is in liquid form in the reactants and neglect its enthalpy; use an enthalpy of reaction of octane at 300 K of −44 820 kJ/kg.

[0.0139; 0.00269; 2200 K]

15 A mixture of one part by volume of vaporised benzene to 50 parts by volume of air is ignited in a cylinder and adiabatic combustion ensues at constant volume. If the initial pressure and temperature of the mixture are 10 bar and 500 K respectively, calculate the maximum pressure and temperature after combustion. (See Chapter 10, Question 1.)

[51.67 bar, 2554 K]

16 A gas engine is operated on a stoichiometric mixture of methane (CH_4) and air. At the end of the compression stroke the pressure and temperature are 10 bar and 500 K respectively. If the combustion process is an adiabatic one at constant volume, calculate the maximum temperature and pressure reached. What has been the effect of dissociation? (See Chapter 10, Question 5).

[59.22 bar, 2960 K]

17 A gas injection system supplies a mixture of propane (C_3H_8) and air to a spark ignition engine, in the ratio of volumes of 1 : 30. The mixture is trapped at 1 bar and 300 K, the volumetric compression ratio is 12:1, and the index of compression is $\kappa = 1.4$. Calculate the equivalence ratio, the maximum pressure and temperature achieved during the cycle, and also the composition (by volume) of the dry exhaust gas. What has been the effect of dissociation? (See Chapter 10, Question 6.)

[0.79334, 119.1 bar, 2883 K;   0.8463, 0.1071, 0.0465]

18 A 10% rich mixture of heptane (C_7H_{16}) and air is trapped in the cylinder of an engine at a pressure of 1 bar and temperature of 400 K. This is compressed and ignited, and at a particular instant during the expansion stroke, when the volume is 20% the trapped volume, the pressure is 27.06 bar. Assuming that the mixture is in chemical equilibrium and contains only CO_2, CO, H_2O, H_2 and N_2, find the temperature and molar fractions of the constituents. This solution presupposes that there is no O_2 in the products; use your results to confirm this.

[2000 K; 0.1037; 0.0293; 0.1434; 0.00868; 0.7148]

19 A turbocharged, intercooled compression ignition engine is operated on octane (C_8H_{18}) and achieves constant pressure combustion. The volumetric compression ratio of the engine is 20 : 1, and the pressure and temperature at the start of compression are 1.5 bar and 350 K respectively. If the air–fuel ratio is 24 : 1 calculate maximum temperature and pressure achieved in the cycle taking into account dissociation of the carbon dioxide and water vapour. Assuming that the combustion gases do not change composition during the expansion stroke, calculate the indicated mean effective pressure (imep, $\bar{p}_i$) of the cycle in bar. What has been the effect of dissociation on the power output of the engine? Assume that the index of compression $\kappa_C = 1.4$, while that of expansion, $\kappa_E = 1.35$. (See Chapter 10, Question 7).

[2487 K; 99.4 bar, 15.10 bar]

20 One method of reducing the maximum temperature in an engine is to run with a rich mixture. A spark ignition engine with a compression ratio of 10:1, operating on the

Otto cycle, runs on a rich mixture of octane and air, with an equivalence ratio of 1.2. The trapped conditions are 1 bar and 300 K, and the index of compression is 1.4. Calculate, taking into account dissociation of the carbon dioxide and water vapour, how much lower the maximum temperature is under this condition than when the engine was operated stoichiometrically. How has dissociation affected the products of combustion? What are the major disadvantages of operating in this mode? (See Chapter 10, Question 8.)

[3020 K, 3029 K]

21 A gas engine with a volumetric compression ratio of 10 : 1 is run on a weak mixture of methane (CH_4) and air, with $\phi = 0.9$. If the initial temperature and pressure at the commencement of compression are 60°C and 1 bar respectively, calculate the maximum temperature and pressure reached during combustion at constant volume, taking into account dissociation of the carbon dioxide and water vapour, under the following assumptions:

(a)   that 10% of the heat released is lost during the combustion period, and

(b)   that compression is isentropic.

Assume the ratio of specific heats, κ, during the compression stroke is 1.4, and the heat of reaction at constant volume for methane is -8.023×10^5 kJ/kmol CH_4. (See Chapter 10, Question 9.)

[2737 K;   71.57 bar]

13

Effect of Dissociation on Combustion Parameters

Dissociation is an equilibrium process by which the products of combustion achieve the minimum Gibbs function for the mixture (see Chapter 12). The effect is to cause the products that would be obtained from complete combustion to break down partially into the original reactants, and other compounds or radicals. This can be depicted on a $U-T$ diagram as shown in Fig 13.1.

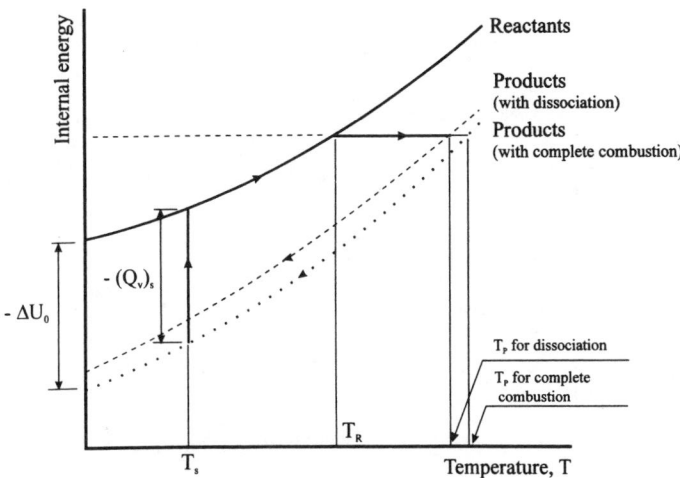

Fig. 13.1 Internal energy–temperature diagram for combustion both with and without dissociation

It can be seen from Fig 13.1 that the effect of dissociation is to reduce the temperature of the products after combustion. This, in turn, reduces the amount of energy that can be drawn from the combustion process and reduces the work output of engines.

The other effect of dissociation is to form pollutants. This is particularly true in combustion in engines, when carbon monoxide (CO) can be formed even when the mixture is weak – that is, there is more than sufficient oxygen in the air to oxidise completely both the carbon and hydrogen in the fuel. If chemical kinetics are considered (see Chapter 14), i.e. account is taken of the finite rate by which equilibrium is achieved, then it is possible to show how some of the major pollutants are formed in the quantities measured at the exit from engines. Chemical kinetic effects and dissociation are the sole causes of the production of NO_x from fuels which do not contain nitrogen (through combining of the nitrogen and oxygen in the air), and they also increase the amount of carbon monoxide (CO) during weak and slightly rich combustion. Processes which include chemical kinetics are *not equilibrium processes*, but they attempt to reach equilibrium if there is sufficient time.

The effect of chemical, or rate, kinetics can be assessed by considering a simple example. Imagine a spark ignition engine in which the mixture is compressed prior to ignition by a spark. If the compression temperature is not too high it can be assumed that the reactants do not react before ignition. After ignition the reactants burn to form products at a high temperature, and these are initially compressed further (as the piston continues to rise to top dead centre [tdc], and the combustion process continues) before being expanded when the piston moves down and extracts work from the gases. As the pressure increases, the temperature of the gas also increases and the products tend to dissociate: they attempt to achieve an equilibrium composition but the speed of the engine is too rapid for this. The effect can be seen in Fig 13.2, which depicts the way in which the actual level of pollutants attempts to follow the equilibrium level, but lags the equilibrium values. This lag means that the maximum level of pollutant (e.g. NO_x) formed does not achieve the maximum equilibrium value, but it also means that the rate of reduction of the pollutants is slower than that of the equilibrium species and the NO_x is usually *frozen* at a level above the equilibrium value of the exhaust gas from the combustion chamber, boiler, gas turbine, diesel engine, etc.

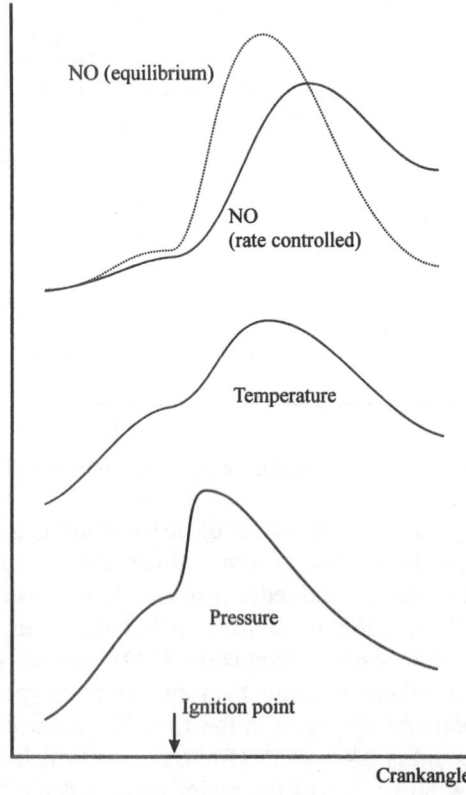

NO (equilibrium)

NO (rate controlled)

Temperature

Pressure

Ignition point

Crankangle

Fig. 13.2 Production of pollutants in equilibrium and chemical kinetically controlled processes

13.1   Calculation of combustion both with and without dissociation

It is possible for the pressure, temperature and equilibrium composition of various mixtures to be calculated using computer programs, based on the principles outlined in Chapter 12. A program written at UMIST and based on the enthalpy coefficients given in Chapter 9 has been used to evaluate the results presented in this chapter. The program, which has a range of fuels programmed into it, can be run in dissociation and non-dissociation modes. A series of calculations was undertaken using octane (C_8H_{18}) and methane (CH_4) as the fuels. The equivalence ratio was varied from a weak value of $\phi = 0.5$ to a rich one of $\phi = 1.25$. It was assumed that the initial pressure and temperature at entry to the combustion chamber were 1 bar and 300 K respectively, and that the fluid was compressed through a volumetric compression ratio of 12, with an index of compression ($\kappa = c_p/c_v$) of 1.4. Combustion then took place at constant volume. This is essentially similar to the combustion that would take place in a reciprocating engine with constant volume combustion at tdc, which is an 'Otto cycle' in which the heat transfer process associated with the air standard cycle is replaced by a realistic combustion process. Combustion in this Otto cycle is adiabatic and occurs at constant volume: this is a constant internal energy process, as shown in Fig 13.1. The principles introduced here can be developed to study the effect of equilibrium on a realistic (finite rate of combustion) cycle.

The computer program was written in a manner that enabled it also to be used for constant pressure combustion, e.g. like that occurring in a gas turbine combustion chamber. In this case the enthalpy of both the reactants and products will be equal.

13.2   The basic reactions

The stoichiometric combustion of methane is defined by the chemical equation

$$CH_4 + 2(O_2 + 3.76N_2) \Rightarrow CO_2 + 2H_2O + 7.52N_2 \tag{13.1}$$

while that for octane is

$$C_8H_{18} + 12.5(O_2 + 3.76N_2) \Rightarrow 8CO_2 + 9H_2O + 47N_2 \tag{13.2}$$

If the mixture is not stoichiometric then the equation for methane becomes

$$CH_4 + \frac{2}{\phi}(O_2 + 3.76N_2) \Rightarrow CO_2 + 2\left(\frac{1}{\phi} - 1\right)O_2 + 2H_2O + \frac{7.52}{\phi}N_2 \tag{13.3}$$

if the mixture is *weak*, i.e. there is less oxygen than required for complete combustion of the fuel. ϕ is called the *equivalence ratio* and in this case $\phi \leqslant 1.0$. For a rich mixture, $\phi \geqslant 1.0$, and if it is assumed that the hydrogen has preferential use of the oxygen, then the equation becomes

$$CH_4 + \frac{2}{\phi}(O_2 + 3.76N_2) \Rightarrow \left(\frac{4 - 3\phi}{\phi}\right)CO_2 + \left(\frac{4\phi - 4}{\phi}\right)CO + 2H_2O + \frac{7.52}{\phi}N_2 \tag{13.4}$$

It will be assumed for the initial examples that the only dissociation is of the products, carbon dioxide and water, which dissociate according to the equations

$$CO_2 \Leftrightarrow CO + \tfrac{1}{2}O_2 \tag{13.5}$$

and

$$H_2O \Leftrightarrow H_2 + \tfrac{1}{2}O_2 \tag{13.6}$$

These equations (eqns (13.5) and (13.6)) have to be added into the basic equations to evaluate the chemical equation with dissociation. The program does this automatically and evaluates the mole fractions of the products. The approach adopted by the program is the same as that introduced in Chapter 12, when the products of combustion were defined by a set of simultaneous equations. Later in the chapter the dissociation of other compounds will be introduced to give a total of 11 species in the products. The method of solving for these species is outlined by Baruah (Chapter 14) in Horlock and Winterbone (1986).

13.3　The effect of dissociation on peak pressure

In general, dissociation will tend to decrease the pressure achieved during the combustion process (when it occurs in a closed system) because it reduces the temperature of the products. This reduction in pressure is always evident with stoichiometric and lean mixtures, although an increase in pressure over the equivalent situation without dissociation can occur in rich mixtures due to the increase in the amount of substance in the products. Figures 13.3 and 13.4 show that the peak pressure is reduced by about 8 bar at the stoichiometric air–fuel ratio ($\phi = 1$) with both methane and octane. Dissociation also changes the equivalence ratio at which the peak pressure is achieved. If there is no dissociation then the peak pressure is always reached at the stoichiometric ratio ($\phi = 1$). However, when dissociation occurs the equivalence ratio at which the peak pressure occurs is moved into the rich region ($\phi > 1$). This is because dissociation tends to increase the amount of substance (n_p) in the products, compared with the non-dissociating case. The peak pressure with dissociation occurs at around $\phi = 1.1$ for methane, and around $\phi = 1.25$ for octane. It is interesting to note from the octane results (Fig 13.4) that the peak pressures achieved with and without dissociation are almost the same – except they occur at a different equivalence ratio.

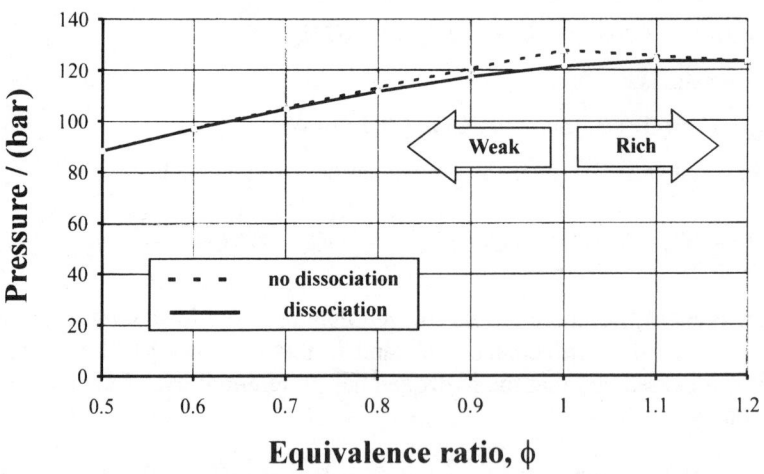

Fig. 13.3 Variation of pressure with equivalence ratio for combustion of methane. Initial pressure: 1 bar; initial temperature: 300 K; compression ratio: 12; compression index: 1.4

13.4　The effect of dissociation on peak temperature

Figures 13.5 and 13.6 show the variation of the peak temperature, produced by an adiabatic combustion process, with equivalence ratio, both with and without dissociation.

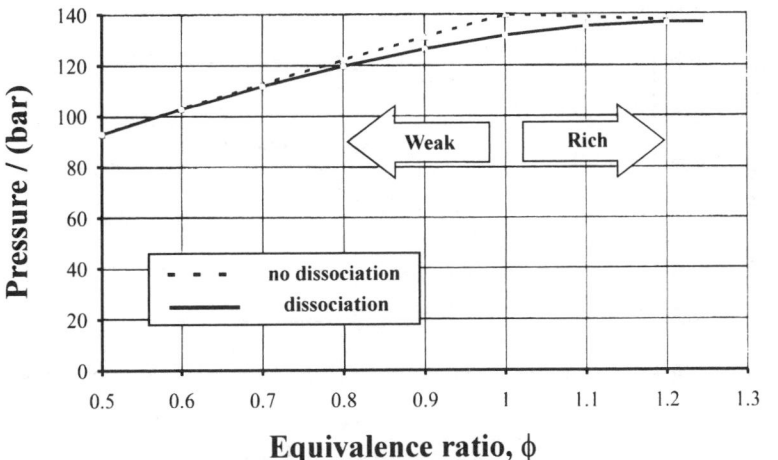

Fig. 13.4 Variation of pressure with equivalence ratio for combustion of octane. Initial pressure: 1 bar; initial temperature: 300 K; compression ratio: 12; compression index: 1.4

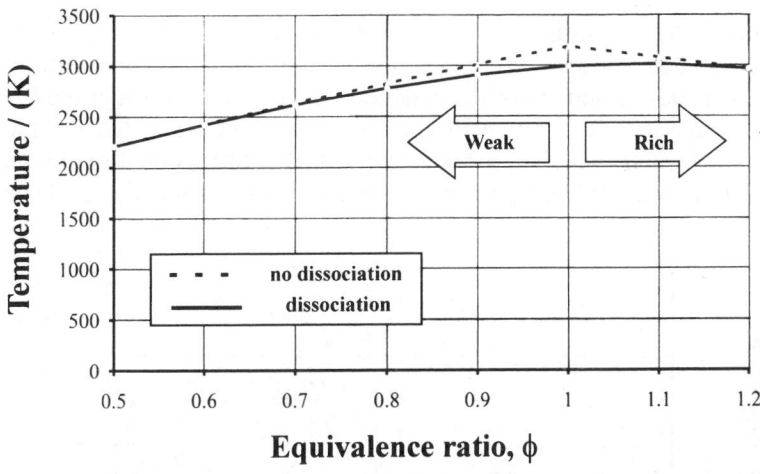

Fig. 13.5 Variation of temperature with equivalence ratio for combustion of methane. Initial pressure: 1 bar; initial temperature: 300 K; compression ratio: 12; compression index: 1.4

Dissociation lowers the temperature for both fuels, and moves the point at which the maximum temperature is achieved. In the case of methane the maximum temperature reduces from 3192 K (at stoichiometric) to 3019 K (at about $\phi = 1.1$ – rich), while for octane the values are 3306 K to 3096 K (at $\phi = 1.2$). The effect of such a reduction in temperature is to lower the efficiency of an engine cycle.

13.5   The effect of dissociation on the composition of the products

The composition of the fuel will affect the composition of the products due to the different stoichiometric air–fuel ratios and the different carbon/hydrogen ratios. The stoichiometric air–fuel ratio (by weight) for methane is 17.16, whilst it is 15.05 for octane. The stoichiometric air–fuel ratio remains in the region 14 to 15 for most of the higher straight hydrocarbon fuels.

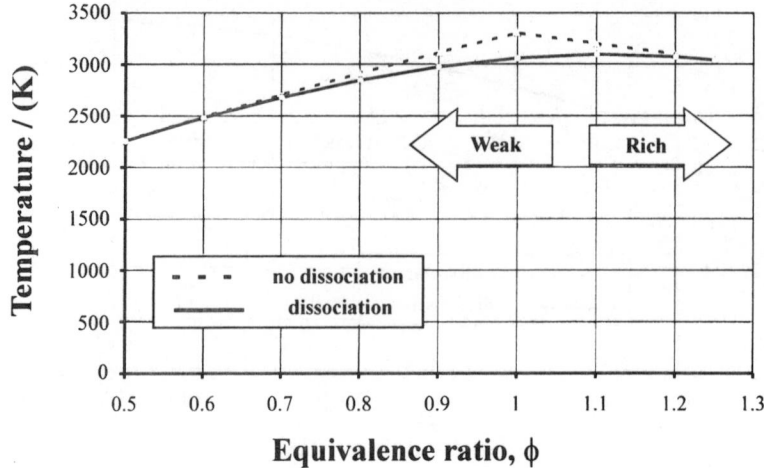

Fig. 13.6 Variation of temperature with equivalence ratio for combustion of octane. Initial pressure: 1 bar; initial temperature: 300 K; compression ratio: 12; compression index: 1.4

The carbon/hydrogen ratio varies from 0.25 to 0.44 for methane and octane respectively, and this affects the composition of the products, both with and without dissociation. The mole fractions of the products for the combustion of methane (taking account of dissociation) are shown in Fig 13.7. It can be seen that the largest mole fraction is that of water, which is more than twice that of carbon dioxide: this reflects the greater proportion of hydrogen atoms in the fuel. (Note: the mole fraction of

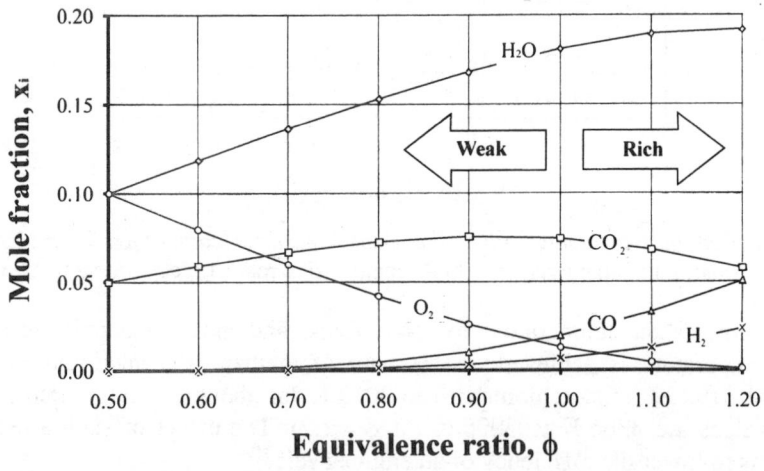

Fig. 13.7 Variation of composition of products with equivalence ratio for combustion of methane. Initial pressure: 1 bar; initial temperature: 300 K; compression ratio: 12; compression index: 1.4

nitrogen, N_2, has been omitted from Figs 13.7, 13.8, 13.9 and 13.10 because it dominates the composition of the products. Its mole fraction remains around 0.79 in all cases.)

The mixture is weak at equivalence ratios of less than unity, and there is excess oxygen for complete combustion up to an equivalence ratio of unity. This can be seen in Figs 13.8

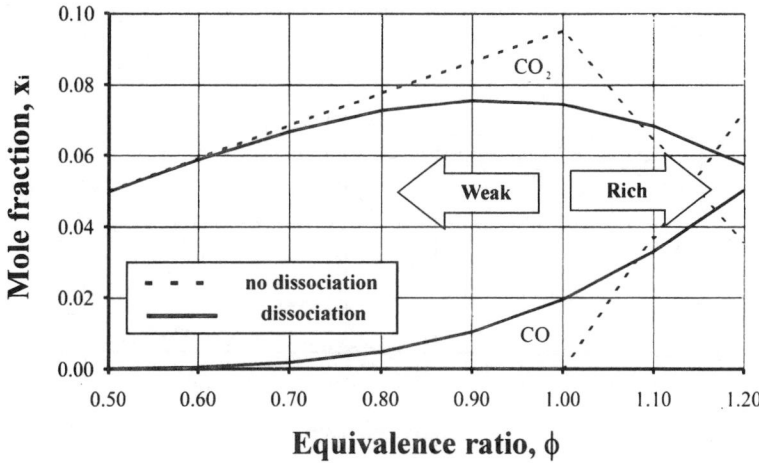

Fig. 13.8 Variation of carbon reactions, with and without dissociation, for combustion of methane

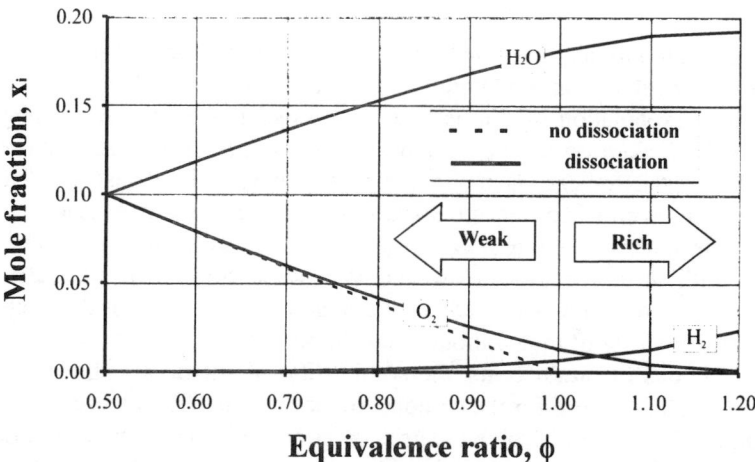

Fig. 13.9 Variation of hydrogen reactions, with and without dissociation, for combustion of methane

and 13.9, where the composition of the products, both with and without dissociation, has been plotted. It can be seen from Fig 13.9, which includes the mole fraction of oxygen, that there is no oxygen in the products at equivalence ratios of greater than unity *if there is no dissociation*. However, when there is dissociation the quantity of oxygen in the products is increased, simply due to the reverse reaction. This reverse reaction also produces some unreacted hydrogen, and dissociation causes this to be produced even with weak mixtures. It should be noted that it was assumed that the hydrogen would be oxidised in preference to the carbon in the reactions without dissociation, and this explains the absence of hydrogen in those cases.

The carbon-related reactions are shown in Fig 13.8. When dissociation is neglected, the carbon is completely oxidised to carbon dioxide in the weak and stoichiometric mixtures, with no carbon monoxide being formed. After the mixture becomes rich, the level of carbon monoxide increases linearly with equivalence ratio. When dissociation is

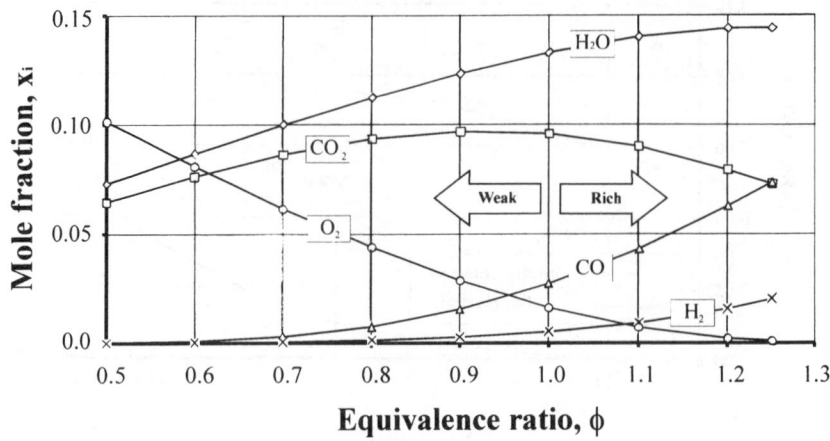

Fig. 13.10 Variation of composition of products with equivalence ratio for combustion of octane. Initial pressure: 1 bar; initial temperature: 300 K; compression ratio: 12; compression index: 1.4

considered, carbon monoxide is formed even in the weak mixture region, but surprisingly there is a lower level of carbon monoxide in the rich region than when there is no dissociation. The explanation of this is that the dissociation of the water releases some oxygen that is then taken up by the carbon. The production of carbon monoxide in lean mixtures is the reason for carbon monoxide in the untreated exhaust gas from internal combustion engines even when they are operating stoichiometrically or lean.

Considering the carbon dioxide, it can be seen that this peaks at an equivalence ratio of unity ($\phi = 1$) without dissociation. However, it peaks at an equivalence ratio of around $\phi = 0.9$ to 1.0 (i.e. weak) when there is dissociation, when the peak level of carbon dioxide is only about 75% of that without dissociation.

The mole fractions of products for the combustion of octane are shown in Fig 13.10. The trends are similar to those for the combustion of methane. The carbon dioxide fraction maximises on the lean side of stoichiometric, at ϕ of about 0.9. It is also noticeable that the mole fractions of water and carbon dioxide are much closer with octane than methane, which simply reflects the higher carbon/hydrogen ratio of octane.

13.6   The effect of fuel on composition of the products

Two different fuels were used in the previous analysis of the effect of dissociation on combustion products. Both are paraffinic hydrocarbons, with a generic structure of C_nH_{2n+2}: for methane the value of $n = 1$, while for octane $n = 8$. This means that the carbon/hydrogen ratio varies from 0.25 to 0.44. This has an effect on the amount of CO_2 produced for each unit of energy released (i.e. kg CO_2/kJ, or more likely kg CO_2/kWh). The effect of this ratio can be seen in Figs 13.7 and 13.10. The maximum mole fraction of CO_2 released with octane is almost 10%, whilst it peaks at around 7.5% with methane: the energy released per unit mass of mixture is almost the same for both fuels, as shown by the similarity of temperatures achieved (see Figs 13.5 and 13.6). Hence, if it is required to reduce the amount of carbon dioxide released into the atmosphere, it is better to burn fuels with a low carbon/hydrogen ratio. This explains, in part, why natural gas is a popular fuel at this time.

13.7 The formation of oxides of nitrogen

It was shown in Chapter 12 that when air is taken up to high temperatures, the nitrogen and oxygen will combine to form oxides of nitrogen, and in particular nitric oxide, NO. It was also stated that it is not enough simply to add the equation

$$N_2 + O_2 \Leftrightarrow 2NO \tag{13.7}$$

but a chain of reactions including 11 species has to be introduced. This is necessary because the formation of atomic oxygen and nitrogen, and the radical OH play a significant part in the amount of NO produced in the equilibrium mixture.

These equations representing equilibrium of the 11 have been used to analyse the equilibrium constituents of the products of combustion for the cases considered above, i.e. the combustion of methane and octane in an engine operating on an 'Otto' cycle. The results are shown in Figs 13.11 and 13.12. It should be noted that the ordinate axis is in logarithmic form to enable all the important constituents to be shown on a single graph. The numerical results for carbon monoxide, carbon dioxide and water are similar to those in Figs 13.9 and 13.10; it is simply the change in scales which has changed the apparent shapes. These results will now be analysed.

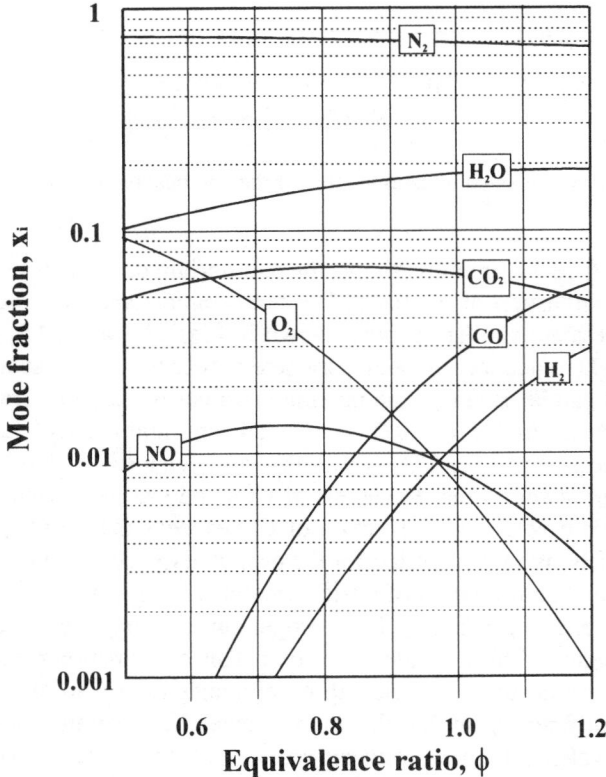

Fig. 13.11 Products of combustion at tdc for the combustion of methane in an Otto cycle

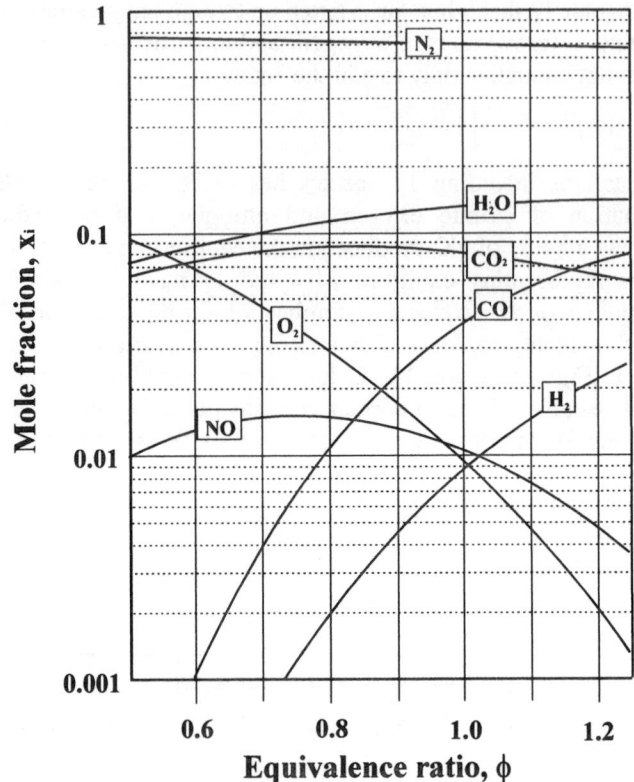

Fig. 13.12 Products of combustion at tdc for the combustion of octane in an Otto cycle

Considering Fig 13.11 first: this shows the equilibrium constituents of the mixture at tdc after adiabatic combustion of methane. The temperatures and pressures achieved by this combustion process are similar to those shown in Figs 13.3 and 13.5, although there will have been a slight reduction in both parameters because the dissociation processes all absorb energy. It can be seen that the maximum values of the mole fraction of both water (≈ 0.19) and carbon dioxide (≈ 0.07) have not been significantly changed by the extra dissociation processes, and the CO_2 still peaks in the vicinity of $\phi = 0.8$ to 0.9. However, detailed examination shows that the levels of CO_2 and CO have been changed in the rich region, and in this case the mole fraction of CO exceeds that of CO_2 at $\phi = 1.2$. This is because more reactions are competing for the available oxygen, and the carbon is the loser.

Now considering the formation of nitric oxide: it was shown in Chapter 12 that nitrogen and oxygen will tend to combine at high temperatures to produce nitric oxide because the enthalpy of formation of nitric oxide is positive. This combination process will be referred to under the general heading of dissociation, although strictly speaking it is the opposite. This example is different from that shown in Chapter 12, when the variation in the amount of NO with equivalence ratio *at constant pressure and temperature* was depicted. In this case the peak pressure and temperature achieved in the combustion process are functions of the equivalence ratio, and the variations are shown in Figs 13.3 and 13.5. In fact, the peak temperature increases from about 2200 K at $\phi = 0.5$, to 3000 K at $\phi \approx 1.1$; the peak pressure goes from about 87 bar to 122 bar over the same range of equivalence ratio. The increase in temperature will tend to increase dissociation effects in all cases, whereas an increase in pressure will tend to decrease the dissociation effects in cases where dissociation increases the total amount of substance (e.g. the CO_2 reaction); this was discussed in Chapter 12. Hence, the changes occurring in the conditions in the cycle tend to run counter to each other to some extent, although the effect of temperature dominates. The value of mole fraction of NO peaks at around $\phi = 0.7$ to 0.8, and obviously does not occur at the highest temperature. This is because the reaction which creates NO is dependent on the amounts of nitrogen (N_2) and oxygen (O_2) present in the mixture, as well as the temperature of the products. At low equivalence ratios there is an abundance of components to react but the temperature is low, whilst at high temperatures the driving force to react is high but there is a deficiency of oxygen. In fact, the nitrogen has to compete with the carbon monoxide and hydrogen for any oxygen that becomes available, and it gains most of the oxygen from the carbon reactions. The maximum level of NO attained is a mole fraction of about 0.13, and this is equivalent to about 13 000 ppm (parts per million [by mass]); this level is typical of that which would be achieved *at equilibrium conditions* in an engine.

Before considering the effect of the rate of reactions on the actual levels of NO achieved, it is worthwhile seeing the effect of fuel on the level of exhaust constituents. The additional reactions do not have a major effect on the exhaust composition with octane as a fuel, and the CO_2 level is still significantly higher than with methane, as would be expected. However, the general level of NO is higher with octane than methane because of the slightly higher temperatures achieved with this fuel, and the larger proportion of CO_2 which can be 'robbed' for oxygen. The maximum levels of NO again occur around $\phi = 0.7$ to 0.8, and result in about 14 000 ppm.

The levels of NO in the exhaust gas of an engine are significantly lower than those calculated above for a number of reasons. First, the reactions occurring in the cylinder are not instantaneous, but are restricted by the rate at which reacting molecules meet; this is known as chemical kinetics or rate kinetics, and is the subject of Chapter 14. The effect of these kinetics is that the instantaneous level of NO in the cylinder does not achieve the equilibrium level, and lags it when it is increasing as shown in Fig 13.2. When the equilibrium concentration is decreasing, the actual level of NO still cannot maintain pace with it, and again lags the level while it decreases down the expansion stroke of the engine, as shown qualitatively in Fig 13.2; in fact these reactions tend to 'freeze' at around 2000 K, and this level then dominates the exhaust value. Secondly, the combustion process does not take place instantaneously at tdc as shown here, but is spread over a significant period of the cycle and the pressures and temperatures are lower than those attained in the 'Otto' cycle: this will reduce the peak equilibrium level of NO.

It is shown in Figs 13.11 and 13.12 that the peak level of NO at equilibrium occurs at around $\phi = 0.7$ to 0.8, but experience shows that the maximum values of NO in engine exhaust systems (around 2000 to 5000 ppm) occur at about $\phi = 0.9$. This can be explained by considering the effect of rate kinetics. It will be shown in Chapter 14 that the rate at which a reaction occurs is exponentially related to temperature. Hence, while the equilibrium level of NO is a maximum at $\phi = 0.7$ to 0.8, the lower temperature at this equivalence ratio limits the amount of NO produced by the reaction. The peak value of NO occurs at $\phi = 0.9$ because the driving force (the equilibrium concentration) and the rate of reaction combine at this equivalence ratio to maximise production of this pollutant.

14

Chemical Kinetics

14.1 Introduction

Up until now it has been assumed that chemical reactions take place very rapidly, and that the equilibrium conditions are reached instantaneously. While most combustion processes are extremely fast, often their speed is not such that combustion can be considered to be instantaneous, compared with the physical processes surrounding them. For example, the combustion in a reciprocating engine running at 6000 rev/min has to be completed in about 5 ms if the engine is to be efficient. Likewise, combustion in the combustion chamber of a gas turbine has to be rapid enough to be completed before the gas leaves the chamber. Such short times mean that it is not possible for all the gases in the combustion chamber to achieve equilibrium – they will be governed by the *chemical kinetics* of the reactions.

Chemical kinetics plays a major role in the formation of pollutants from combustion processes. For example, oxygen and nitrogen will coexist in a stable state at atmospheric conditions, and the level of oxides of nitrogen (NO_x) will be negligible. However, if the oxygen and nitrogen are involved in a combustion process then they will join together at the high temperature to form NO_x which might well be *frozen* into the products as the temperature drops. This NO_x is a pollutant which is limited by legislation because of its irritant effects. NO_x is formed in all combustion processes, including boilers, gas turbines, diesel and petrol engines; it can be removed in some cases by the use of catalytic converters.

It was shown in Chapters 12 and 13 that significant dissociation of the normal products of combustion, carbon dioxide and water can occur at high temperature. The values shown were based on the equilibrium amounts of the substances, and would only be achieved after infinite time; however, the rates at which chemical reactions occur are usually fast and hence some reactions get close to equilibrium even in the short time the gases are in the combustion zone. An analysis of the kinetics of reactions will now be presented.

14.2 Reaction rates

Reaction rates are governed by the movement and breakdown of the atoms or molecules in the gas mixture: reactions will occur if the participating 'particles' collide. The number of collisions occurring in a mixture will be closely related to the number densities (number per unit volume) of the 'particles'. The number density can be defined by the molar concentration, c, which is the amount of substance *per unit volume*. This is obviously a measure of the number of particles per unit volume since each amount of substance is proportional to the number of molecules; this is illustrated in Fig 14.1. The molar concentration will be denoted by enclosing the reactant or product symbol in [ ].

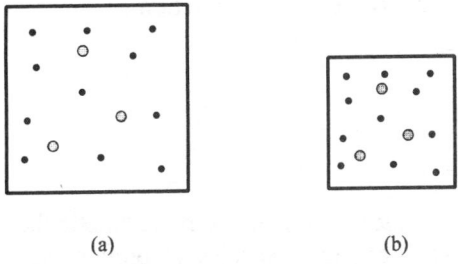

(a)   (b)

Fig. 14.1 Diagrammatic representation of molar concentration. Molar concentration in (a) is approximately half that in (b)

Reactions occur when two, or more, reactants are capable of reacting. Many simple chemical reactions are second order, e.g.

$$A + B \underset{k_b}{\overset{k_f}{\rightleftharpoons}} C + D$$

(14.1)

$$CO_2 + H_2 \underset{k_b}{\overset{k_f}{\rightleftharpoons}} H_2O + CO$$

where the first reaction is a general one and the second is an example based on the water gas reaction (chosen because the same number of reactants and products exist on both sides).

These reactions can be written as

$$\nu_1' A_1 + \nu_2' A_2 + \cdots \underset{k_b}{\overset{k_f}{\rightleftharpoons}} \nu_1'' A_1 + \nu_2'' A_2 + \cdots$$

(14.2)

where $A_1, A_2, \ldots$ etc are the elements or compounds involved in the reaction, and $\nu_1, \nu_2, \ldots$ etc are the respective stoichiometric coefficients. Equation (14.2) can be further generalised to

$$\sum_{i=1}^{q} \nu_i' A_i \underset{k_b}{\overset{k_f}{\rightleftharpoons}} \sum_{i=1}^{q} \nu_i'' A_i$$

(14.3)

where q = total number of species being considered, ν' represents the stoichiometric coefficients of the reactants, and ν'' that of the products.

In this case q has been taken as the same on both sides of the equation because it is assumed that the same species *can* exist on both sides. If a species does not exist on one of the sides it is represented by a stoichiometric coefficient of zero ($\nu = 0$). For example,

$$CO_2 + H_2 \Leftrightarrow H_2O + CO$$

(14.4)

can be represented by

$$\sum_{i=1}^{4} v'_{ij} A_i \underset{k_b}{\overset{k_f}{\rightleftharpoons}} \sum_{i=1}^{4} v''_{ij} A_i \tag{14.5}$$

where $q = 4$

v_{ij} = stoichiometric coefficient of i in reaction j

$A_1 = CO_2$	$v'_1 = 1$	$v''_1 = 0$
$A_2 = H_2$	$v'_2 = 1$	$v''_2 = 0$
$A_3 = H_2O$	$v'_3 = 0$	$v''_3 = 1$
$A_4 = CO$	$v'_4 = 0$	$v''_4 = 1$

The law of mass action, which is derived from the kinetic theory of gases, states that the rate of formation, or depletion, of a species is proportional to the product of molar concentrations of the reactants, each raised to the power of its stoichiometric coefficient. Hence the rate of formation of species j, for an elementary reaction, is

$$R_j \propto [A_1]^{v_{1j}} [A_2]^{v_{2j}} \cdots [A_n]^{v_{nj}}$$

or $\tag{14.6}$

$$R_j \propto \prod_{i=1}^{n} [A_i]^{v_{ij}} \quad \text{for } (j = 1, \dots s)$$

and s = total number of simultaneous reactions.
Equation (14.6) may be written as

$$R_j = k_j \prod_{i=1}^{n} [A_i]^{v_{ij}} \quad \text{for } (j = 1, \dots s) \tag{14.7}$$

where k_j is the *rate constant for the reaction*. In eqns (14.1) to (14.5) the reaction was described as having a forward and backward direction, with rates k_f and k_b. The reaction shown in eqn (14.7) can be described in this way as

$$R_{fj} = k_{fj} \prod_{i=1}^{n} [A_i]^{v'_{ij}} \quad \text{for } (j = 1, \dots s) \tag{14.8}$$

and

$$R_{bj} = k_{bj} \prod_{i=1}^{n} [A_i]^{v''_{ij}} \quad \text{for } (j = 1, \dots s) \tag{14.9}$$

The rate of change with time of species i is proportional to the change of the stoichiometric coefficients of A_i in the reaction equation (the system is effectively a first order one attempting to achieve the equilibrium state). Thus, for the forward direction

$$\left. \frac{d[A_i]_j}{dt} \right|_f = (v'_{ij} - v''_{ij}) k_{fj} \prod_{i=1}^{n} [A_i]^{v'_{ij}} = \Delta v_{ij} k_{fj} \prod_{i=1}^{n} [A_i]^{v'_{ij}} \tag{14.10}$$

and, for the backward direction

$$\left.\frac{d[A_i]_j}{dt}\right|_b = (v''_{ij} - v'_{ij})k_{bj}\prod_{i=1}^{n}[A_i]^{v''_{ij}} = -\Delta v_{ij}k_{bj}\prod_{i=1}^{n}[A_i]^{v''_{ij}} \tag{14.11}$$

Hence the *net rate of formation of* A_i is

$$\left.\frac{d[A_i]_j}{dt}\right|_{net} = \Delta v_{ij}\left[k_{fj}\prod_{i=1}^{n}[A_i]^{v'_{ij}} - k_{bj}\prod_{i=1}^{n}[A_i]^{v''_{ij}}\right] \tag{14.12}$$

Consider the following rate controlled reaction equation:

$$v_a A + v_b B \underset{k_b}{\overset{k_f}{\rightleftharpoons}} v_c C + v_d D \tag{14.13}$$

The net rate of generation of species C is given by

$$\frac{d[C]}{dt} = [k_f[A]^{v_a}[B]^{v_b} - k_b[C]^{v_c}[D]^{v_d}] \tag{14.14}$$

Using the notation $[A]/[A]_e = \alpha$, $[B]/[B]_e = \beta$, $[C]/[C]_e = \gamma$ and $[D]/[D]_e = \delta$, where the suffix e represents equilibrium concentrations, gives

$$d[C]/dt = k_f \alpha^{v_a}\beta^{v_b}[A]_e^{v_a}[B]_e^{v_b} - k_b\gamma^{v_c}\delta^{v_d}[C]_e^{v_c}[D]_e^{v_d} \tag{14.15}$$

At equilibrium

$$k_f[A]_e^{v_a}[B]_e^{v_b} = k_b[C]_e^{v_c}[D]_e^{v_d} = R_j \tag{14.16}$$

Therefore the net rate is

$$d[C]/dt = R_j[\alpha^{v_a}\beta^{v_b} - \gamma^{v_c}\delta^{v_d}] \tag{14.17}$$

14.3 Rate constant for reaction, k

The rate constant for the reaction, k, is related to the ability of atoms or ions to combine. In combustion engineering this will usually occur when two or more particles collide. Obviously the collision rate is a function of the number of particles per unit volume, and their velocity of movement, i.e. their concentration and temperature. Most chemical reactions take place between two or three constituents because the probability of more than three particles colliding simultaneously is too small. It has been found experimentally that most reactions obey a law like that shown in Fig 14.2.

This means that the rate constant for the reaction can be defined by an equation of the form

$$k = Ae^{-E/RT} \tag{14.18}$$

This equation is called the *Arrhenius equation*. The factor A is called the pre-exponential factor, or the frequency factor, and is dependent on the rate at which collisions with the required molecular orientation occur. A sometimes contains a temperature term, indicating that the number of collisions is related to the temperature in those cases. The term E in the exponent is referred to as the *activation energy*. The values of A and E are dependent on the reactions being considered, and some values are

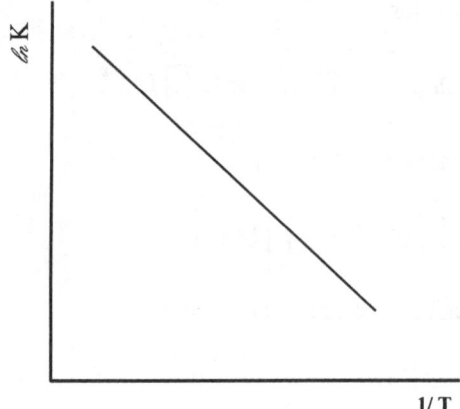

Fig. 14.2 Relationship between reaction rate and temperature

introduced below. The significance of the activation energy, E, is shown in Fig 14.3, where it can be seen to be the energy required to ionise a particular molecule and make it receptive to reacting. It appears in the exponential term because not all activated molecules will find conditions favourable for reaction.

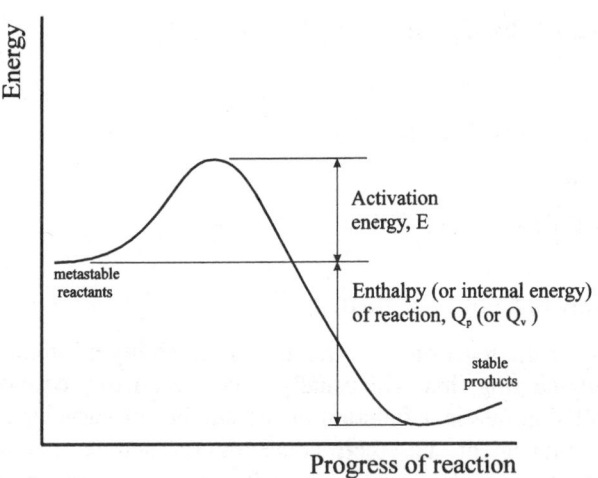

Fig. 14.3 Schematic interpretation of activation energy

14.4 Chemical kinetics of NO

The chemical kinetics for the formation of NO are relatively well understood, and will be developed here. The chemical kinetics for other pollutants can be derived in a similar way if the necessary reaction rates are available, although it should be recognised that most other pollutants are produced by reactions between the oxygen in the air and a constituent of the fuel (e.g. carbon or sulfur). The formation of NO in combustion processes can occur from two sources: *thermal NO* and *prompt NO*. Thermal NO is formed by the combination of the oxygen and nitrogen in the air, and will be produced even if there is no nitrogen in the fuel itself. This section will restrict itself to considering thermal NO. Prompt NO is thought to be formed in the flame as a result of the combination of the *nitrogen in the fuel* with the oxygen in the air. The amount of nitrogen in most conventional hydrocarbon fuels is usually very low.

The governing equations for the mechanism of NO formation are (Lavoie *et al.* 1970)

(1) $N + NO \Leftrightarrow N_2 + O$, $\qquad k_{f1} = 3.1 \times 10^{10} \times e^{(-160/T)}$ m³/kmol s $\qquad$ (14.19)

(2) $N + O_2 \Leftrightarrow NO + O$, $\qquad k_{f2} = 6.4 \times 10^6 \times T \times e^{(-3125/T)}$ m³/kmol s (14.20)

(3) $N + OH \Leftrightarrow NO + H$, $\qquad k_{f3} = 4.2 \times 10^{10}$ m³/kmol s $\qquad\qquad$ (14.21)

(4) $H + N_2O \Leftrightarrow N_2 + OH$, $\qquad k_{f4} = 3.0 \times 10^{10} \times e^{(-5350/T)}$ m³/kmol s $\qquad$ (14.22)

(5) $O + N_2O \Leftrightarrow N_2 + O_2$, $\qquad k_{f5} = 3.2 \times 10^{12} \times e^{(-18900/T)}$ m³/kmol s $\qquad$ (14.23)

(6) $O + N_2O \Leftrightarrow NO + NO$, $\qquad k_{f6} = k_{f5}$ $\qquad\qquad\qquad\qquad\qquad$ (14.24)

(7) $N_2O + M \Leftrightarrow N_2 + O + M$, $\quad k_{f7} = 10^{12} \times e^{(-30500/T)}$ m³/kmol s $\qquad$ (14.25)

In these equations the rate constants (k_{fi}) are all in m³/kmol s. M is a third body which may be involved in the reactions, but is assumed to be unchanged by the reactions. M can be assumed to be N_2. These equations can be applied to the zone containing 'burned' products, which exists after the passage of the flame through the unburned mixture. It will be assumed that H and OH, and O and O_2 are in equilibrium with each other; these values can be calculated by the methods described in Chapter 12.

14.4.1 RATE EQUATIONS FOR NITRIC OXIDE

The rate of formation of nitric oxide can be derived in the following manner. Consider reaction *j*: let k_{fj} be the forward reaction rate, k_{bj} the backward rate, and R_j the 'one way' equilibrium rate. Also let

$$[NO]/[NO]_e = \alpha, \qquad [N]/[N]_e = \beta, \qquad [N_2O]/[N_2O]_e = \gamma$$

where suffix e denotes equilibrium values. Then the following expressions are obtained for the burned gas, or product, volume.

Expression for nitric oxide, NO
From eqn (14.19) the net rate is

$$-k_{f1}[N][NO] + k_{b1}[N_2][O] = -\alpha\beta k_{f1}[N]_e[NO]_e + k_{b1}[N_2]_e[O]_e$$

But

$$k_{f1}[N]_e[NO]_e = k_{b1}[N_2]_e[O]_e = R_1,$$

so the net rate for eqn (14.19) becomes $-\alpha\beta R_1 + R_1$. Using a similar procedure for eqns (14.20), (14.21) and (14.24) involving NO gives the following expression, which also allows for the change in volume over a time step:

$$\frac{1}{V}\frac{d}{dt}([NO]V) = -\alpha(\beta R_1 + R_2 + R_3 + 2\alpha R_6) + R_1 + \beta(R_2 + R_3) + 2\gamma R_6 \qquad (14.26)$$

where V is the volume of the products zone.

Expression for atomic nitrogen, N
Equations (14.19), (14.20) and (14.21), which all involve N, give

$$\frac{1}{V}\frac{d}{dt}([N]V) = -\beta(\alpha R_1 + R_2 + R_3) + R_1 + \alpha(R_2 + R_3) \tag{14.27}$$

Expression for nitrous oxide N$_2$O
Equations (14.22) to (14.25) all involve N$_2$O and can be combined to give

$$\frac{1}{V}\frac{d}{dt}([N_2O]V) = -\gamma(R_4 + R_5 + R_6 + R_7) + R_4 + R_5 + \alpha^2 R_6 + R_7 \tag{14.28}$$

A finite time is required for the reactions to reach their equilibrium values; this is called the *relaxation time*. It has been found (Lavoie *et al.* 1970) that the relaxation times of reactions (14.27) and (14.28) are several orders of magnitude shorter than those of reaction (14.26), and hence it can be assumed that the [N] and [N$_2$O] values are at steady state, which means that the right-hand sides of eqns (14.27) and (14.28) are zero. Then, from eqn (14.27)

$$\beta = \frac{R_1 + \alpha(R_2 + R_3)}{(\alpha R_1 + R_2 + R_3)} \tag{14.29}$$

and from eqn (14.28)

$$\gamma = \frac{R_4 + R_5 + \alpha^2 R_6 + R_7}{(R_4 + R_5 + R_6 + R_7)} \tag{14.30}$$

These values can be substituted into eqn (14.26) to give

$$\frac{1}{V}\frac{d}{dt}([NO]V) = 2(1 - \alpha^2)\left[\frac{R_1}{1 + \alpha[R_1/(R_2 + R_3)]} + \frac{R_6}{1 + [R_6/(R_4 + R_5 + R_7)]}\right] \tag{14.31}$$

This is the rate equation for NO that is solved in computer programs to evaluate the level of NO in the products of combustion. It should be noted that the variation in the *molar concentration* is a first-order differential equation in time and relates the rate of change of [NO] to the instantaneous ratio of the actual concentration of NO to the equilibrium value (i.e. α). When the actual level of [NO] is at the equilibrium level then $\alpha = 1$ and the rate of change of [NO] $= 0$.

14.4.2   *INITIAL RATE OF FORMATION OF NO*

Heywood *et al.* (1971) derived the initial rate of formation of NO in the following way. It can be shown that eqn (14.31) is the dominant equation for the initial formation of NO. When the nitric oxide (NO) starts to form, the value of $\alpha = 0$, and eqn (14.31) can be written

$$\frac{1}{V}\frac{d}{dt}([NO]V) = 2R_1 \tag{14.32}$$

$$R_1 = k_{f1}[O]_e[N_2]_e \tag{14.33}$$

Heywood (1988) stated that, using a three equation set for the formation of NO, the value of the rate constant for the reaction in the forward direction is

$$k_{f1} = 7.6 \times 10^{13} \times e^{(-38000/T)} \text{ cm}^3/\text{mol s} \tag{14.34}$$

It is necessary to note two factors about eqn (14.34): the rate constant for the reaction is in cm³/mol s, and the exponential term is significantly different from that in eqn (14.19). Hence

$$\frac{1}{V} \frac{d}{dt} ([NO]V) = 2k_{f1}[O]_e[N_2]_e \tag{14.35}$$

Heywood (1988) also showed that

$$[O]_e = \frac{K_{p(O)}[O_2]_e^{1/2}}{(\Re T)^{1/2}}$$

where

$$K_{p(O)} = \text{equilibrium constant for the reaction } \tfrac{1}{2}O_2 \Leftrightarrow O$$
$$= 3.6 \times 10^3 \times e^{(-31090/T)} \text{ atm}^{1/2} \tag{14.36}$$

Substituting this value into eqn (14.34) gives

$$\frac{d[NO]}{dt} = \frac{6 \times 10^{16}}{T^{1/2}} e^{(-69090/T)}[O_2]_e^{1/2}[N_2]_e \text{ cm}^3/\text{mol s} \tag{14.37}$$

The values of $[O_2]_e$ and $[N_2]_e$, which should be in mol/cm³, can be obtained from an equilibrium analysis of the mixture. The value of $[O_2]_e$ can be calculated from the perfect gas law, because

$$[O_2]_e = \frac{n_{[O_2]_e}}{V} = \frac{p_{[O_2]_e}}{\Re T} = \frac{x_{[O_2]_e}p}{\Re T} \tag{14.38}$$

Likewise

$$[N_2]_e = \frac{x_{[N_2]_e}p}{\Re T} \tag{14.39}$$

Calculating the initial rate of change of NO with time gives the results shown in Fig 14.4. The conditions used to evaluate Fig 14.4 were: pressure, 15 bar; fuel, octane, which are based on those in Heywood (1988). Using these conditions the equilibrium concentrations of all the reactants were calculated using the techniques described in Chapter 12. This gave the following mole fractions for the condition $T = 2000$ K; $\phi = 0.6$:

$$x_O = 4.7725 \times 10^{-5}; \qquad x_{N_2} = 7.5319 \times 10^{-1}; \qquad x_{O_2} = 7.7957 \times 10^{-2}$$

The rate kinetic values, which depict the rate of change of mole fraction of NO, were then calculated in the following way. Equation (14.37) is based on cm³, mol and seconds rather than SI units of m³, kmol and seconds and this must be taken into account.

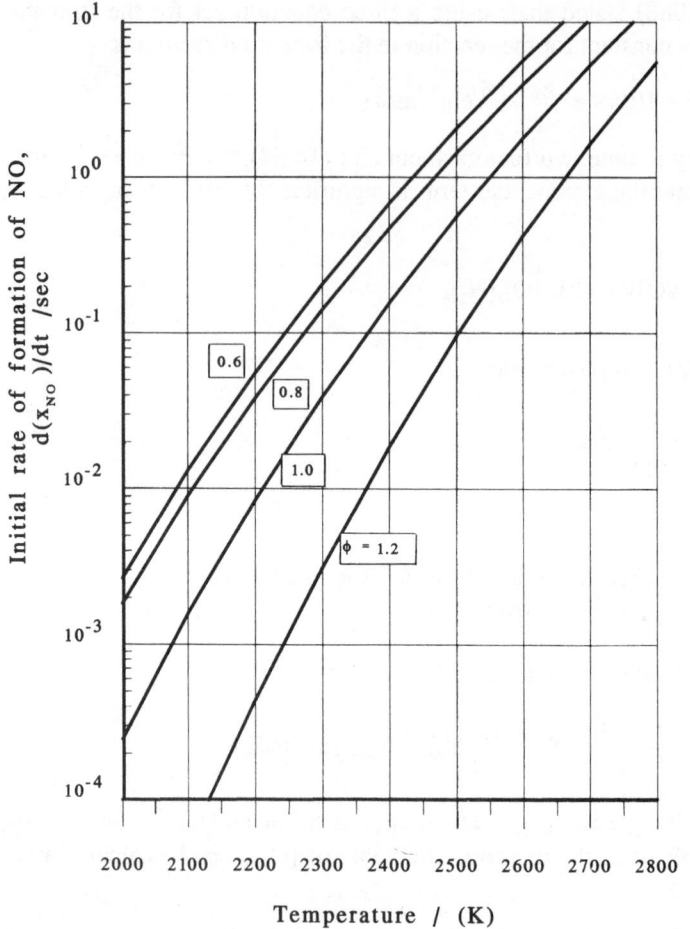

Fig. 14.4 Initial rate of formation of nitric oxide for combustion of octane at a pressure of 15 bar

The molar concentration is defined as

$$[M] = \frac{x_M n}{V} = \frac{x_M}{v_m}, \quad \text{where } M \text{ is a general substance} \tag{14.40}$$

$$v_m = \frac{\mathfrak{R}T}{p} = \frac{8.3143 \times 2000}{15 \times 10^5} \times 10^6 = 11086 \text{ cm}^3/\text{mol} \tag{14.41}$$

$$[O]_e = \frac{x_O}{11086} = \frac{4.7725 \times 10^{-5}}{11086} = 4.3051 \times 10^{-9} \text{ mol}/\text{cm}^3$$

$$[N_2]_e = \frac{x_{N_2}}{11086} = \frac{7.5319 \times 10^{-1}}{11086} = 6.7941 \times 10^{-5} \text{ mol}/\text{cm}^3$$

$$[O_2]_e = \frac{x_{O_2}}{11086} = \frac{7.7957 \times 10^{-2}}{11086} = 7.0320 \times 10^{-6} \text{ mol}/\text{cm}^3$$

Using a combination of eqns (14.34) and (14.35) gives

$$\frac{d[NO]}{dt} = 2 \times 7.6 \times 10^{13} \times e^{(-38000/T)}[O]_e[N_2]_e$$

$$= 15.2 \times 10^{13} \times e^{(-38000/2000)} \times 4.3051 \times 10^{-9} \times 6.7941 \times 10^{-5}$$

$$= 2.491 \times 10^{-7} \text{ mol/cm}^3 \text{ s} \tag{14.42}$$

A similar value can be obtained from eqn (14.37), which gives

$$\frac{d[NO]}{dt} = \frac{6 \times 10^{16}}{T^{1/2}} \times e^{(-69090/T)}[O_2]_e[N_2]_e$$

$$= \frac{6 \times 10^{16}}{2000^{1/2}} \times e^{(-69090/2000)} \times (7.0320 \times 10^{-6})^{1/2} \times 6.7941 \times 10^{-5}$$

$$= 2.4022 \times 10^{-7} \text{ mol/cm}^3 \text{ s} \tag{14.43}$$

These values can be converted from molar concentrations to more useful parameters and, in this case, they will be depicted as rate of formation of mole fraction. Rearranging eqn (14.40) gives

$$x_M = [M]v_m, \text{ where } M \text{ is a general substance} \tag{14.44}$$

Hence, the rate of production of NO in terms of mole fraction is given by

$$\frac{d(x_{NO})}{dt} = v_m \frac{d[NO]}{dt} \tag{14.45}$$

For this condition

$$\frac{d(x_{NO})}{dt} = 11086 \times 2.491 \times 10^{-7} = 2.762 \times 10^{-3} \text{ s}^{-1} \tag{14.46}$$

The curves in Fig 14.4 were calculated in this manner. It can be seen that either eqns (14.34) and (14.35), or eqn (14.37) give similar results. This indicates that the dissociation equation, eqn (14.36), for oxygen is similar to that used in the program for calculating the values of mole fraction of atomic oxygen.

Other kinetics-controlled pollutants

Carbon monoxide (CO) is formed in processes in which a hydrocarbon is burned in the presence of oxygen. If the mixture is rich (i.e. there is more fuel than oxygen available to oxidise it) then there is bound to be some carbon monoxide formed. However, even if the mixture is lean there will be some carbon monoxide due to the dissociation of the carbon dioxide; this was discussed in Chapters 12 and 13. The amount of carbon monoxide reduces as the mixture gets leaner, until misfire occurs in the combustion device.

Sulfur dioxide (SO$_2$) is formed by oxidation of the sulfur in the fuel.

Carbon particulates are formed during diffusion combustion because the oxygen in the air is not able to reach all the carbon particles formed during the pyrolysis process while they are at a suitable condition to burn.

14.5   The effect of pollutants formed through chemical kinetics

14.5.1   PHOTOCHEMICAL SMOG

Photochemical smogs are formed by the action of sunlight on oxides of nitrogen, and the subsequent reactions with hydrocarbons. Photochemical smogs were first identified in Los Angeles in the mid-1940s, and brought about the stringent legislation introduced in California to control the emissions from automobiles. These smogs act as bronchial irritants and can also irritate the eyes.

Among the pollutants involved in photochemical smogs are ozone, nitrogen dioxide and peroxyacyl nitrate (PAN). The nitrogen dioxide, and other oxides of nitrogen, are primary pollutants produced by dissociation in combustion reactions, and both 'prompt' and 'thermal' NO_X can be involved in the reactions. Ozone and PAN are secondary pollutants produced by the action of sunlight on the primary pollutants and the atmosphere. Glassman (1986) described in some detail the chemical chain that results in photochemical smog.

Oxides of nitrogen can be measured using non-dispersive infra-red (NDIR) equipment and the values are often quoted in parts per million (ppm), g/m^3 (at normal temperature and pressure), g/kWh, or g/mile (if applied to a vehicle engine).

14.5.2   SULFUR DIOXIDE (SO_x) EMISSIONS

Sulfur emissions can only occur if there is sulfur in the fuel. The levels of sulfur in diesel fuel and petrol are being reduced at the refinery in the developed world, and the sulfur levels are often around 0.5% or lower. SO_x emissions from vehicles are not a major problem. In the developing world, the levels of sulfur might be significantly higher and then the SO_x emissions from vehicles become a more serious problem. A further advantage of reducing the sulfur content of diesel fuels is that this reduces the synergy between the sulfur and the particulates, and reduces the quantity of particulates produced. The levels of sulfur dioxide produced in some power plant can be extremely high. Diesel engines running on 'heavy fuel', which has a high percentage of sulfur, will produce significant quantities of sulfur dioxide. Power stations running on heavy fuels or some coals will produce large quantities of sulfur fumes, which smell noxious, cause corrosion and form acid rain. Many coal burning power stations are now being fitted with flue gas desulfurisation, a process by which the sulfur dioxide reacts with limestone to form gypsum. These processes are suitable for stationary plant, but do consume copious amounts of limestone.

Sulfur in fuel has been a problem for a long time and caused high levels of acid rain in UK until recently. The move away from coal to gas for electricity generation, and the imposition of smokeless zones in urban areas have reduced the problem to a more acceptable level. Sulfur can be removed from crude oil by catalytic hydrodesulfurisation, but is often left in the residual oils which are used for marine propulsion, power generation, etc. The use of orimulsion, a fuel obtained by water extraction of low-grade oils, also brings worries of high sulfur fuels.

Sulfur in the fuel tends to be oxidised into SO_2 during the combustion process. This can then form sulfurous (H_2SO_3) and sulfuric (H_2SO_4) acids, which have corrosive effects on exhaust stacks, chimneys, etc., before being released into the surrounding atmosphere to form acid rain. It is possible to use flue gas desulfurisation to reduce the levels of SO_X emitted to the atmosphere, but such systems are expensive to install and run. It has been stated that if all the SO_2 emitted from the UK power stations were converted to sulfuric acid the supply of this acid would exceed the UK demand.

14.5.3 PARTICULATE EMISSIONS

These are generated mainly by devices which rely on diffusion burning (see Chapter 15), in which a rich zone can produce carbon particles prior to subsequent oxidation. Hence, particulate emissions tend to come from diesel engines, gas turbines and boiler plant. The carbon formed during combustion can be an asset in a boiler because it increases the radiant heat transfer between the flame and the tubes of the boiler; however, it is necessary to ensure that the soot emissions from the stack are not excessive. The carbon formed in diesel engine and gas turbine combustion is not beneficial and it increases the heat transfer to the components, increasing the need to provide cooling. The words *particulates* and *soot* are often used synonymously, but there is a difference in nature between these emissions. Dry soot is usually taken to mean the carbon that is collected on a filter paper in the exhaust of an engine. The unit of measurement of soot is usually the Bosch Smoke Number, which is assessed by the 'reflectance' of a filter paper on which the soot has been collected. Particulates contain more than simply the dry soot; they are soot particles on which other compounds (often polycyclic aromatic hydrocarbons [PAH]) have condensed. The PAH compounds have a tendency to be carcinogenic. The level of particulates is strongly affected by the amount of sulfur in the fuel, and increases with sulfur content. Particulates are measured by trapping the particles on glass-fibre filter papers placed in a dilution tunnel, and then weighing the quantity; they might be quoted in g/kW h.

14.5.4 GREENHOUSE EFFECT

The greenhouse effect is the name given to the tendency for the carbon dioxide in the atmosphere to permit the passage of visible (short wavelength) light, while absorbing the long wavelength infra-red transmission from objects on Earth. The effect of this is to cause the temperature of the Earth to increase. The evidence for global warming is not conclusive because the temperature of the Earth has never been constant, as shown by the Ice Ages of the past. There have been significant variations in temperature through the last few centuries, with warm summers at times and the freezing over of the Thames at others. It will take some time to prove conclusively whether global warming is occurring. What is indisputable is that mankind has released a large amount of carbon from its repository in hydrocarbons back into the atmosphere as CO_2. The most convenient form of energy available to man is that found as gas, oil or coal. These are all hydrocarbon fuels with the quantity of hydrogen decreasing as the fuel becomes heavier. The combustion of any of these fuels will produce carbon dioxide.

Obviously, the policy that should be adopted is for the amount of CO_2/kW or CO_2/mile to be reduced. This can be achieved by using more efficient engines, or changing the fuel to, say, hydrogen. The most efficient power station, using conventional fuels, in terms of CO_2/kW is the combined cycle gas turbine plant (CCGT) running on methane. Such plant can achieve thermal efficiencies higher than 50% and claims of efficiencies as high as 60% have been made. Large marine diesel engines also achieve thermal efficiencies of greater than 50%, and other smaller diesel engines can achieve around 50% thermal efficiencies when operating at close to full-load. It is presaged that the inter-cooled regenerated gas turbine will also have efficiencies of this order, and it is claimed, will compete with turbo-charged diesel engines down to the 5 MW size range.

The hydrogen powered engine produces no carbon dioxide, but does form NO_x. However, the major problem with hydrogen is its production and storage, particularly for mobile applications. California has now introduced the requirement for the zero emissions vehicle (ZEV) which must produce no emissions. At present the only way to achieve this is by an electric vehicle. It has been said that a ZEV is an electric car running in California on electricity produced in Arizona! This is the major problem that engineers have to explain to politicians and legislators: the Second Law states that you cannot get something for nothing, or break-even. Many ideas simply move the source of pollution to somewhere else.

14.5.5   CLEAN-UP METHODS

The clean-up methods to be adopted depend upon the pollutant and the application. As stated above, the exhaust of a power station can be cleaned up using a desulfurisation plant to remove the sulfur compounds. It is also possible to remove the grit from the power station boilers by cyclone and electro-static precipitators. These plants tend to be large and would be inappropriate for a vehicle, although investigations to adapt these principles to diesel engines are continuing.

All petrol-driven cars being sold in the USA, Japan and Europe are fitted with *catalytic converters* to clean up the gaseous emissions. This currently requires that the engine is operated at stoichiometric mixture strength so that there is sufficient oxygen to oxidise the unburned hydrocarbons (uHCs) while enabling the carbon monoxide (CO) and NO_x to be reduced. This means that the engine has to be operated under closed loop control of the air–fuel ratio, and this is achieved by fitting a λ sensor (which senses the fuel–air equivalence ratio) in the exhaust system. The error signal from the sensor is fed back to the fuel injection system to change the mixture strength. The need to operate the engine at stoichimetric conditions at all times has a detrimental effect on the fuel consumption, and investigations into lean operation (reducing) catalysts are being undertaken.

At present it is not possible to use catalytic converters on diesel engines because they always operate in the lean burn regime. This causes a problem because there is no mechanism for removing the NO_x produced in the diesel engine combustion process, and it has to be controlled in the cylinder itself. Lean burn catalysts, based on zeolites, are currently under research and will have a major impact on engine operation – both spark-ignition and diesel.

Another pollutant from diesel engines is the particulates. These are a significant problem and attempts have been made to control them by traps. Cyclone traps are being investigated, as well as gauze-based devices. A major problem is that after a short time the trap becomes full and the carbon has to be removed. Attempts have been made to achieve this by burning the particulates, either by self-heating or external heating. At present this area is still being investigated.

14.6   Other methods of producing power from hydrocarbon fuels

All of the devices discussed above, i.e. reciprocating engines, gas turbines, steam turbines, etc., produce power from hydrocarbon fuels by using the energy released from the combustion process to heat up a working fluid to be used in a 'heat engine'. Hence, all of these devices are limited by the Second Law efficiency of a heat engine, which is itself constrained by the maximum and minimum temperatures of the working cycle. It was shown in Chapters 2 and 4 that the maximum work obtainable from the combustion of a hydrocarbon fuel was equal to the change in Gibbs energy of the fuel as it was transformed from reactants to products: if there was some way of releasing all of this energy to produce power then a Second Law efficiency of 100% would be achievable. A device which can perform this conversion is known as a *fuel cell*; this obeys the laws of thermodynamics but, because it is not a heat engine, it is not constrained by the Second Law. Fuel cells are discussed in Chapter 17.

14.7 Concluding remarks

This is the first time non-equilibrium thermodynamics has been introduced. It has been shown that, while most thermodynamics processes are extremely rapid, it is not always possible for them to reach equilibrium in the time available.

The effect of rate kinetics on the combustion reactions is that some of the reactions do not achieve equilibrium and this can be a major contributor to pollutants. The rate equations for nitric oxide have been developed.

Finally, the pollutants caused by combustion have been introduced, and their effects have been described.

PROBLEMS

1 A reaction in which the pre-exponential term is independent of temperature is found to be a hundred times faster at 200°C than it is at 25°C. Calculate the activation energy of the reaction. What will be the reaction rate at 1000°C?

[30 840 kJ/kmol; 13 814]

2 A chemical reaction is found to be 15 times faster at 100°C than at 25°C. Measurements show that the pre-exponential term contains temperature to the power of 0.7. Calculate the activation energy of the reaction. What will be the reaction rate at 700°C?

[31 433 kJ/kmol; 15 204]

3 The rate of formation of nitric oxide (NO) is controlled by the three reversible chemical reactions:

$$O + N_2 \underset{k_1^-}{\overset{k_1^+}{\rightleftharpoons}} NO + N$$

$$N + O_2 \underset{k_2^-}{\overset{k_2^+}{\rightleftharpoons}} NO + O$$

$$N + OH \underset{k_3^-}{\overset{k_3^+}{\rightleftharpoons}} NO + H$$

Use the steady state approximation for the nitrogen atom concentration and the assumption of partial equilibrium for the reactions governing the concentrations of O, O_2, H and OH, show that

$$\beta = \frac{\delta R + \alpha}{\alpha R + 1}$$

where $\beta = [N]/[N]_e$, $\delta = [N_2]/[N_2]_e$, $\alpha = [NO]/[NO]_e$, $R = R_1/(R_2 + R_3)$, and R_j is the equilibrium reaction rate of reaction j, and [ ] denotes the molar concentration, and [ ]$_e$ is the equilibrium molar concentration. Derive an expression for $d[NO]/dt$ in terms of R_1, β, α and δ.

At a particular stage in the formation of nitric oxide the values of R and α are 0.26 and 0.1 respectively. Why is $\delta = 1$ likely to be a good approximation in this case? What is the error if the rate of formation of NO is evaluated from the larger approximation $d[NO]/dt = 2R_1$, rather than the equation derived in this question?

[3.64%]

4 The rate of change of mole concentration of constituent A in a chemical reaction is expressed as

$$\frac{d[A]}{dt} = -k[A]^n$$

While mole concentration is the dominant property in the reaction it is much more usual for engineers to deal in mole fractions of the constituents. Show that the rate of change of mole fraction of constituent A is given by

$$\frac{dx_A}{dt} \propto -kx_A^n \rho^{n-1}$$

where ρ = density. Also show how the rate of change of mole fraction is affected by pressure.

$[dx_A/dt \propto p^{n-1}]$

15
Combustion and flames

15.1 Introduction

Combustion is the mechanism by which the chemical (bond) energy in a 'fuel' can be converted into thermal energy, and possibly mechanical, power. Most combustion processes require at least two components in the *reactants* – usually a *fuel* and an *oxidant*. The chemical bonds of these reactants are rearranged to produce other compounds referred to as *products*. The reaction takes place in a *flame*. There are three parameters which have a strong influence on combustion: *temperature, turbulence* and *time*. In designing combustion systems attention must be paid to optimising these parameters to ensure that the desired results are achieved. In reciprocating engines the time available for combustion is limited by the operating cycle of the engine, and it is often necessary to increase the turbulence to counterbalance this effect. In furnaces the time available for combustion can be increased by lengthening the path taken by the burning gases as they traverse the chamber.

There are two basically different types of flame: *premixed* and *diffusion*. An example of premixed flames occurs in conventional spark-ignition (petrol, natural gas, hydrogen) engines. In these engines the fuel and air are mixed (often homogeneously) during the admission process to the engine, either by a carburettor or low pressure inlet manifold fuel injection system. Ignition is initiated by means of a spark, which ignites a small volume of the charge in the vicinity of the spark plug; this burning region then spreads through the remaining charge as a flame front. This type of combustion mechanism can be termed *flame traverses charge* (FTC), and once combustion has commenced it is very difficult to influence its progress. The diffusion flame occurs in situations of heterogeneous mixing of the fuel and air, when fuel-rich and fuel-lean regions of mixture exist at various places in the combustion chamber. In this case the progress of combustion is controlled by the ability of the fuel and air to mix to form a combustible mixture – it is controlled by the *diffusion* of the fuel and air. An example of this type of combustion is met in the diesel engine, where the fuel is injected into the combustion chamber late in the compression stroke. The momentum of the fuel jet entrains air into itself, and at a suitable temperature and pressure part of the mixture spontaneously ignites. A number of ignition sites may exist in this type of engine, and the fuel–air mixture then burns as the local mixture strength approaches a stoichiometric value. This type of combustion is controlled by the mixing (or diffusion) processes of the fuel and air. Other examples of diffusion combustion are gas turbine combustion chambers, and boilers and furnaces.

These different combustion mechanisms have an effect on how the energy output of the combustion process can be controlled. In the homogeneous, premixed, combustion process the range of air–fuel ratios over which combustion will occur is limited by the flammability of the mixture. In old petrol engines the mixture strength varied little, and was close to stoichiometric at all operating conditions (the mixture was often enriched at high load to reduce combustion temperatures). In modern engines, attempts are made to run the engines with weaker mixtures (i.e. *lean-burn*). At present, all spark-ignition engines operate with a throttle to limit the air flow at low loads. This is known as *quantitative governing* of the engine power because the *quantity* of charge entering the engine is controlled, and this controls the power output. Throttling the engine reduces the pressure in the inlet manifold, and effectively the engine has to *pump* air from the inlet manifold to the exhaust system, giving a negative pumping work of up to 1 bar mep. The advantage of lean-burn engines is that the load control can be partly by making the mixture leaner, and partly by throttling. Honda have an engine which will run as lean as 24:1 air–fuel ratio on part load; this reduces the combustion temperatures, and hence NO_x, and also reduces the pumping work over a substantial part of the operating range. The power output of a diesel engine is controlled by changing the amount of fuel injected into the cylinder – the air quantity is not controlled. This means that the overall air–fuel ratio of the diesel engine changes with load, and the *quality* of the charge is controlled. This is referred to as *qualitative governing* and has the benefit of not requiring throttling of the intake charge. The diesel engine operates over a broad range of air–fuel ratios. However, the richest operating regime of a diesel engine is usually at an air–fuel ratio of more than 20:1, whereas the petrol engine can achieve the stoichiometric ratio of about 15:1. This means that, even at the same engine speed, the diesel engine will only produce about 70% of the power output of the petrol engine. However, the lack of throttling, higher compression ratio, and shorter combustion period mean that the diesel engine has a higher thermal efficiency than the petrol engine. (Note: the efficiency of engines should be compared on the basis of the specific fuel consumption [kg/kW h] rather than miles per gallon because the calorific values [kJ/kg] of diesel fuel and petrol are approximately the same, but the calorific value of diesel fuel in kJ/m^3 is about 15% greater than that of petrol because the fuel density is about 15% greater. A diesel vehicle which achieves only 15% more miles/gallon than a petrol engine is not more efficient!) Gas turbine combustion chambers and boilers also control their output in a qualitative manner, by varying the air–fuel ratio.

An important feature of a flame is its rate of progress through the mixture, referred to as the *flame speed*. Flame speed is easiest understood when related to a premixed laminar flame, but the concept of flame speed applies in all premixed combustion.

15.2   Thermodynamics of combustion

When fuel and air are mixed together they are in metastable equilibrium, as depicted schematically in Fig 15.1. This means that the Gibbs energy of the reactants is not at the minimum, or equilibrium, value but is at a higher level. Fortunately it is still possible to apply the laws of thermodynamics to the metastable state, and hence the energies of reactants can be calculated in the usual way. Even though the reactants are not at the minimum Gibbs energy, spontaneous change does not usually occur: it is necessary to provide a certain amount of energy to the mixture to initiate the combustion – basically to 'ionise' the fuel constituents. Once the fuel constituents (usually carbon and hydrogen) have been 'ionised' they will combine with the oxygen in the air to form compounds with a lower

Gibbs energy than the reactants; these are called the *products* (usually carbon dioxide, water, etc). In the case of the spark-ignition engine the ionisation energy is provided by the spark, which ignites a small kernel of the charge from which the flame spreads. In the diesel engine part of the mixture produced in the cylinder cannot exist in the metastable state: it will spontaneously ignite. Such a mixture is termed a *hypergolic mixture*.

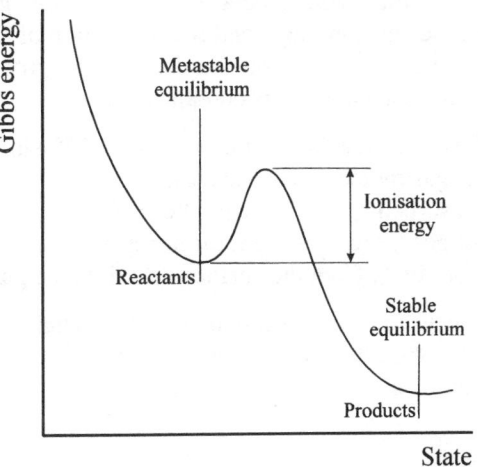

Fig. 15.1 Energy states associated with combustion

Some mixtures are unstable at room temperature, and their constituents spontaneously ignite. An example of such a mixture is hydrogen and fluorine. This concept of spontaneous ignition will be returned to later.

All of these thermodynamic processes take place somewhere in the combustion zone, and many of them occur in the flame. Before passing on to the detailed discussion of flames it is worthwhile introducing some definitions and concepts.

15.2.1   REACTION ORDER

The overall order of a reaction is defined as $n = \sum_{i=1}^{q} v_i$, summed over the q species in the reactants.

First-order reactions

These are reactions in which there is spontaneous disintegration of the reactants. These reactions do not usually occur, except in the presence of an 'inert' molecule.

Second-order reactions

These are the most common reactions because they have the highest likelihood of a successful collision occurring.

Third-order reactions

These are less likely to occur then second-order ones but can be important in combustion. An example is when OH and H combine to produce an H_2O molecule. This H_2O molecule will tend to dissociate almost immediately unless it can pass on its excess energy – usually to a nitrogen molecule in the form of increased thermal energy.

15.2.2　PROCESSES OCCURRING IN COMBUSTION

The thermodynamics of combustion generally relates to the gas in isolation from its surroundings. However, the surroundings, and the interaction of the gas with the walls of a container, etc, can have a major effect on the combustion process. The mechanisms by which the combusting gas interacts with its container are:

(i)　transport of gaseous reactant to the surface – diffusion;
(ii)　adsorption of gas molecules on surface;
(iii)　reaction of adsorbed molecules with the surface;
(iv)　desorption of gas molecules from the surface;
(v)　transport of products from the surface back into the gas stream – diffusion.

　These effects might occur in a simple container, say an engine cylinder, or in a catalytic converter. In the first case the interaction might stop the reaction, while in the second one it might enhance the reactions.

15.3　Explosion limits

The kinetics of reactions was introduced previously, and it was stated that reactions would, in general, only occur when the atoms or ions of two constituents collided. The reaction rates were derived from this approach, and the *Arrhenius equations* were introduced. The tendency for a mixture spontaneously to explode is affected by the conditions in which it is stored. A mixture of hydrogen and oxygen at 1 bar and 500°C will remain in a metastable state, and will only explode if ignited. However, if the pressure of that mixture is reduced to around 10 mm Hg (about 0.01 bar) there will be a spontaneous explosion. Likewise, if the pressure increased to about 2 bar there would also be an explosion. It is interesting to examine the mechanisms which make the mixture become hypergolic. The variation of explosion limits with state for the hydrogen–oxygen mixture is shown in Fig 15.2.

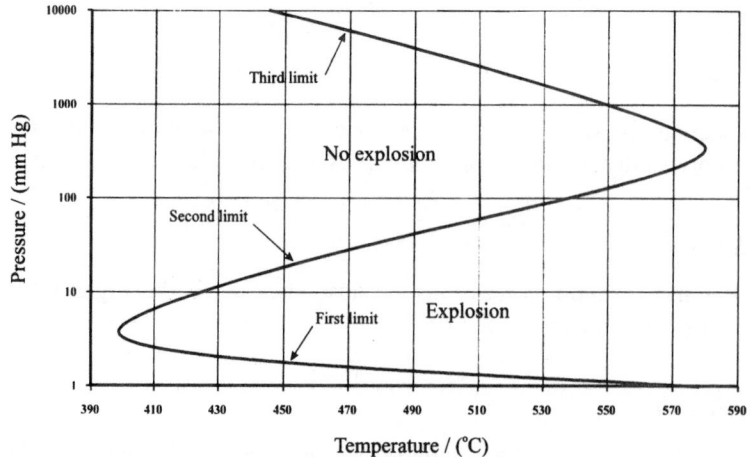

Fig. 15.2 Explosion limits for a mixture of hydrogen and oxygen (from Lewis and von Elbe, 1961)

The kinetic processes involved in the H_2–O_2 reaction are:

$$H + O_2 \rightarrow OH$$

which leads to the further branching steps

$$O + H_2 \rightarrow OH + H \qquad (15.1)$$
$$OH + H_2 \rightarrow H_2O + H$$

The first two of the three steps are called branching steps and produce two *radicals* (highly reactive ions) for each one consumed. The third step does not increase the number of radicals. Since all steps are necessary for the reaction to occur, the multiplication factor (i.e. the number of radicals produced by the chain) is between 1 and 2. The first step is highly *endothermic* (it requires energy to be *supplied* to achieve the reaction), and will be slow at low temperatures. This means that an H atom can survive a lot of collisions without reacting, and can be destroyed at the wall of the container. Hence, H_2–O_2 mixtures can exist in the metastable state at room temperatures, and explosions will only occur at high temperatures, where the first step proceeds more rapidly.

15.3.1 THE EFFECT OF MULTIPLICATION FACTOR ON THE TENDENCY TO EXPLODE

The effect of the multiplication factor can be examined in the following way. Assume a straight chain reaction has 10^8 collisions/s, and there is 1 chain particle/cm³, with 10^{19} molecules/cm³. Then all the molecules will be consumed in 10^{11} seconds, which is approximately 30 years. However, if the multiplication factor is now 2 then $2^N = 10^{19}$, giving $N = 62$. This means that all the reactants' molecules will be consumed in 62 generations of collisions, giving a total reaction time of 62×0^{-8} seconds, or 0.62 μs: an extremely fast reaction! If the multiplication factor is only 1.01 then the total reaction time is still only 10 ms. Hence, the speed of the reaction is very dependent on the multiplication factor of the reactions, but the overall multiplication factors do not have to be very high to achieve rapid combustion.

A general branched chain reaction may be written

$$M \xrightarrow{k_1} R \qquad \text{initiation}$$

$$R + M \xrightarrow{k_2} \alpha R + M \qquad \text{chain branching, } \alpha > 1$$

$$R + M \xrightarrow{k_3} P \qquad \text{product formation removes radical} \qquad (15.2)$$

$$\left. \begin{array}{l} R \xrightarrow[\text{wall}]{k_4} \text{destruction} \\[2mm] R \xrightarrow[\text{gas}]{k_5} \text{destruction} \end{array} \right\} \quad \text{chain termination}$$

where M is a molecule, R is a radical, and P is a product. α is the multiplication factor. The value of α necessary to achieve an explosion can be evaluated. The rate of formation of the product, P, is given by

$$\frac{d[P]}{dt} = k_3[R][M] \qquad (15.3)$$

The steady state condition for the formation of radicals is

$$\frac{d[R]}{dt} = 0 = k_1[M] + k_2(\alpha - 1)[R][M] - k_3[R][M] - k_4[R] - k_5[R] \qquad (15.4)$$

Solving eqn (15.4) for R and substituting into eqn (15.3) gives

$$\frac{d[P]}{dt} = \frac{k_1 k_2 [M]^2}{\{k_3[M] + k_4 + k_5 - k_2(\alpha - 1)[M]\}} \qquad (15.5)$$

The rate of production of the product P becomes infinite when the denominator is zero, giving

$$\alpha_{crit} = 1 + \frac{k_3[M] + k_4 + k_5}{k_2[M]} = \left(1 + \frac{k_3}{k_2}\right) + \frac{k_4 + k_5}{k_2[M]} \qquad (15.6)$$

Thus, if $\alpha_{react} > \alpha_{crit}$ the reaction is explosive; if the $\alpha_{react} < \alpha_{crit}$ then the combustion is non-explosive and progresses at a finite rate. A few explosion limits, together with flammability limits are listed in Table 15.1. This text will concentrate on non-explosive mixtures from now on.

Table 15.1 Flammability and explosion limits (mixtures defined in % volume) at ambient temperature and pressure

	Lean		Rich		
Mixture	Flammability	Explosion	Flammability	Explosion	Stoichiometric
H_2–air	4	18	74	59	29.8
CO–O_2	16	38	94	90	66.7
CO–air	12.5		74		29.8
NH_3–O_2	15	25	79	75	36.4
C_3H_8–O_2	2	3	55	37	16.6
CH_4–air	5.3		15		9.51
C_2H_6–air	3.0		12.5		5.66
C_3H_8–air	2.2		9.5		4.03
C_4H_{10}–air	1.9		8.5		3.13

15.4   Flames

A flame is the usual mechanism by which combustion of hydrocarbons takes place in air. It is the region where the initial breakdown of the fuel molecules occurs. There are two different types of flame, as described above: premixed flames and diffusion flames. Premixed flames will be dealt with first because it is easier to understand their mechanism.

15.4.1   *PREMIXED FLAMES*

Premixed flames occur in any homogeneous mixture where the fuel and the oxidant are mixed prior to the reaction. Examples are the Bunsen burner flame and the flame in most spark-ignited engines. Premixed flames can progress either as *deflagration* or *detonation* processes. This text will consider only deflagration processes, in which the flame progresses subsonically. Detonation processes do occur in some premixed, spark-ignited engines, when the 'end gas' explodes spontaneously making both an audible knock and causing damage to the combustion chamber components.

When considering laminar flame speed, it is useful to start with a qualitative analysis of a Bunsen burner flame, such as depicted in Fig 15.3. If the flow velocity at the exit of the tube is low then the flow of mixture in the pipe will be laminar. The resulting flame speed will be the *laminar flame speed*. While most flames are not laminar, the laminar flame speed is a good indication of the velocity of combustion under other circumstances. It can be seen from Fig 15.3(b) that the shape (or angle) of the inner luminous cone is defined by the ratio of the *laminar flame speed* (or *burning velocity*), u_f, to the flow velocity of the mixture. In fact, the laminar flame speed $u_f = u_g \sin \alpha$. While this is a relatively simple procedure to perform, it is not a very accurate method of measuring laminar flame speed because of the difficulty of achieving a straight-sided cone, and also defining the edge of the luminous region. Other methods are used to measure the laminar flame speed, including the rate of propagation of a flame along a horizontal tube and flat burners.

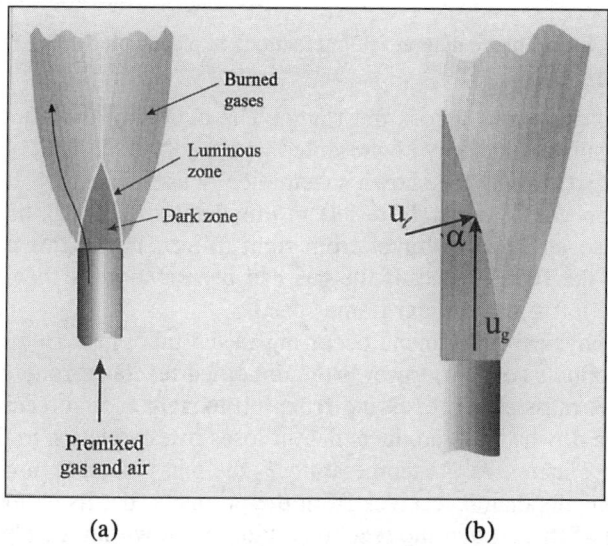

(a)   (b)

Fig. 15.3 Schematic diagram of Bunsen burner flame: (a) general arrangement; (b) velocity vectors

15.4.2 *LAMINAR FLAME SPEED*

There are a number of theories relating to laminar flame speed. These can be classified as:

- thermal theories
- diffusion theories
- comprehensive theories.

The original theory for laminar flame speed was developed by Mallard and le Chatelier (1883), based on a thermal model. This has been replaced by the Zel'dovitch and Frank-Kamenetsky (1938), and Zel'dovitch and Semenov (1940) model which includes both

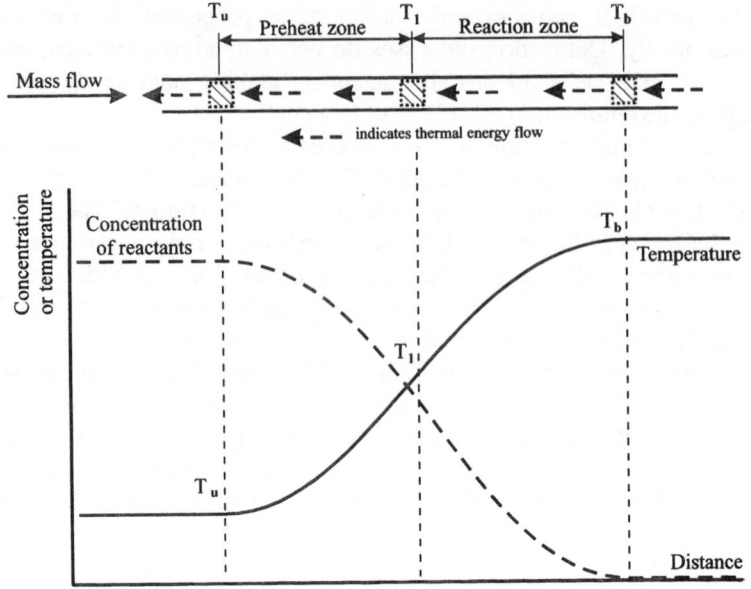

Fig. 15.4 Schematic diagram of interactions in plane combustion flame

thermal and species diffusions across the flame. The details of these models will not be discussed but the results will simply be presented.

A plane flame in a tube may be shown schematically as in Fig 15.4. It can be seen that the thermal diffusion goes from right to left in this diagram, i.e. against the direction of flow. The flame also attempts to travel from right to left, but in this case the gas flow velocity is equal to the flame speed. If the gas had been stationary then the flame would have travelled to the left at the laminar flame speed.

Considering the physical phenomena occurring in the tube: heat flows, by conduction, from the burned products zone (b) towards the unburned reactants zone (u), while the gas flows from u to b. A mass element passing from left to right at first receives more heat by conduction from the downstream products than it loses by conduction to the reactants, and hence its temperature increases. At temperature T_1 the mass element now loses more heat to the upstream elements than it receives from the products, but its temperature continues to increase because of the exothermic reaction taking place within the element. At the end of the reaction, defined by T_b, the chemical reaction is complete and there is no further change in temperature.

The Zel'dovitch *et al.* (1938, 1940) analysis results in the following equation for laminar flame speed

$$u_1 = \left(\frac{2k}{c_p \rho_u c_u} \frac{Z' e^{-E/RT_b}}{(T_b - T_u)} \frac{RT_b^2}{E} \right)^{1/2} \tag{15.7}$$

In obtaining eqn (15.7) the assumption had been made that the Lewis number

$$Le = k/\rho c_p D = 1$$

where $k =$ thermal conductivity, $\rho =$ density, $c_p =$ specific heat at constant pressure, and $D =$ mass diffusivity Hence, Le is the ratio between thermal and mass diffusivities, and this obviously has a major effect on the transport of properties through the reaction zone. The assumption $Le = 1$ can be removed to give the following results for first- and second-order reactions.

For first-order reactions

$$u_1 = \left\{ \frac{2k_b c_{p_b} Z'}{\rho_u \bar{c}_p^2} \frac{T_u}{T_b} \frac{n_R}{n_P} \frac{A}{B} \left(\frac{RT_b^2}{E}\right)^2 \frac{e^{-E/RT_b}}{(T_b - T_u)^2} \right\}^{1/2} \tag{15.8a}$$

and for second-order reactions

$$u_1 = \left\{ \frac{2k_b c_{p_b}^2 Z' c_u}{\rho_u \bar{c}_p^3} \left(\frac{T_u}{T_b}\right)^2 \left(\frac{n_R}{n_P}\right)^2 \left(\frac{A}{B}\right)^2 \left(\frac{RT_b^2}{E}\right) \frac{e^{-E/RT_b}}{(T_b - T_u)^3} \right\}^{1/2} \tag{15.8b}$$

where Z' is the pre-exponential term in the Arrhenius equation and c_u is the initial volumetric concentration of reactants.

Equations (15.8a) and (15.8b) can be simplified to

$$u_1 \approx \left\{ \frac{k c_{p_b}^2}{\rho_u \bar{c}_p^3} c_u Z e^{-E/RT_b} \right\}^{1/2} \approx \left\{ \frac{k}{\rho_u c_p} R \right\}^{1/2} \approx (\alpha R)^{1/2} \tag{15.9}$$

Hence the laminar flame speed is proportional to the square root of the product of thermal diffusivity, α, and the rate of reaction, R. Glassman (1986) shows that the flame speed can be written as

$$u_1 = \left\{ \frac{k_b}{\rho_u c_p} \frac{(T_b - T_{ig})}{(T_{ig} - T_u)} \frac{m_{w_b}}{\rho_u} R \right\}^{1/2} \tag{15.10}$$

which is essentially the same as eqn (15.9), where $R = Z e^{-E/RT_b}$. Obviously the laminar flame speed is very dependent on the temperature of the products, T_b, which appears in the rate equation. This means that the laminar flame speed, u_1, will be higher if the reactants temperature is high, because the products temperature will also be higher. It can also be shown that $u_1 \propto p^{(n-2)/2}$, where n is the *order of the reaction*, and $n \approx 2$ for a reaction of hydrocarbon with oxygen. This means that the effect of pressure on u_1 is small. Figure 15.5 (from Lewis and von Elbe, 1961) shows the variation of u_1 with reactant and mixture strength for a number of fundamental 'fuels'. It can be seen that, in general, the maximum value of u_1 occurs at close to the stoichiometric ratio, except for hydrogen and carbon monoxide which have slightly more complex reaction kinetics. It is also apparent that the laminar flame speed is a function both of the reactant and the mixture strength. The effect of the reactant comes through its molecular weight, m_w. This appears in more than one term in eqn (15.10) because density and thermal conductivity are both functions of m_w. The net effect is that $u_1 \propto 1/m_w$. This explains the ranking order of flame speeds shown in Fig 15.5, with the laminar flame speed for hydrogen being much higher than the others shown. While the molecular weight is a guide to the flame speed of a fuel, other more complex matters, such as the reaction rates, included as Z' in these equations, also have a big influence on the results obtained. Figure 15.6, from Metgalchi and Keck (1980, 1982) shows a similar curve, but for fuels which are more typical of those used in spark-ignition engines.

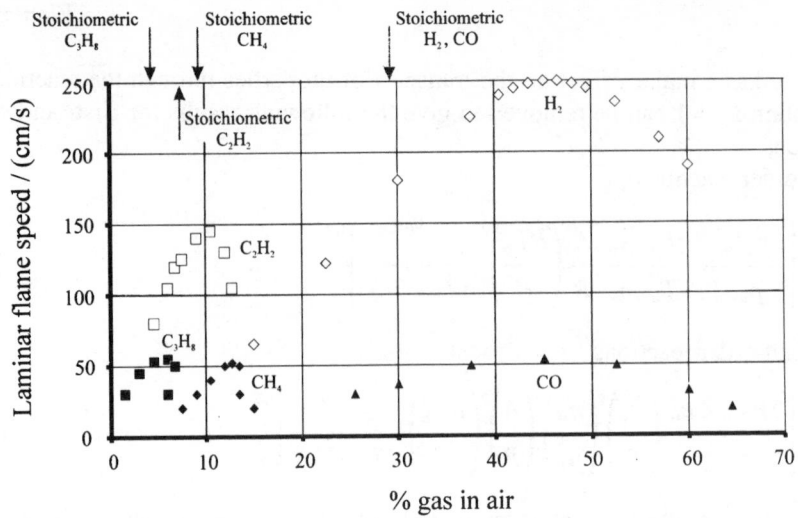

Fig. 15.5 Variation of laminar flame speed with reactants and mixture strength. $p = 1$ bar; $T_u = 298$ K

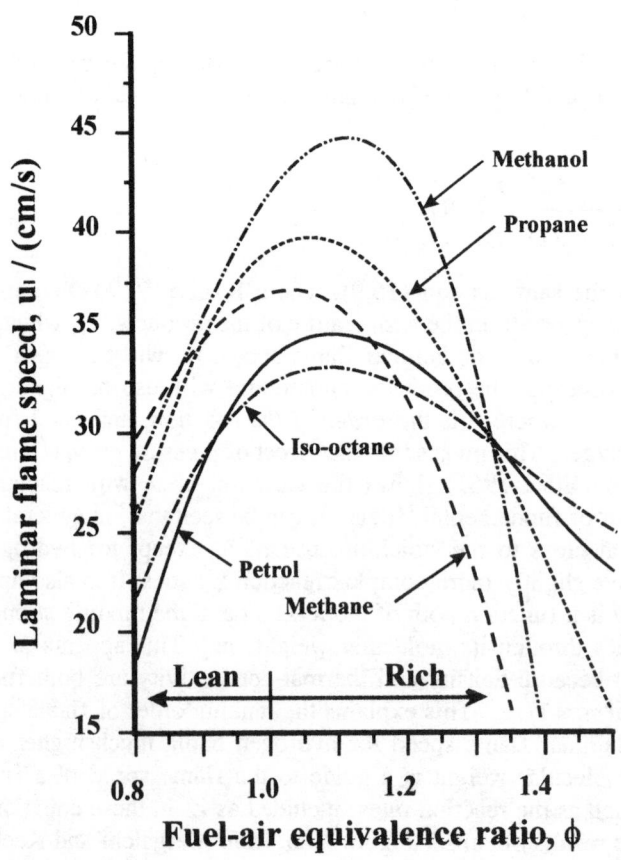

Fig. 15.6 Variation of laminar flame speed with mixture strength for typical fuels (based on 1 atm, 300 K)

It can be seen that the laminar flame speed is dependent on mixture strength, and this has a major influence on the design of engines operating with lean mixtures. The other feature to notice is that the laminar flame speed would remain approximately constant in an engine operating over a speed range of probably 800 to 6000 rev/min. If the combustion process depended on laminar burning then the combustion period in terms of crank angle would change by a factor of more than 7 : 1. Consider an engine operating with methane (CH_4) at stoichiometric conditions: the laminar flame speed is about 50 cm/s (0.5 m/s). If the engine bore is 100 mm then the combustion period will be 0.1 s. At 800 rev/min this is equivalent to 480° crankangle – longer than the compression and expansion periods! Obviously the data in Figs 15.5 and 15.6 are not directly applicable to an engine; this is for two reasons.

First, the initial conditions of the reactants, at temperatures of 298 K or 300 K, are much cooler than in an operating engine. Kuehl (1962) derived an expression for the combustion of propane in air, which gave

$$u_1 = \frac{0.78 \times 10^4}{\left(\dfrac{10^4}{T_b} + \dfrac{900}{T_u}\right)^{4.938}} p^{-0.09876} \qquad (15.11)$$

where

p = pressure (bar)

T = temperature (K)

u_1 = laminar flame speed (m/s)

It can be seen from eqn (15.11) that the effect of pressure on flame speed is very small, as suggested above. Figure 15.7 shows how the laminar flame speed increases with reactants temperature. It has been assumed that the adiabatic temperature rise remains constant at 2000 K, which is approximately correct for a stoichimetric mixture. It can be seen that the speed increases rapidly, and reaches a value of around 4 m/s when the reactants temperature

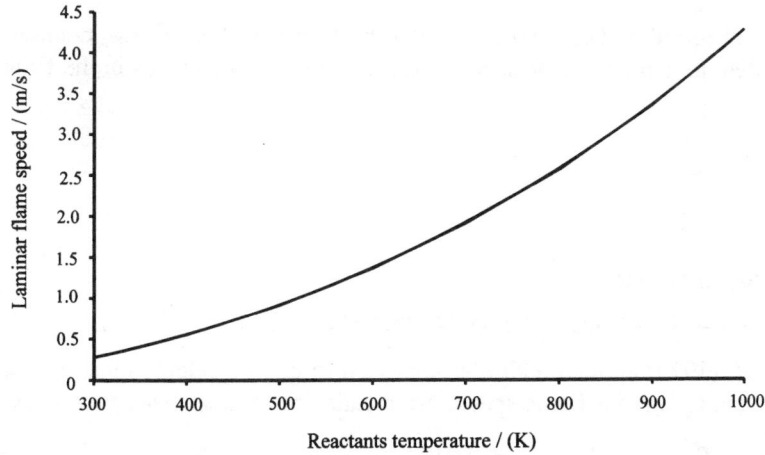

Fig. 15.7 Variation of laminar flame speed with reactants' temperature (predicted by Kuehl's equation with $p = 1$ bar)

is 1000 K. This is an increase of about a factor of 8 on the previous value which would reduce the combustion duration to about 60° crankangle. This is still quite a long combustion duration, especially since it has been evaluated at only 800 rev/min; some other feature must operate on the combustion process to speed it up. This, second, parameter is *turbulence*, which enhances the laminar flame speed as described below.

15.5.3   TURBULENT FLAME SPEED

It was shown previously that the laminar flame speed is too low to enable engines to operate efficiently, particularly if they are required to work over a broad speed range. The laminar flame speed must be enhanced in some way, and turbulence in the flow can do this. A popular model describing how turbulence increases the flame speed is the *wrinkled laminar flame model*, which is shown diagrammatically in Fig 15.8.

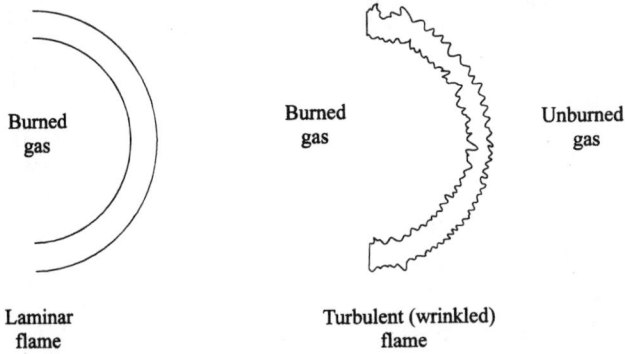

| Burned gas | Burned gas | Unburned gas |

Laminar flame   Turbulent (wrinkled) flame

Fig. 15.8 Comparison of laminar and turbulent flames

The effect of turbulence on the flame is threefold:

- the turbulent flow distorts the flame so that the surface area is increased;
- the turbulence may increase the transport of heat and active species;
- the turbulence may mix the burned and unburned gases more rapidly.

The theory of turbulent flames was initiated by Damkohler (1940) who showed that the ratio of turbulent to laminar flame speeds, based on large scale eddies in the flow, is

$$\frac{u_t}{u_l} = \sqrt{\frac{\varepsilon}{v}},\tag{15.12}$$

where

ε = eddy diffusivity

v = kinematic viscosity of the unburned gas

Heikal *et al.* (1979) applied a similar approach to engine calculations, and defined the ratio of turbulent to laminar flame speed, often called the *flame speed factor*, as

$$f_f = \frac{u_t}{u_l} = \left[1 + \frac{bv}{aP_r} \left(\frac{r(u_l + c\bar{V}_p^d)}{v} \right)^a \right]^{1/2}\tag{15.13}$$

where

α = molecular thermal diffusivity

P_r = Prandtl number

$\overline{V}_p$ = mean piston speed

a, b, c, d are all empirical constants

Experience shows that in spark ignition engines the flame speed factor is a strong function of engine speed but is not greatly affected by load. Experiments by Lancaster *et al.* (1976) have also shown that the level of turbulence intensity in an engine cylinder increases with engine speed, but is not quite proportional to it. This means that while turbulence increases the flame speed significantly, the length of the burning period increases as engine speed is increased, which explains why the ignition timing has to be advanced.

15.5 Flammability limits

Flammability limits were introduced in Table 15.1, where they were listed with the explosion limits. The flammability limit of a mixture is defined as the mixture strength beyond which, lean or rich, it is not possible to sustain a flame. The flammability limit, in practice, is related to the situation in which the flame is found. If the flame is moving in a confined space it will be extinguished more easily because of the increase in the interaction of the molecules with the walls; this is known as *quenching*. It is also possible for the flame to be extinguished if the level of turbulence is too high, when the flame is *stretched* until it breaks. The lean flammability limit is approximately 50% of stoichiometric, while the rich limit is around three times stoichiometric fuel–air ratio.

Bradley *et al.* (1987, 1992) investigated the way in which turbulence 'stretches' the flame and causes it to be extinguished. The results are summarised in Fig 15.9, which is a graph of the ratio of turbulent to laminar flame speed, i.e. the flame speed factor defined in eqn (15.13), against the ratio of turbulence intensity to laminar flame speed. The abscissa is closely related to the Karlovitz number, K, which is a measure of the flame stretch:

$$K = \left(\frac{1}{F_1} \frac{dF_1}{dt} \right) \tau_1 = \left(\frac{u'}{l_T} \right) \left(\frac{\delta_1}{u_1} \right) \tag{15.14}$$

where

F_1 = area of laminar flame;

τ_1 = transit time for flow through laminar flame;

u' = turbulence intensity;

l_T = Taylor microscale;

δ_1 = laminar flame thickness;

u_1 = laminar flame speed.

Also shown in Fig 15.9 are lines of constant KLe, which is the product of Karlovitz and Lewis numbers, and lines of constant Re/Le^2, which is the ratio of Reynolds number to the square of Lewis number. If the value of KLe is low (e.g. 0.005) then the flame is a wrinkled laminar one, while if the value of KLe is high (e.g. 6) the flame is stretched sufficiently to quench it. This shows that the design of a combustion chamber must be a compromise between a high enough value turbulence intensity to get a satisfactory flame speed, and one which does not cause extinction of the flame.

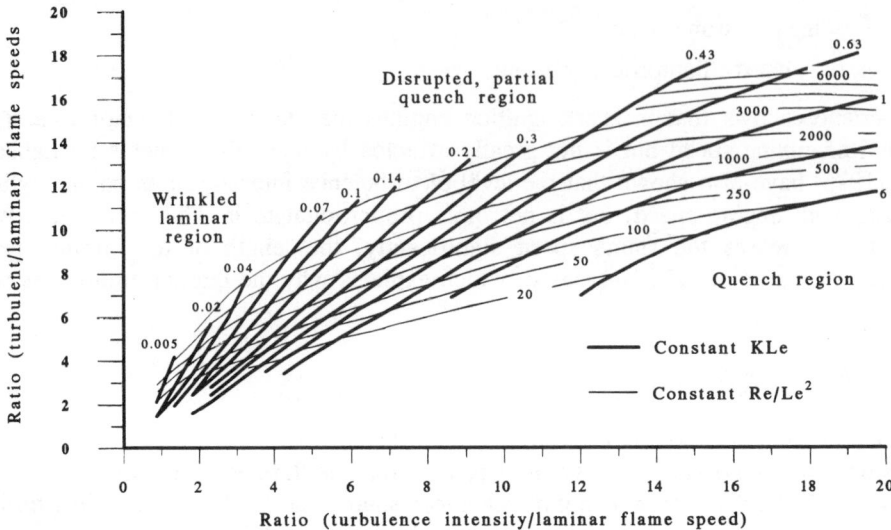

Fig. 15.9 The effect of turbulence on the turbulent flame speed and tendency to quench for a premixed charge

It can be seen from Table 15.1 that the spread of flammability for hydrogen is much higher than the other fuels This makes it an attractive fuel for homogeneous charge engines because it might be possible to control the load over a wide range of operation by qualitative governing rather than throttling. This would enable the hydrogen powered engine to achieve brake thermal efficiencies similar to those of the diesel engine. The restricted range of flammability limits for hydrocarbon fuels limits the amount of power reduction that can be achieved by *lean-burn* running; it also restricts the ability of operating the engine lean to control NO_x. Honda quote one of their engines operating as lean as 24 : 1 air–fuel ratio, which enables both the engine power to be reduced and the emissions of NO_x to be controlled. Such lean-burn operation is achieved through careful design of the intake system and the combustion chamber. In practice, in a car engine the lean limit is set by the driveability of the vehicle and the tendency to misfire. A small percentage of misfires from the engine will make the uHC emissions unacceptable – these misfires might not be perceptible to the average driver.

The rich and lean flammability limits come closer together as the quantity of inert gas added to a mixture is increased. Fortunately, the rich limit of flammability is more affected than the lean one, and basically the lean limit, which is usually the important one for engine operation, is not much changed.

15.6　Ignition

The ignition process is an extremely important one in the homogeneous charge engine because it has to be initiated by an external source of energy – usually a spark-plug. It can be shown that the minimum energy for ignition, based on supplying sufficient energy to the volume of mixture in the vicinity of the spark-gap to cause a stable flame, is

$Q_{p,min} \propto k^3/p^2u_1^3$. This means that the minimum energy is increased as the inverse of the laminar flame speed cubed. In other words, much more energy is required when u_1 is low. The typical level of energy required to ignite the mixture in a spark-ignition engine is around 30 mJ. As shown above, u_t is dependent on the initial temperature of the reactants, and the equivalence ratio of the charge (because this affects the adiabatic temperature rise, and hence T_b). This means that the strongest spark is needed when the engine is cold, and also if the engine is running lean. This dependence also explains why it is attractive to develop *stratified charge engines*, with a rich mixture zone around the spark-plug. The typical variation of minimum energy to ignite a mixture is shown in Fig 15.10.

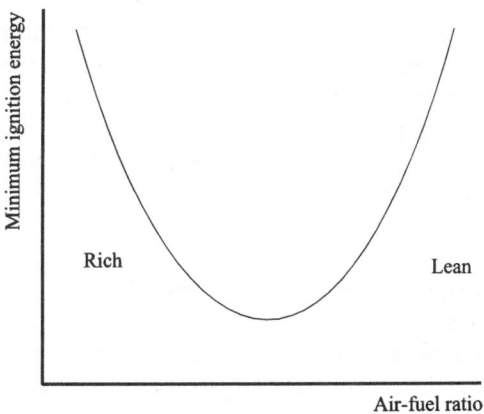

Fig. 15.10 Minimum ignition energy against air–fuel ratio

15.7 Diffusion flames

The previous sections all related to premixed flames of the type found in Bunsen burners, or spark-ignition engines. The other major class of flames is called *diffusion flames*; in these flames the rates of reaction are not controlled by the laminar flame speed but the rate at which the fuel and air can be brought together to form a combustible mixture. This type of combustion occurs in

- open flames, when mixing with secondary air enables combustion of a rich premixed core to continue to completion;
- gas turbine combustion chambers, when the liquid fuel sprays are mixed with the air in the combustion chamber;
- diesel engines, when the injected fuel has to mix with the air in the chamber before combustion can take place.

A typical arrangement of a diffusion flame might be that shown in Fig 15.11. This is a simple example of a jet of hydrogen passing into an oxygen atmosphere. The principle is the same if the fuel is a more complex gaseous one passing into air; the major difference will be that the products of combustion will be more complicated. The three sections across the jet (at A, B and C) show the way in which the oxygen diffuses into the jet, usually by turbulent mixing brought about by the jet entraining the surrounding oxygen. At section A the hydrogen and oxygen are completely separate, as indicated by the mass fraction curves. By section B, some way from the end of the nozzle, the hydrogen has been mixed with the oxygen just outside the jet diameter due to turbulent entrainment.

There has not yet been any mixing of hydrogen and oxygen in the potential core. Some combustion has also taken place by this section, as indicated by the mass fraction of water. Section C is located almost at the end of the mixing zone, after the end of the potential core, and there is no pure hydrogen left in the jet. The edge of the mixing zone is now well within the diameter of the jet.

Considering the concentration of hydrogen and oxygen along the centre-line of the jet, it can be seen that there is pure hydrogen right up to the end of the potential core. After that the hydrogen and oxygen on the centre-line combine to form water, and it is not until the end of the mixing zone that the oxygen concentration starts to rise again, as the water and oxygen mix and dilute each other.

In this example it has been assumed that the oxygen and hydrogen burn as soon as they come intimately in contact. This presupposes that the chemical reaction rate is much faster than the diffusion rates; this is usually a reasonable assumption.

A similar, but more complex analysis may be made of the injection of diesel fuel into the cylinder of a diesel engine. In this case the entrainment of the fuel and air does not take place in the gaseous phase, but occurs because the droplets of fuel leaving the nozzle impart their momentum on the surrounding air by aerodynamic drag. Once the fuel and air are mixed it is possible for the droplets to evaporate to create a combustible mixture. The details of diesel combustion are beyond this course, but the principles are similar to the combustion of gaseous jets. Gas turbine combustion chambers work in a similar way, with a spray of fuel entraining primary air to initiate the combustion, and subsequent entrainment of secondary air to complete the process. Individual combustion systems will now be considered.

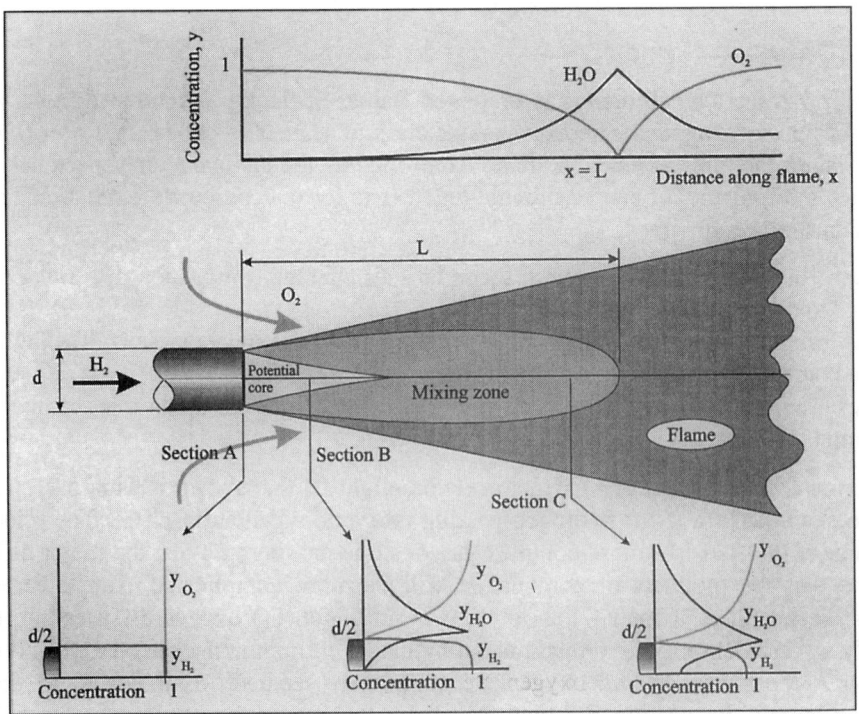

Fig. 15.11 Combustion of a jet of hydrogen in an oxygen atmosphere

15.8 Engine combustion systems

Different engines have different combustion systems which reflect the way in which the fuel and air are brought together. These systems will be described briefly.

15.8.1 SPARK-IGNITION ENGINES

In spark-ignition engines the fuel and air are usually premixed prior to admission to the engine cylinder. This is done either in a carburettor, or more commonly now by a port or manifold fuel injection system. [Some of the two- and four-stroke engines being investigated for use in cars do have direct injection into the engine cylinder using an air-blast injector. These engines will not be considered here, but are a hybrid between the fully premixed spark-ignition engine and the diesel engine in their combustion mechanism.] All these systems prepare the charge prior to it entering the cylinder, although it is probable that the fuel enters the cylinder with a large proportion in the liquid phase. Under fully warmed-up conditions this fuel will have evaporated by the time of ignition. At start-up this will not be the case, and enrichment, beyond stoichiometric, is done to ensure that the light fractions of the fuel give a combustible mixture; the remaining liquid fuel causes high levels of unburned hydrocarbons (uHCs). It was stated above that a high level of *turbulent* gas motion in the cylinder will increase the flame speed, and this can be achieved by various mechanisms. In older engines the shape of the piston and cylinder head produced a *squish* motion as the piston approached tdc, which enhanced the turbulence in the region of the spark plug, and increased the flame speed (see Fig 15.12).

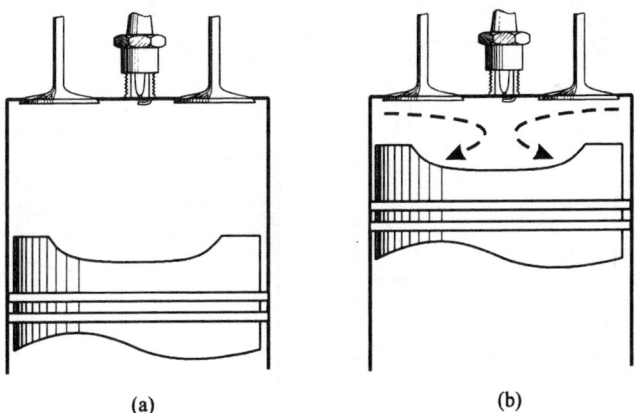

(a)   (b)

Fig. 15.12 Squish flow in 'bath tub' combustion chamber: (a) piston at mid-stroke; (b) piston approaching tdc and squishing gas from top land region

Other designs have been proposed, including the May combustion chamber (Figs 15.13(a) and (b)) which produces a high level of turbulence by 'squeezing' the gas into a small combustion chamber under either the intake or exhaust valve. While this system produces high turbulence it occurs too late to achieve its aims. A May chamber was fitted to a Jaguar engine and produced better fuel economy by enabling leaner mixtures to be used. More modern engines attempt to increase the turbulence levels around the spark plug by the break-up of *barrel swirl* or *tumble*. The gas entering the engine has a combination of swirl (vortex motion in a horizontal plane) and barrel swirl (vortex motion in the vertical plane) (see Figs 15.13 (c) and (d)). Swirl momentum is preserved during compression, but is not very useful for spark-ignition engine combustion. Barrel swirl cannot be preserved as ordered motion because the shape of the vortex is destroyed as the aspect ratio of the combustion chamber changes; barrel swirl is broken down into smaller scale motion which enhances the flame speed.

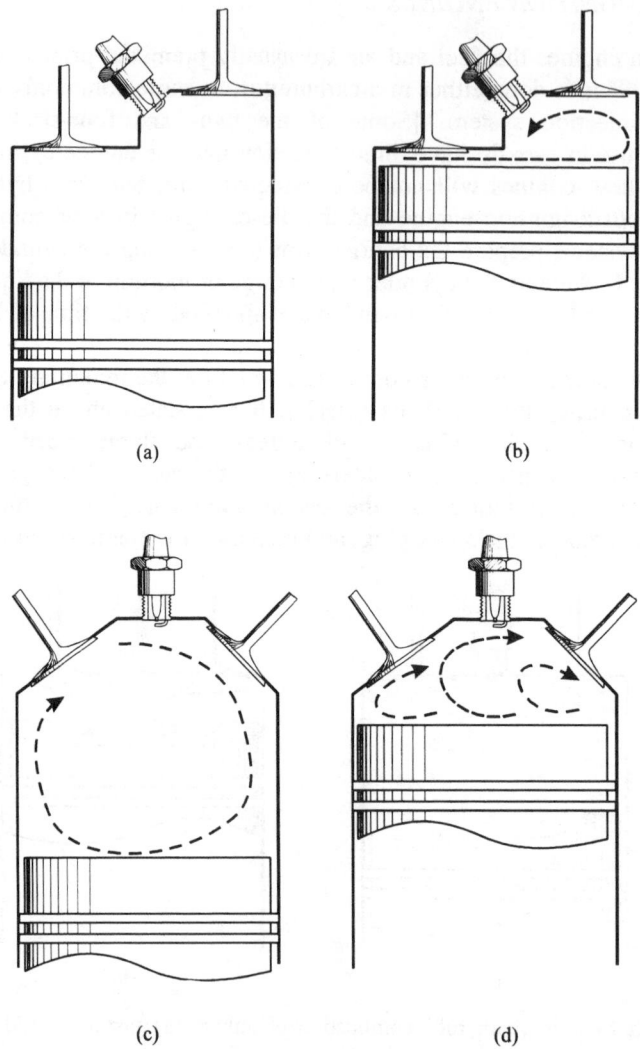

(a)   (b)

(c)   (d)

Fig. 15.13 High activity cylinder heads for spark-ignition engines: (a) May combustion chamber – note combustion chamber under valve; (b) May combustion chamber with piston near tdc – note squishing of charge into compact chamber; (c) modern pent-roof (4 valve) head – note barrel swirl set up in cylinder; (d) pent-roof chamber with piston near tdc – note high activity of gas around plug

Further features which must be borne in mind with spark-ignition engine combustion chambers are:

    (i)   reducing zones where combustion can be quenched (see top land region in Fig 15.11);

(ii)   limiting zones which might trap the end gases and cause *detonation*;

(iii)   reducing crevice volumes where uHCs might be trapped, e.g. around top ring
        groove.

15.8.2   DIESEL ENGINES

The design of diesel engine combustion chambers is different from that of spark-ignition
engines because of the nature of the diesel process. Fuel is injected into the diesel engine
cylinder, in liquid form, through a high pressure injector. This fuel enters the engine as a
jet, or jets, which has to entrain air to enable evaporation of the fuel and subsequent
mixing to a point where hypergolic combustion occurs. The mixing and combustion
processes are similar to those shown in Fig 15.11 for a gaseous jet. The droplet size of the
fuel varies but is of the order of 20 μm; the size depends on the injection nozzle hole size
and the fuel injection pressure, which might be 0.20 mm and 700 bar in a small high-speed
direct injection diesel engine. The prime considerations in the design of a combustion
chamber for a diesel engine are to obtain efficient mixing and preparation of the fuel and
air in the time available in the cycle.

The majority of small diesel engines for passenger cars use the *indirect injection* (*idi*)
process. In these engines, see Fig 15.14(b), the fuel is injected not directly into the main
combustion chamber but into a small swirl chamber (e.g. Ricardo Comet V) connected to
the main chamber by a throat. This approach was adopted because it was possible to
achieve good mixing of the fuel and air in the pre-chamber due to the high swirl velocities
that could be generated there as air was forced into the pre-chamber by the piston
travelling towards tdc. This approach had a further advantage that relatively simple fuel
injection equipment could be used, with a single hole nozzle and low pressure injection
pump. Combustion of the rich mixture commences in the pre-chamber, and the burning
gases enter the main chamber (which contains pure air) and generate high turbulence
which ensures good mixing of the burning plume and the 'secondary' air. Such combustion
systems are very tolerant of fuel quality, need relatively simple fuel injection equipment,

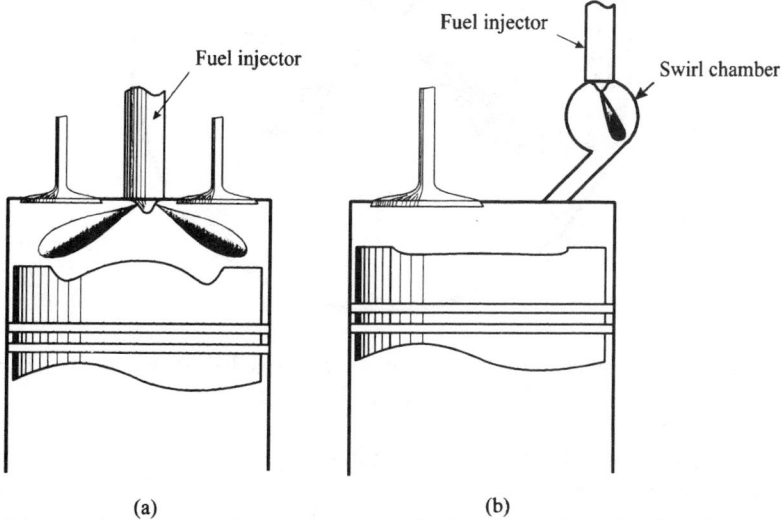

(a)   (b)

Fig. 15.14 Some combustion chamber arrangements for diesel engines: (a) basic schematic of
direct injection diesel engine; (b) basic schematic of indirect injection diesel engine

and can be run at relatively low air–fuel ratios before producing black smoke. Their disadvantage is that their fuel economy is about 10% or more worse than their direct injection counterparts.

Direct injection (di) diesel engines (shown schematically in Fig 15.14(a)) dominate the larger size range. Nearly all truck diesel engines are direct injection, as are those for rail and marine applications. The bottom limit of size for di diesel engines is reducing all the time, and four cylinder engines as small as 2 litres are available for van and car applications. The largest di diesel engines have *quiescent* combustion chambers, in which there is no organised air motion. The mixing of the fuel and air is achieved by the multiple fuel jets entraining air into themselves and bringing about the necessary mixing. More than six holes might be used in the fuel injector nozzle to give good utilisation of the air in the chamber. The relatively low engine speed of large engines allows sufficient time for combustion to occur. As the engine bore size reduces so the engine speed increases, and the time available for combustion becomes shorter. Also, the appropriate size of injector hole reduces, until it reaches the limit that can be achieved by production techniques (around 0.18 mm diameter). The increased engine speed, and reduced time for mixing, requires that the rate of mixing of the fuel and air is enhanced above that which can be achieved using a quiescent combustion chamber. The mixing rate can be increased by imparting air motion to the charge, and this is done in the form of *swirl* in the diesel engine. The fuel jets injected into a swirling flow have the peripheral fuel stripped off to form a combustible mixture, which is where ignition is initiated. Figure 15.15 shows the ignition points located by high speed photography in a high speed di engine.

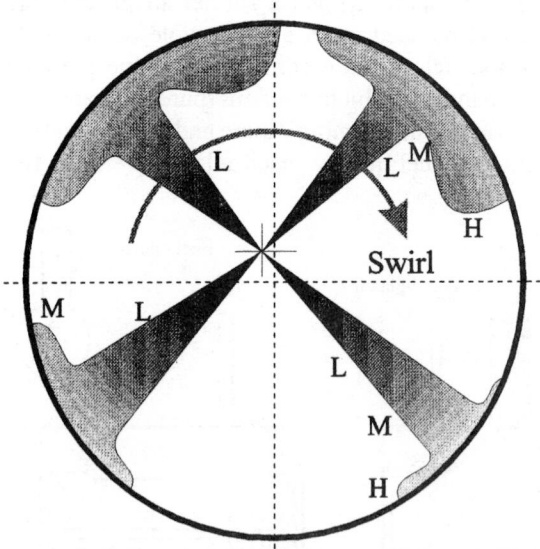

Fig. 15.15 Sprays and combustion initiation points in a small di diesel engine: H = high swirl; M = medium swirl; L = low swirl

Figure 15.15 also shows that the fuel jets impact upon the wall of the piston bowl. This also plays an important contribution in the mixing of the fuel and air, and is necessary at present because high enough mixing rates cannot be achieved in the centre of the bowl itself.

The design of combustion systems for diesel engines must aim at 'optimising' the overall combustion process. This takes place in two distinct modes: pre-mixed and diffusion combustion. The first fuel that is injected into the cylinder is not ready to ignite until it has evaporated and produced the necessary conditions for hypergolic combustion. This takes a finite time, referred to as the *ignition delay period*, during which the fuel is prepared but not yet ignited. At the end of this period there is rapid combustion of the premixed fuel and air, which gives rise to a high rate of heat release and produces high temperatures in the combustion chamber. This period has a major effect on the amount of NO_x produced in the engine. Typical equations for the ignition delay period are

$$t_{ig} = \frac{0.446e^{(4650/T_{ig})}}{p_{ig}^{1.19}} \text{ ms} \tag{15.15a}$$

or

$$t_{ig} = (0.36 + 0.22\bar{V}_p)e^{[E_A(1/RT - 1/17190)(21.2/(p - 12.4))^{0.63}]} \text{ deg crankangle} \tag{15.15b}$$

where

$\bar{V}_p$ = mean piston speed (m/s)

p = pressure (bar)

T = temperature (K)

E_A = activation energy = $618840/(CN + 25)$

CN = cetane number

Both eqns (15.15a) and (15.15b) have a similar form, and are related to the Arrhenius equation introduced in Chapter 14 to define the rate equations. Equation (15.15b) is a more recent formulation than eqn (15.15a), and has a more complex structure. Both equations are the result of experimental tests on engines with a range of fuels, and cannot be extended far beyond the regime under which they were evaluated, but they do give a basic structure for ignition delay. It should be noted that eqn (15.15b) contains a term for the cetane number of the fuel. The value of E_A reduces as the cetane number increases and this means that the ignition delay is inversely related to cetane number. A mechanical method for limiting the overall ignition delay is to use two-stage or split injection. In this type of system a small quantity of pilot fuel is injected into the cylinder some time prior to the main injection process. The pilot charge is prepared and ready to ignite before the main charge enters the chamber, and in this way the premixed combustion is limited to the pilot charge.

After the premixed period is over, the main combustion period commences, and this is dominated by diffusion burning, controlled by the mixing of the fuel and air. Whitehouse and Way (1970) attempted to model both periods by the two equations given below.

Reaction rate:

$$R = \frac{K}{N} \frac{p_{O_2}}{\sqrt{T}} e^{(-E_A/T)} \int_{\alpha_{inj}}^{\alpha} (P - R) \, d\alpha \tag{15.16a}$$

and preparation rate:

$$P = K' m_i^{(1-x)} m_u^x (p_{O_2})^z \tag{15.16b}$$

where

p_{O_2} = partial pressure of oxygen

P = rate of preparation of fuel by mixing

R = rate of reaction

m_i = mass of fuel injected

m_u = mass of fuel unburned

These equations result in instantaneous heat release patterns of the form shown in Fig 15.16. One of the diagrams has a short ignition delay, and it can be seen that the instantaneous rate of heat release does not reach such a high level as for the long delay. This is because the time for the physical and chemical processes to enable the fuel to reach a hypergolic state is less, and consequently less fuel is available for spontaneous ignition. The long delay results in a large amount of fuel burning spontaneously, with high temperature rises, high rates of pressure rise $(dp/d\alpha)$ and a high level of noise generation. This initial period is governed by the rate of reaction, R. After the premixed phase has taken place, the temperatures inside the combustion chamber are high and the rate of reaction is much faster than the rate of preparation, P. At this stage the combustion process is governed by eqn (15.16b), which models a diffusion process. During this process the rate of combustion is controlled by the rate at which the fuel and air mix, and in this phase the hydrocarbon fuel in the centre of the jet is burning in insufficient oxygen. The fuel pyrolyses and forms the precursors of the carbon particles produced in the exhaust system. It is important to mix the burning fuel with the air at the appropriate rate to ensure that the carbon produced during the combustion process is consumed before the exhaust valve opens.

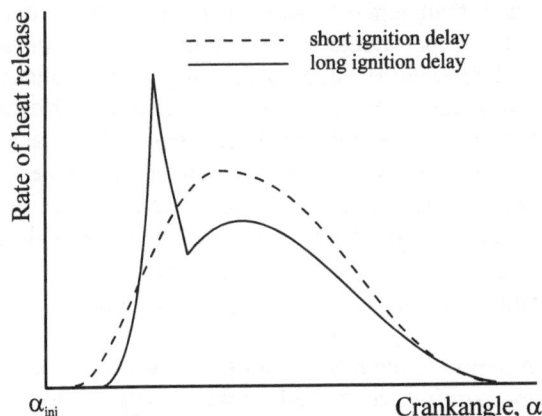

Fig. 15.16 Effect of ignition delay on rate of heat release diagram

15.8.3   GAS TURBINES

Combustion in gas turbines takes place at essentially constant pressure, although there is a small pressure drop through the chamber which should be taken into account when undertaking design. The *overall air–fuel ratio* in the combustion chamber will be around 40 : 1, although there will be regions in which much richer mixtures exist. Figure 15.17 is a schematic section of an *annular combustion chamber*, which will be located circumferentially around the body of the gas turbine. Annular combustion chambers have advantages over the older types of tubular combustors in terms of pressure loss and compactness.

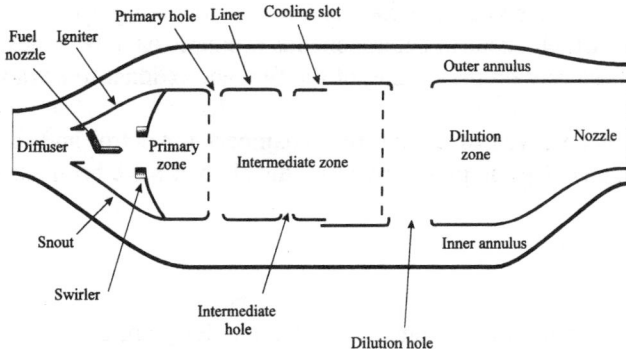

Fig. 15.17 Section through aero gas turbine annular combustion chamber

The combustion in the chamber takes place in a number of regions. When considering the combustion process it is useful to remember that the three parameters – *temperature, turbulence, time* – have to be optimised to achieve good, clean combustion. The air enters the chamber through the *diffuser*, where its velocity is reduced and the air-flow is distributed between the primary air entering the snout and the secondary air entering the inner and outer annuli. The *snout* imparts a high swirl into the flow to induce a strong recirculation region in the *primary zone*; this swirl enables the residence time in the primary zone to be lengthened without making this region excessively long. It also promotes good mixing through turbulence. The fuel injector will be either a high pressure spray atomiser or an air-blast injector. Sometimes a smaller pilot injector is used to assist ignition. Ignition is achieved by means of a powerful plasma jet igniter. In the *primary combustion zone* the air–fuel ratio will be approximately stoichiometric (15 : 1), and the adiabatic temperature rise will take the temperature up to about 2000 K, at which temperature NO_x will be formed. The gas will leave the primary zone and then be diluted with *secondary air* bleeding through the side walls of the chamber: the charge is then in the *intermediate zone*. This air will induce further turbulence and supply more oxygen to complete the burning of the soot formed during the rich primary combustion. The secondary air will also lower the gas temperature to about 1800 K, and reduce the amount of NO_x produced. Next, the gases will enter the *dilution zone* where they will be further diluted, and their temperature will be reduced to about 1300–1400 K. The design of the dilution zone will also reduce the circumferential temperature distribution. Finally the cooling air which has been passed down the outside of the combustion chamber will be mixed in with the hot gases, and the products will be passed to the turbine through the *nozzle*, which matches the flow from the combustion chamber to the requirements of the turbine.

15.9 Concluding remarks

The physical phenomena which affect combustion have been introduced. The differences between homogeneous, pre-mixed reactants and heterogeneous reactants, and their effect on combustion, have been described. It has been shown that pre-mixed combustible mixtures exhibit a characteristic combustion velocity called the laminar flame speed. The laminar flame speeed is related to the mixture strength and the temperature of the reactants, but is too slow for most engineering applications and must be enhanced by turbulence to achieve levels appropriate for powerplant. The level of turbulence must be balanced against the other combustion properties to ensure that the flame is not extinguished.

Diffusion flames which occur in heterogeneous mixtures have been described, and it has been shown that these rely on the mixing of the fuel and oxidant to achieve combustion of the mixture.

Examples have been given of combustion chambers for petrol and diesel engines, and gas turbines. The general principles derived in this chapter have been used to analyse their operation.

PROBLEMS

1 (a) One of the main problems encountered in the design of a diesel engine combustion system is the mixing of the air and fuel sufficiently rapidly to ensure complete combustion. Explain, using diagrams, how these problems are catered for in the design of

    (i)   large automotive diesel engines;
    (ii) the smallest automotive diesel engines.

Give two relative advantages of each type of combustion system.

(b) Compare and contrast the combustion systems of diesel and spark ignition engines in the forms they are applied to passenger cars.

2 (a) What is meant by the terms

    (i)   a global reaction;
    (ii)   an elementary reaction;
    (iii) a reaction mechanism.

(b) Describe the steps required to form a chain reaction and explain why chain reactions are important in combustion.

(c) A reaction is found to be 25 times faster at 400 K than at 300 K. Measurements of the temperature exponent yielded a value of 0.7. Calculate the activation energy of the reaction. How much faster will the reaction be at 1000 K?

3 A combustible mixture of gas and air is contained in a well insulated combustion bomb. It is ignited at a point and a thin flame propagates through the mixture, completely burning the reactants. This mechanism produces multiple zones of products. Prove that the temperature of an element of gas mixture which burned at pressure p_b has a temperature $T(p, p_b)$ at a pressure $p > p_b$:

$$T(p, p_b) = \left[T_1 \left(\frac{p_b}{p_1} \right)^{(\kappa - 1)/\kappa} + \frac{Q'_p}{c_p} \right] \left(\frac{p}{p_b} \right)^{(\kappa - 1)/\kappa}$$

where

> κ = ratio of specific heats,
>
> Q'_p = calorific value of fuel,
>
> and suffix 1 defines the conditions before ignition.

Calculate the final pressure, p_2, in terms of p_1, T_1, Q'_p, c_v, and κ. What is the difference between the final temperature and that of the first gas to burn if $T_1 = 300$ K, $\kappa = 1.3$ and $Q'_p/c_v = 1500$ K.

4 The structure of ethylene is $H_2C = CH_2$. Estimate the enthalpy of reaction when 1 kmol of ethylene is completely oxidised. Compare the value obtained with the tabulated value of -1323.2 MJ/kmol. Give reasons for the difference between the values.

Neglecting dissociation, find the temperature reached after constant pressure combustion of ethylene with 50% excess air if the initial temperature of the reactants is 400 K. The specific heat at constant pressure, c_p, of ethylene is approximately 1.71 kJ/kg K over the temperature range of the reactants.

[1302.3 MJ/kmol; 2028 K]

5 A method of improving engine fuel consumption and reducing the emissions of NO_x in a spark ignition engine is to run it lean, i.e. with a weak mixture. Discuss the problems encountered when running engines with weak mixtures, and explain how these can be overcome by design of the engine combustion chamber.

6 Describe the construction of a boiler for burning pulverised coal. Explain how this design optimises the temperature, turbulence and time required for good combustion. What are the main emissions from this type of plant, and how can they be reduced?

16
Irreversible Thermodynamics

16.1 Introduction

Classical thermodynamics deals with transitions from one equilibrium state to another and since it does not analyse the changes between state points it could be called *thermostatics*. The term *thermodynamics* will be reserved, in this chapter, for dynamic non-equilibrium processes.

In previous work, *phenomenological* laws have been given which describe irreversible processes in the form of proportionalities, e.g. Fourier's law of heat conduction, Ohm's law relating electrical current and potential gradient, Fick's law relating flow of matter and concentration gradient etc. When two of these phenomena occur simultaneously they interfere, or couple, and give rise to new effects. One such cross-coupling is the reciprocal effect of thermoelectricity and electrical conduction: the Peltier effect (evolution or absorption of heat at a junction due to the flow of electrical current) and thermoelectric force (due to maintenance of the junctions at different temperatures). It is necessary to formulate coupled equations to deal with these phenomena, which are 'phenomenological' inasmuch as they are experimentally verified laws but are not a part of the comprehensive theory of irreversible processes.

It is possible to examine irreversible phenomena by statistical mechanics and the kinetic theory but these methods are on a molecular scale and do not give a good macroscopic theory of the processes. Another method of considering non-equilibrium processes is based on 'pseudo-thermostatic theories'. Here, the laws of thermostatics are applied to a part of the irreversible process that is considered to be reversible and the rest of the process is considered as irreversible and not taken into account. Thomson applied the second law of thermostatics to thermoelectricity by considering the Thomson and Peltier effects to be reversible and the conduction effects to be irreversible. The method was successful as the predictions were confirmed by experiment but it has not been possible to justify Thomson's hypothesis from general considerations.

Systematic macroscopic and general thermodynamics of irreversible processes can be obtained from a theorem published by Onsager (1931a,b). This was developed from statistical mechanics and the derivation will not be shown but the results will be used. The theory, based on Onsager's theorem, also shows why the incorrect thermostatic methods give correct results in a number of cases.

16.2 Definition of irreversible or steady state thermodynamics

All previous work on macroscopic 'thermodynamics' has been related to equilibrium. A system was said to be in equilibrium when no spontaneous process took place and all the thermodynamic properties remained unchanged. The macroscopic properties of the system were spatially and temporally invariant.

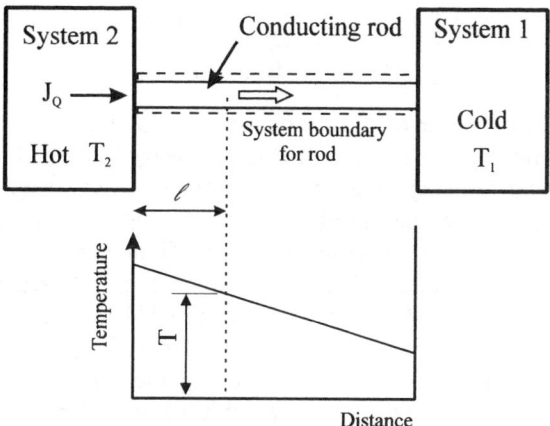

Fig. 16.1 Steady state conduction of heat along a bar

Consider the system shown in Fig 16.1, in which a thermally insulated rod connects two reservoirs at temperatures T_2 and T_1 respectively. Heat flows between the two reservoirs by conduction along the rod. If the reservoirs are very large the conduction of energy out of, or into, them can be considered not to affect them. A state will be achieved when the rate of heat flow, dQ/dt, entering the rod equals the rate of heat flow, $-dQ/dt$, leaving the rod. If a thermometer were inserted at any point in the rod the reading would not change with time, but it would be dependent on position. If the rod were of uniform cross-section the temperature gradient would be linear. (This is the basis of the Searle's bar conduction experiment.)

The temperature in the bar is therefore a function of position but is independent of time. The overall system is in a 'stationary' or 'steady' state but not in 'equilibrium', for that requires that the temperature be uniform throughout the system. If the metal bar were isolated from all the influences of the surroundings and from the heat sources, i.e. if it is made an *isolated system*, the difference between the *steady state* and *equilibrium* becomes obvious. In the case where the system was in *steady state*, processes would occur after the isolation (equalisation of temperature throughout the bar); where the system was already in *equilibrium*, they would not.

16.3 Entropy flow and entropy production

Still considering the conduction example given above. If the heat flows into the left-hand end of the bar due to an infinitesimal temperature difference, i.e. the process is reversible, the left-hand reservoir loses entropy at the rate

$$\frac{dS_2}{dt} = -\frac{1}{T_2}\frac{dQ}{dt} \tag{16.1}$$

Similarly the right-hand reservoir gains entropy at the rate

$$\frac{dS_1}{dt} = -\frac{1}{T_1} \frac{dQ}{dt} \tag{16.2}$$

Thus, the total change of entropy for the whole system is

$$\frac{dS}{dt} = \frac{dS_1}{dt} + \frac{dS_2}{dt} = \frac{dQ}{dt} \left[\frac{1}{T_1} - \frac{1}{T_2} \right] = \frac{dQ}{dt} \left[\frac{T_2 - T_1}{T_2 T_1} \right] \tag{16.3}$$

Now $T_2 > T_1$ and therefore the rate of change of entropy, $dS/dt > 0$.

To understand the meaning of this result it is necessary to consider a point in the bar. At the point l from the left-hand end the thermometer reading is T. This reading is independent of time and is the reading obtained on the thermometer *in equilibrium* with the particular volume of the rod in contact with it. Hence, the thermometer indicates the 'temperature' of that volume of the rod. Since the temperature is constant the system is in a 'steady state', and at each point in the rod the entropy is invariant with time. However, there is a net transfer of entropy from the left-hand reservoir to the right-hand reservoir, i.e. entropy is 'flowing' along the rod. The total entropy of the composite system is increasing with time and this phenomenon is known as 'entropy production'.

16.4   Thermodynamic forces and thermodynamic velocities

It has been suggested that for systems not far removed from equilibrium the development of the relations used in the thermodynamics of the steady state should proceed along analogous lines to the study of the dynamics of particles, i.e. the laws should be of the form

$$J = LX \tag{16.4}$$

where J is the *thermodynamic velocity* or *flow*;
$\quad\quad\;$ X is the *thermodynamic force*;
$\quad\quad\;$ L is a coefficient independent of X and J and is scalar in form, while both J and X are vector quantities.

The following simple relationships illustrate how this law may be applied.
Fourier's equation for one-dimensional conduction of heat along a bar is

$$\frac{dQ}{dt} = -kA \frac{dT}{dl} \tag{16.5}$$

where Q = quantity of energy (heat); T = temperature; A = area of cross-section; l = length; k = thermal conductivity.
Ohm's Law for flow of electricity along a wire, which is also one-dimensional, is

$$I = \frac{dq_l}{dt} = -\lambda A \frac{de}{dl} \tag{16.6}$$

where I = current; q_l = charge (coulomb); e = potential difference (voltage); A = area of cross-section of wire; l = length; λ = electrical conductivity.

Fick's Law for the diffusion due to a concentration gradient is, in one dimension

$$\frac{dn_i}{dt} = -k\frac{dC_i}{dl} \tag{16.7}$$

where n_i = amount of substance i, C_i = concentration of component i, and k = diffusion coefficient. This law was derived in relation to biophysics by analogy with the laws of thermal conduction to explain the flow of matter in living organisms. It will be shown later that this equation is not as accurate as one proposed by Hartley, in which the gradient of the ratio of chemical potential to temperature is used as the driving potential.

Other similar relationships occur in physics and chemistry but will not be given here. The three equations given above relate the flow of one quantity to a difference in potential: hence, there is a flow term and a force term as suggested by eqn (16.4). It will be shown that although eqns (16.5), (16.6) and (16.7) appear to have the correct form, they are not the most appropriate relationships for some problems. Equations (16.5), (16.6) and (16.7) also define the relationship between individual fluxes and potentials, whereas in many situations the effects can be coupled.

16.5   Onsager's reciprocal relation

If two transport processes are such that one has an effect on the other, e.g. heat conduction and electricity in thermoelectricity, heat conduction and diffusion of gases, etc., then *the two processes are said to be coupled*. The equations of coupled processes may be written

$$\left.\begin{array}{l} J_1 = L_{11}X_1 + L_{12}X_2 \\ J_2 = L_{21}X_1 + L_{22}X_2 \end{array}\right\} \tag{16.8}$$

Equation (16.8) may also be written in matrix form as

$$\begin{bmatrix} J_1 \\ J_2 \end{bmatrix} = \begin{bmatrix} L_{11} & L_{12} \\ L_{21} & L_{22} \end{bmatrix}\begin{bmatrix} X_1 \\ X_2 \end{bmatrix} \tag{16.8a}$$

It is obvious that in this equation the basic processes are defined by the diagonal coefficients in the matrix, while the other processes are defined by the off-diagonal terms.

Consider the situation where diffusion of matter is occurring with a simultaneous conduction of heat. Both of these processes are capable of transferring energy through a system. The diffusion process achieves this by mass transfer, i.e. each molecule of matter carries some energy with it. The thermal conductivity process achieves the transfer of heat by the molecular vibration of the matter transmitting energy through the system. Both achieve a similar result of redistributing energy but by different methods. The diffusion process also has the effect of redistributing the matter throughout the system, in an attempt to achieve the equilibrium state in which the matter is evenly distributed with the minimum of order (i.e. the maximum entropy or minimum chemical potential). It can be shown, by a more complex argument, that thermal conduction will also have an effect on diffusion. First, if each process is considered in isolation the equations can be written

$J_1 = L_{11}X_1$ the equation of conduction without any effect due to diffusion

$J_2 = L_{22}X_2$ the equation of diffusion without any effect due to conduction

Now, the diffusion of matter has an effect on the flow of energy because the individual, diffusing, molecules carry energy with them, and hence an effect for diffusion must be included in the term for thermal flux, J_1. Hence

$$J_1 = L_{11}X_1 + L_{12}X_2 \tag{16.9}$$

where L_{12} is the *coupling coefficient* showing the effect of mass transfer (diffusion) on energy transfer.

In a similar manner, because conduction has an effect on diffusion, the equation for mass transfer can be written

$$J_2 = L_{21}X_1 + L_{22}X_2 \tag{16.10}$$

where L_{21} is the coupling coefficient for these phenomena.

A general set of coupled linear equations is

$$J_i = \sum_k L_{ik}X_k \tag{16.11}$$

The equations are of little use unless more is known about the forces X_k and the coefficients L_{ik}. This information can be obtained from *Onsager's reciprocal relation*. There is considerable latitude in the choice of the forces X, but Onsager's relation chooses the forces in such a way that when each flow J_i is multiplied by the appropriate force X_i, the sum of these products is equal to the rate of creation of entropy per unit volume of the system, θ, multiplied by the temperature, T. Thus

$$T\theta = J_1X_1 + J_2X_2 + \cdots = \sum_i J_iX_i \tag{16.12}$$

Equation (16.12) may be rewritten

$$\theta = \sum_i J_ix_i \tag{16.13}$$

where

$$x_i = \frac{X_i}{T}$$

Onsager further showed that if the above mentioned condition was obeyed, then, in general

$$L_{ik} = L_{ki} \tag{16.14}$$

This means that the *coupling matrix* in eqn (16.8a) is *symmetric*, i.e. $L_{12} = L_{21}$ for the particular case given above. The significance of this is that the effects of parameters on each other are equivalent irrespective of which is judged to be the most, or least, significant parameter. Consideration will show that if this were not true then it would be possible to construct a system which disobeyed the laws of thermodynamics. It is not proposed to derive Onsager's relation, which is obtained from molecular considerations – it will be assumed to be true.

In summary, the thermodynamic theory of an irreversible process consists of first finding the *conjugate fluxes and forces*, J_i and x_i, from eqn (16.13) by calculating the entropy production. Then a study is made of the phenomenological equations (16.11) and

Onsager's reciprocal relation (16.14) is used to solve these. The whole procedure can be performed within the realm of macroscopic theory and is valid for any process.

16.6 The calculation of entropy production or entropy flow

In section 16.3 the concept of entropy flow was introduced. At any point in the bar, l, the entropy flux, J_S, may be defined as the entropy flow rate per unit cross-section area. At l the entropy flow rate will be

$$\frac{dS}{dt} = \frac{d}{dt}\left(\frac{Q}{T}\right) = \frac{1}{T}\frac{dQ}{dt} \tag{16.15}$$

and the entropy flux is

$$J_S = \frac{1}{A}\frac{dS}{dt} = \frac{1}{A}\left(\frac{1}{T}\frac{dQ}{dT}\right) \tag{16.16}$$

The heat flow rate J_Q is defined as

$$J_Q = \frac{dQ/dt}{A} \tag{16.17}$$

Hence

$$J_S = \frac{J_Q}{T} \tag{16.18}$$

The rate of production of entropy per unit volume, θ, is

$$\theta = \frac{d}{dV}\left(\frac{dS}{dt}\right) = \frac{d}{dl}\frac{(dS/dt)}{A} = \frac{d}{dl}\left(\frac{1}{A}\frac{dS}{dt}\right)$$

i.e.

$$\theta = \frac{dJ_S}{dl} \qquad \text{from eqn (16.16)} \tag{16.19}$$

Substituting from eqn (16.18) for J_S gives

$$\theta = \frac{d}{dl}\left(\frac{J_Q}{T}\right) = -\frac{J_Q}{T^2}\frac{dT}{dl} \tag{16.20}$$

Equation (16.20) is the rate of production of entropy per unit volume at point l in the rod where the temperature is T. This equation was derived for thermal conduction *only*.

If the rod was at a uniform temperature, i.e. in equilibrium, $dT/dl = 0$ and $\theta = 0$. For conduction $dT/dl < 0$, thus θ is positive, i.e. entropy is produced not dissipated.

A similar calculation may be made for the flow of electricity along a wire. The wire may be considered to be in contact along its length with a reservoir at temperature T. If an electric current density, J_l $(=I/A)$, flows due to a potential difference $d\varepsilon$, and since the wire is at constant temperature T because of its contact with the reservoir, then the electrical work must be equal to the heat transferred from the wire.

The rate of doing electrical work (power) is $-J_1A\,d\varepsilon$, and the total rate of heat production is $\dot{Q}$. Hence

$$\dot{Q} = -J_1A\,d\varepsilon \qquad (16.21)$$

The total change of entropy in volume dV is

$$\theta\,dV = -\frac{J_1A\,d\varepsilon}{T} \qquad (16.22)$$

Hence, the rate of production of entropy per unit volume is, because $dV = A\,dl$,

$$\theta = -\frac{J_1}{T}\frac{d\varepsilon}{dl} \qquad (16.23)$$

Equation (16.23) applies for electrical flow with *no* heat conduction.

16.7   Thermoelectricity – the application of irreversible thermodynamics to a thermocouple

16.7.1   *THERMOELECTRIC PHENOMENA*

As an aid to understanding the analysis which follows, a summary will be given of the effects involved in a thermocouple. A thermocouple consists of two junctions of dissimilar metals, one held at a high temperature the other at a low temperature, as shown in Fig 16.2.

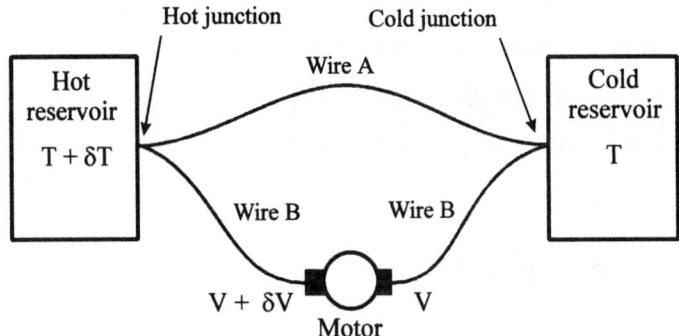

Fig. 16.2 Schematic of a thermocouple being used to drive an electric motor

Two different wires A and B are joined to form a loop with two junctions, each placed in a heat reservoir. A perfectly reversible motor is inserted in wire B. If the temperatures of the two reservoirs are different then not only will heat be transported along the wires A and B but also a flow of electrical current will occur. The following phenomena may be isolated:

(i)   If the entire system is kept at a uniform temperature and the motor is driven as a generator then a current will flow around the circuit. It is found that under these circumstances heating occurs at one junction and cooling at the other. This is known as the *Peltier effect* and the heating (or cooling) is equal to πI, where π = Peltier coefficient and I is the current. Because of the resistance of the wires it is necessary to do work to drive the generator. This work is equal to the product of current and potential difference across the wire and is called *Joulean heating*; this is the I^2R loss, which is dissipated in the reservoirs.

(ii)   If the reservoirs are put at different temperatures a potential difference is set up even when no electric current flows,. This is called the *Seebeck effect. Fourier conduction* will also occur along the wires.

(iii)   If both a temperature difference and electric current occur simultaneously then a third effect occurs. This is called the *Thomson heat* and occurs in a wire carrying a current of electricity along a temperature gradient. Consider a wire in a temperature field in the absence of an electric current. At each point in the wire there will exist a temperature as shown in Fig 16.1. Now suppose an electrical current flows. It is found that a flow of heat is required to keep the wire at the same temperature as previously. This effect is additional to Joulean heating and is proportional to the temperature gradient. The heat flux is given by

$$\frac{\mathrm{d}\dot{Q}}{\mathrm{d}l} = \sigma I \frac{\mathrm{d}T}{\mathrm{d}l} \tag{16.24}$$

where $\dot{Q}$ = rate of heat transfer; I = current; $\mathrm{d}T/\mathrm{d}l$ = temperature gradient in direction of I; and σ = the Thomson Coefficient.

The above laws are purely phenomenological. The various coefficients defined in them can be used in engineering practice. It is impossible to measure some of these effects in isolation. For example, if it is attempted to measure the potential difference between two points not at the same temperature, the measuring instrument and the electrical wires used to make the connections will constitute part of a thermocouple circuit. When making connections it is necessary that the measuring points are in an isothermal region. These connections will still give rise to contact and thermally induced voltages but these are usually negligibly small.

It is not possible to measure the thermoelectric characteristics of a single material, except for the Thomson effect, because they are junction effects. Hence, the *Seebeck and Peltier coefficients are the properties of pairs of metals*, not single metals.

16.7.2   UNCOUPLED EFFECTS IN THERMOELECTRICITY

Consider the two possible types of uncoupled flow. Let suffix 1 refer to heat flow processes and suffix 2 refer to electrical flow processes.

Heat flow

For uncoupled heat flow the rate of entropy production per unit volume is, from eqns (16.13) and (16.20),

$$\theta = \frac{J_1 X_1}{T} = \frac{J_Q X_1}{T} = -\frac{J_Q}{T^2} \frac{\mathrm{d}T}{\mathrm{d}l} \tag{16.25}$$

where

$$X_1 = -\frac{1}{T} \frac{\mathrm{d}T}{\mathrm{d}l} \tag{16.26}$$

Electrical flow

For uncoupled electrical flow, from eqns (16.13) and (16.23)

$$\theta = \frac{J_2 X_2}{T} = \frac{J_I X_2}{T} = -\frac{J_I}{T^2}\frac{d\varepsilon}{dl} \tag{16.27}$$

$$X_2 = -\frac{d\varepsilon}{dl} \tag{16.28}$$

16.7.3   THE COUPLED EQUATIONS OF THERMOELECTRICITY

The Onsager relations as given by eqn (16.8) may be applied, i.e.

$$\left.\begin{array}{l}J_1 = L_{11}X_1 + L_{12}X_2 \\ J_2 = L_{21}X_1 + L_{22}X_2\end{array}\right]$$

These become, in this case, for heat flow

$$J_Q = -\frac{L_{11}}{T}\frac{dT}{dl} - L_{12}\frac{d\varepsilon}{dl} \tag{16.29}$$

for electrical flow

$$J_I = -\frac{L_{21}}{T}\frac{dT}{dl} - L_{22}\frac{d\varepsilon}{dl} \tag{16.30}$$

From eqn (16.18), the entropy flux J_S is given by $J_S = J_Q/T$, and hence eqn (16.29) may be written

$$J_S = -\frac{L_{11}}{T^2}\frac{dT}{dl} - \frac{L_{12}}{T}\frac{d\varepsilon}{dl} \tag{16.31}$$

It has been assumed that both the electrical and heat flow phenomena may be represented by empirical laws of the form $J = LX$.

At constant temperature, Ohm's law states $I = -\lambda A\, d\varepsilon/dl$, giving

$$J_I = \frac{I}{A} = -\lambda\frac{d\varepsilon}{dl} \tag{16.32}$$

where λ is the electrical conductivity of the wire at constant temperature.

If dT is set to zero in eqn (16.30), i.e. the electrical current is flowing in the absence of a temperature gradient, then

$$-\lambda\frac{d\varepsilon}{dl} = -L_{22}\frac{d\varepsilon}{dl} \quad \Rightarrow L_{22} = \lambda \tag{16.33}$$

If the *entropy flux* in the wire is divided by the *electrical current flowing at constant temperature*, and this ratio is called *the entropy of transport, S^**, then

$$S^* = \left(\frac{J_S}{J_I}\right)_T = \left(-\frac{L_{12}}{T}\frac{d\varepsilon}{dl}\right)\bigg/\left(-\lambda\frac{d\varepsilon}{dl}\right) = \frac{L_{12}}{TL_{22}} \tag{16.34}$$

The entropy of transport is basically the rate at which entropy is generated per unit energy flux in an *uncoupled process*. It is a useful method for defining the cross-coupling terms in the coupled equations. Thus

$$L_{12} = TL_{22}S^* = \lambda TS^* \qquad (16.35)$$

By Onsager's reciprocal relation

$$L_{21} = L_{12} = \lambda TS^* \qquad (16.36)$$

Substituting for L_{22}, L_{12} and L_{21} in eqns (16.29) and (16.30) gives

$$J_Q = -\frac{L_{11}}{T}\frac{dT}{dl} - \lambda TS^*\frac{d\varepsilon}{dl} \qquad (16.37)$$

and

$$J_I = -\frac{\lambda TS^*}{T}\frac{dT}{dl} - \lambda\frac{d\varepsilon}{dl} \qquad (16.38)$$

Now if there is zero current flow (i.e. $J_I = 0$), Fourier's law of heat conduction may be applied, giving

$$J_Q = \frac{dQ/dt}{A} = -k\frac{dT}{dl} \qquad (16.39)$$

However, if $J_I = 0$ then eqn (16.38) gives

$$\frac{d\varepsilon}{dl} = -S^*\frac{dT}{dl} \qquad (16.40)$$

Substituting this value for $d\varepsilon/dl$ in eqn (16.37) gives

$$J_Q = -\frac{kdT}{dl} = -\frac{L_{11}}{T}\frac{dT}{dl} + \lambda S^{*2}T\frac{dT}{dl} \qquad (16.41)$$

Hence

$$L_{11} = (k + \lambda TS^{*2})T \qquad (16.42)$$

Substituting the coefficients into eqns (16.29) and (16.30) gives:
for heat flow

$$J_Q = -(k + \lambda TS^{*2})\frac{dT}{dl} - \lambda TS^*\frac{d\varepsilon}{dl} \qquad (16.43)$$

for electrical flow

$$J_I = -\lambda S^*\frac{dT}{dl} - \lambda\frac{d\varepsilon}{dl} \qquad (16.44)$$

The entropy flux is obtained from eqn (16.31).

$$J_S = -\frac{(k + \lambda TS^{*2})}{T}\frac{dT}{dl} - \lambda S^*\frac{d\varepsilon}{dl} \qquad (16.45)$$

It is possible to define another parameter, the heat of transport Q^*, where $Q^* = (J_Q/J_I)$. This is basically the 'thermal energy' which is transported by a flow of electrical energy when there is no temperature gradient, and indicates the magnitude of the off-diagonal terms in the cross-coupling matrix in eqn (16.8a).

From eqns (16.43) and (16.44), when $dT = 0$

$$\left(\frac{J_Q}{J_I}\right)_T = \left(-\lambda TS^* \frac{d\varepsilon}{dl}\right) \Big/ \left(-\lambda \frac{d\varepsilon}{dl}\right) = TS^* \tag{16.46}$$

Hence

$$Q^* = TS^* \tag{16.47}$$

Equation (16.47) is interesting because it retains the basic generic form relating 'heat', temperature and 'entropy', namely that entropy is evaluated by dividing a heat transfer term by a temperature term.

It is now possible to relate the equations derived above to the various physical phenomena observed in experiments. Previously, the Seebeck effect was defined as the potential difference set up in a wire due to a temperature gradient without any current flow (i.e. $(d\varepsilon/dT)_{J_I = 0}$). From eqn (16.44), if there is zero current flow

$$-S^* \frac{dT}{dl} = \frac{d\varepsilon}{dl} \quad \text{or} \quad \frac{d\varepsilon}{dT} = -S^* \tag{16.48}$$

Hence, S^* is a measure of the magnitude of the Seebeck effect, and the value of S^* for most materials is non-zero.

If the wire is kept at a constant temperature, a flow of heat (thermal energy) will occur due to the electrical potential difference. From eqn (16.43)

$$J_Q = -\lambda TS^* \frac{d\varepsilon}{dl} \tag{16.49}$$

This transport of thermal energy due to an electrical field is known as the Thomson effect.

16.7.4   THE THERMOCOUPLE

A thermocouple is a device for recording temperature at a point. It can be represented diagrammatically as shown in Fig 16.3.

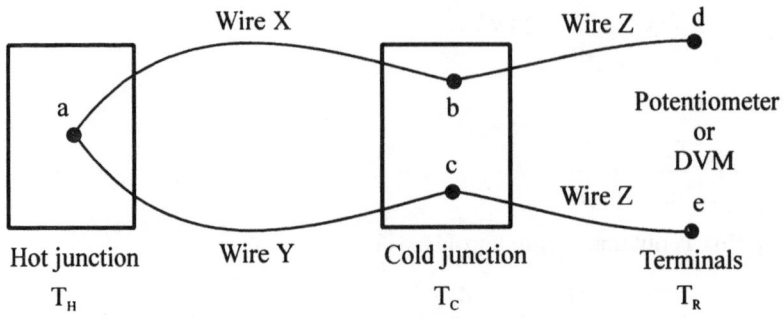

Fig. 16.3 Schematic diagram of a thermocouple

The thermocouple consists of two wires X and Y of dissimilar metals forming a junction at a. The ends b and c of the wires are immersed in an ice bath to form the cold junction and leads of material Z are connected to materials X and Y at points b and c, respectively, and these connections are inserted into a cold junction. These leads of material Z are then connected to a potentiometer or DVM at d and e.

The voltage representing temperature T_H is obtained due to the thermoelectric effect in the leads when zero current flows, i.e. the potentiometer is balanced and $J_l = 0$. At zero current, as previously discussed, the emf at the potentiometer (or digital voltmeter [DVM]) equals the thermal emf generated by the thermocouple.

From eqn (16.48) the potential difference between each end of a wire of material M is related to the temperature difference between the points by the equation.

$$d\varepsilon)_{J_l=0} = -S_M^* \, dT)_{J_l=0} \tag{16.50}$$

where S_M^* is the value of S^* for material M.

Equation (16.50) can be applied to each wire in the system shown above, to give the following:

wire ec

$$\varepsilon_c - \varepsilon_e = -\int_{T_R}^{T_C} S_Z^* \, dT$$

wire ca

$$\varepsilon_a - \varepsilon_c = -\int_{T_C}^{T_H} S_Y^* \, dT$$

wire ab

$$\varepsilon_b - \varepsilon_a = -\int_{T_H}^{T_C} S_X^* \, dT$$

wire bd

$$\varepsilon_d - \varepsilon_b = -\int_{T_C}^{T_R} S_Z^* \, dT$$

Adding these equations to obtain the potential difference between points e and d gives the potential difference measured by the potentiometer or DVM:

$$\varepsilon_d - \varepsilon_e = -\int_{T_R}^{T_C} S_Z^* \, dT - \int_{T_C}^{T_H} S_Y^* \, dT - \int_{T_H}^{T_C} S_X^* \, dT - \int_{T_C}^{T_R} S_Z^* \, dT$$

$$= -\int_{T_C}^{T_H} (S_Y^* - S_X^*) \, dT \tag{16.51}$$

This can be written as

$$\varepsilon_{X,Y} = \varepsilon_d - \varepsilon_e = \int_{T_C}^{T_H} (S_X^* - S_Y^*) \, dT \tag{16.52}$$

Thus, the emf of a thermocouple at any particular temperature T_H is dependent only on the materials between the *hot* and *cold* junctions if the leads bd and ce are both made of the same material. Since S^* is defined in terms of J_S and J_l it is independent of the length of the constituent wires, and hence the *emf generated is also independent of the length of the wires*.

Thermocouples are normally bought as pairs of wires which have been calibrated by experimental tests. For low temperatures (up to 150°C) Cu-Ni thermocouples may be used, for high temperatures platinum-rhodium is used. The calibration of the thermocouple is dependent only on the wires used to make the junctions; the length of thermocouples is not a parameter in the calibration, see eqns (16.50) and (16.52).

Thermocouple with junctions at T_C and T_H and connections in one of the measuring wires

Figure 16.3 showed an idealised thermocouple in which the connections between the thermoelectric pair of wires and those connecting the thermocouple to the measuring device were maintained at the cold junction temperature. It might be inconvenient to set up the thermocouple in this way, and a more convenient arrangement is shown in Fig 16.4.

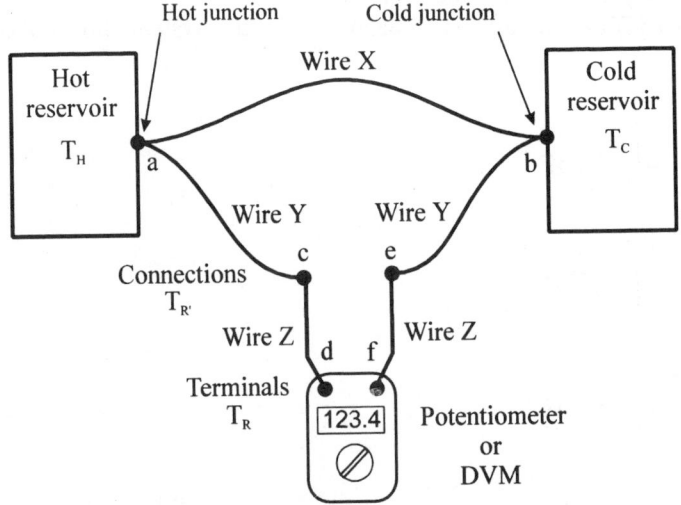

Fig. 16.4 Conventional layout of a thermocouple

It can be seen from Fig 16.4 that there are now four junctions between dissimilar materials, and each of these has the capability of generating an emf. This system will be analysed to determine the potential difference at the DVM.

Applying $d\varepsilon)_{J_l=0} = -S^* \, dT)_{J_l=0}$ to the system gives the following results:

Wire dc

$$\varepsilon_c - \varepsilon_d = -\int_{T^R}^{T^{R'}} S_Z^* \, dT$$

Wire cb

$$\varepsilon_b - \varepsilon_c = -\int_{T_{R'}}^{T_C} S_Y^* \, dT$$

Wire ba

$$\varepsilon_a - \varepsilon_b = -\int_{T_C}^{T_H} S_X^* \, dT$$

Wire *ae*

$$\varepsilon_e - \varepsilon_a = -\int_{T_H}^{T_{R'}} S_Y^* \, dT$$

Wire *ef*

$$\varepsilon_f - \varepsilon_e = -\int_{T_{R'}}^{T_R} S_Y^* \, dT$$

The emf at the potentiometer is given by

$$\varepsilon_f - \varepsilon_d = -\int_{T_{R'}}^{T_R} S_Z^* \, dT - \int_{T_H}^{T_{R'}} S_Y^* \, dT - \int_{T_C}^{T_H} S_X^* \, dT - \int_{T_{R'}}^{T_C} S_Y^* \, dT - \int_{T_R}^{T_{R'}} S_Z^* \, dT$$

$$= -\int_{T_C}^{T_H} S_X^* \, dT - \int_{T_H}^{T_{R'}} S_Y^* \, dT - \int_{T_{R'}}^{T_C} S_Y^* \, dT$$

$$= -\int_{T_C}^{T_H} S_X^* \, dT + \int_{T_C}^{T_H} S_Y^* \, dT = -\int_{T_C}^{T_H} (S_X^* - S_Y^*) \, dT \qquad (16.53)$$

This is the same result given before in eqn (16.51). There is one important point which is apparent from this result and that is that the two junctions between wires Y and Z, points *c* and *e*, must be at the same temperature, i.e. $(T_{R'})_c = (T_{R'})_e$. If these junctions are not at the same temperature then each of them will act as another junction and generate an electrical potential. Hence, care must be taken when setting up a thermocouple wired in a manner similar to Fig 16.4 to ensure that points *c* and *e* are close together and at the same temperature.

16.7.5 OTHER EFFECTS IN THERMOCOUPLES

Peltier effect or Peltier heating

As defined in section 16.7.1, this effect occurs when the entire system is kept at uniform temperature and a current flows around the circuit. The temperature of the system tends to rise due to both Ohmic and Peltier heating. If the Ohmic heating is neglected initially then the Peltier coefficient, defined below in eqn (16.54), can be evaluated

$$J_{Q\text{Peltier}} = \pi J_I \qquad (16.54)$$

where π = Peltier coefficient.

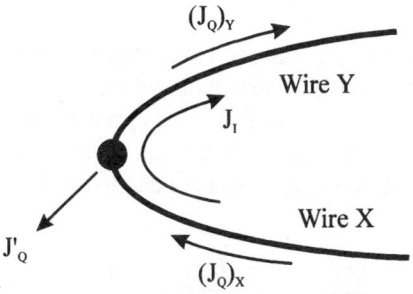

Fig. 16.5 Peltier heating at a junction

The Peltier heating effect at the junction in Fig 16.5 is given by

$$J_Q' = (J_Q)_X - (J_Q)_Y \tag{16.55}$$

where J_Q' is the heat that must be transferred to a reservoir to maintain the temperature of the junction at T. From eqn (16.54)

$$J_Q' = \pi_{X,Y} J_I \tag{16.56}$$

Hence, the heat transfer from the junction to maintain the temperature constant is

$$\pi_{X,Y} J_I = (J_Q)_X - (J_Q)_Y \tag{16.57}$$

This is the definition of the Peltier effect, and hence the Peltier coefficient is given by

$$\pi_{X,Y} = \left(\frac{J_Q}{J_I}\right)_X - \left(\frac{J_Q}{J_I}\right)_Y \tag{16.58}$$

The ratio $(J_Q/J_I)_{T=\text{constant}}$ was defined previously and termed heat of transport, Q^*. It was shown (eqn (16.47)) that $Q^* = TS^*$. Thus

$$\pi_{X,Y} = TS_X^* - TS_Y^* = T(S_X^* - S_Y^*) \tag{16.59}$$

It is apparent from the form of eqn (16.59) that the Peltier heating is reversible, i.e. if the current is reversed the direction of heat flow will be reversed.

Thomson effect or Thomson heating

This was defined previously in eqn (16.24). Rewriting this equation in the nomenclature now employed and applying it to the small element of wire, Δl, shown in Fig 16.6,

$$(J_Q)_{\text{Th}} = \sigma J_I \Delta T \tag{16.60}$$

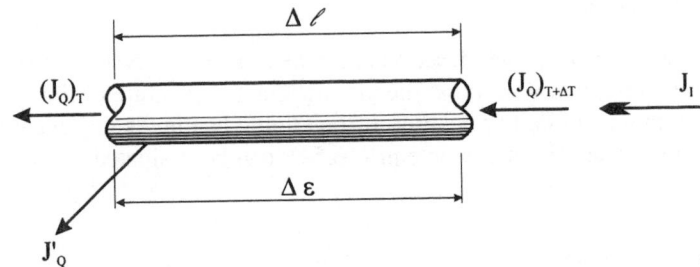

Fig. 16.6 Small element of a wire subject to current flow

Assume that initially there is no current flow, but that the heat flow gives rise to a temperature drop ΔT. Now if an electric current is switched on it is found that a flow of heat is required to keep the wire at the same temperatures as previously; this is in addition to Joulean heating (see section 16.7.1 (iii)). Let the required heat transfer, to a series of reservoirs, be J_Q'. Under these conditions the heat flow due to heat conduction is the same as previously.
Then

$$J_Q' = \underbrace{(J_Q)_{T+\Delta T}}_{\substack{\text{Heat flow rate, } dQ/dt, \\ \text{at } T+\Delta T}} - \underbrace{(J_Q)_T}_{\substack{\text{Heat flow rate,} \\ dQ/dt, \text{ at } T}} + \underbrace{J_I \Delta \varepsilon}_{\text{Ohmic heating}} \tag{16.61}$$

At any given temperature, T, the heat of transport, Q^*, is (eqn (16.47))

$$Q^* = \left(\frac{J_Q}{J_I}\right)_T = TS^*$$

and hence

$$(J_Q)_T = J_I TS^* \tag{16.62}$$

At a temperature, $T + \Delta T$

$$(J_Q)_{T+\Delta T} = J_I(T + \Delta T)\left(S^* + \frac{\mathrm{d}S^*}{\mathrm{d}T}\Delta T\right)$$

$$= J_I\left(TS^* + \frac{T\,\mathrm{d}S^*}{\mathrm{d}T}\Delta T + S^*\,\Delta T\right) \tag{16.63}$$

if multiples of small terms are negected. Now, for the element

$$\Delta\varepsilon = -\frac{\mathrm{d}\varepsilon}{\mathrm{d}l}\Delta l \tag{16.64}$$

and

$$\Delta T = -\frac{\mathrm{d}T}{\mathrm{d}l}\Delta l \tag{16.65}$$

Equation (16.38) can be rearranged to give

$$\frac{\mathrm{d}\varepsilon}{\mathrm{d}l} = -\frac{J_I}{\lambda} - S^*\frac{\mathrm{d}T}{\mathrm{d}l} \tag{16.66}$$

Substituting the above relations into eqn (16.61) gives

$$J_Q' = J_I\left(TS^* + \frac{T\,\mathrm{d}S^*}{\mathrm{d}T}\Delta T + S^*\,\Delta T\right) - J_I TS^* + J_I\left(\frac{J_I}{\lambda}\Delta l - S^*\,\Delta T\right)$$

$$= \underbrace{J_I T\frac{\mathrm{d}S^*}{\mathrm{d}T}\Delta T}_{\substack{\text{Thomson heat extracted}\\\text{to maintain temperature}}} + \underbrace{J_I^2\frac{\Delta l}{\lambda}}_{\text{Ohmic heat generated}} \tag{16.67}$$

Now the Thomson heat was defined in eqn (16.60) as

$$J_Q = \sigma J_I\,\Delta T$$

Thus

$$\sigma = -T\frac{\mathrm{d}S^*}{\mathrm{d}T} \tag{16.68}$$

For the thermocouple shown in Fig 16.3, the difference in Thomson coefficients for the two wires is

$$\sigma_X - \sigma_Y = -T \frac{\mathrm{d}}{\mathrm{d}T} (S_X^* - S_Y^*) \qquad (16.69)$$

16.7.6 SUMMARY

The equations for thermocouple phenomena for materials X and Y acting between temperature limits of T_H and T_C are as follows:

Seebeck effect
Seebeck coefficient:

$$\varepsilon_{X,Y} = \int_{T_C}^{T_H} (S_X^* - S_Y^*)\mathrm{d}T$$

Peltier effect
Peltier coefficient:

$$\pi_{X,Y} = T(S_X^* - S_Y^*)$$

Thomson effect
Thomson coefficient:

$$\sigma_X - \sigma_Y = -T \frac{\mathrm{d}}{\mathrm{d}t} (S_X^* - S_Y^*)$$

16.8 Diffusion and heat transfer

16.8.1 BASIC PHENOMENA INVOLVED

The classical law of diffusion is due to Fick (1856). This states that the diffusion rate is proportional to the concentration gradient, and is the mass transfer analogy of the thermal conduction law. The constant of proportionality in Fick's law is called the diffusion coefficient. It was found from experimental evidence that the diffusion coefficient tended to vary with conditions and that better proportionality could be obtained if the rate of diffusion was related to the gradient of chemical potential; this law is due to Hartley. Figure 16.7 shows an adiabatic system made up of two parts connected by a porous membrane, or a pipe with a bore that is small compared with the mean free path of the molecules.

The Soret effect

This is a thermal diffusion effect. It is characterised by the setting up of a concentration gradient as a result of a temperature gradient.

The Dufour effect

This is the inverse phenomena to the Soret effect and is the non-uniformity of temperature encountered due to concentration gradients.

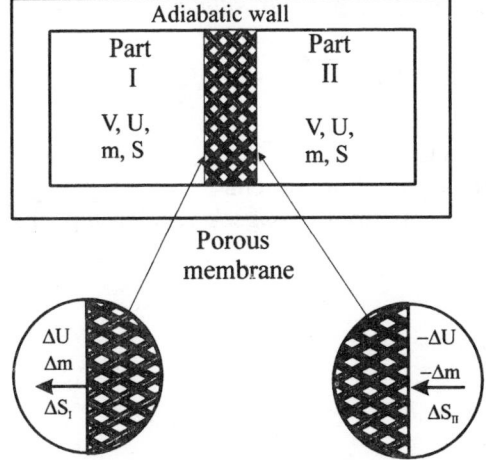

Fig. 16.7 Schematic diagram of two containers connected by a porous membrane, or small bore pipe

16.8.2 DEFINING THE FORCES AND FLUXES

To choose the forces and fluxes it is necessary to consider the rate of entropy generation (see section 16.5, and de Groot (1951)). Suppose that a system comprising two parts, I and II connected by a hole, is enclosed in a reservoir. It will be assumed that both parts are the same volume, V, and when in thermostatic equilibrium the energy, U, and mass, m, in each part is equal, and there is an entropy, S, associated with each part of the system. (U and m were chosen as parameters because these obey conservation laws.)

The system is isolated, hence variations ΔU and Δm in part I give rise to variations $-\Delta U$ and $-\Delta m$ in part II. The variation in entropy due to these changes may be found from Taylor's series:

$$\Delta S_{\text{I}} = \left(\frac{\partial S}{\partial U}\right)_m \Delta U + \left(\frac{\partial S}{\partial m}\right)_U \Delta m + \frac{1}{2}\left(\frac{\partial^2 S}{\partial U^2}\right)\Delta U^2 + \left(\frac{\partial^2 S}{\partial U\, \partial m}\right)\Delta U\, \Delta m + \frac{1}{2}\left(\frac{\partial^2 S}{\partial m^2}\right)\Delta m^2$$

+ higher order terms $\qquad\qquad$ (16.70)

ΔS_{II} may be found by a similar expansion, but the linear terms in ΔU and Δm are negative. Thus, the change of entropy of the Universe is

$$\Delta S = \Delta S_{\text{I}} + \Delta S_{\text{II}} = \left(\frac{\partial^2 S}{\partial U^2}\right)_m \Delta U^2 + 2\frac{\partial^2 S}{(\partial U\, \partial m)}\Delta U\, \Delta m + \left(\frac{\partial^2 S}{\partial m^2}\right)_U \Delta m^2$$

+ higher order terms ... $\qquad\qquad$ (16.71)

The time rate of change of ΔS, i.e. the rate of generation of entropy, is given by

$$\frac{d}{dt}(\Delta S) = \left(\frac{\partial^2 S}{\partial U^2}\right)(2\,\Delta U)\Delta \dot U + 2\left(\frac{\partial^2 S}{\partial U\, \partial m}\right)\Delta m\, \Delta \dot U + 2\left(\frac{\partial^2 S}{\partial U\, \partial m}\right)\Delta U\, \Delta \dot m + \left(\frac{\partial^2 S}{\partial m^2}\right)(2\,\Delta m)\,\Delta \dot m$$

$$= 2\,\Delta \dot U\left\{\left(\frac{\partial^2 S}{\partial U^2}\right)\Delta U + \left(\frac{\partial^2 S}{\partial U\, \partial m}\right)\Delta m\right\} + 2\,\Delta \dot m\left\{\left(\frac{\partial^2 S}{\partial m^2}\right)\Delta m + \left(\frac{\partial^2 S}{\partial U\, \partial m}\right)\Delta U\right\}$$

$$= 2\,\Delta \dot U\, \Delta\left(\frac{\partial S}{\partial U}\right)_m + 2\,\Delta \dot m\, \Delta\left(\frac{\partial S}{\partial m}\right)_U \qquad\qquad (16.72)$$

and the rate of generation of entropy *per unit volume* is

$$\frac{d}{dt}(\Delta s) = \frac{1}{2}\frac{d}{dt}(\Delta S) = \Delta\dot{U}\,\Delta\left(\frac{\partial S}{\partial U}\right)_m + \Delta\dot{m}\,\Delta\left(\frac{\partial S}{\partial m}\right)_U \tag{16.73}$$

because the combined volume of systems I and II is $2V$, and the original terms were defined in relation to a single part of the system.

It was stated in eqn (16.12), from Onsager's relationship, that for a two-component system

$$T\theta = J_1 X_1 + J_2 X_2$$

which may be written, for this system, as

$$T\theta = \frac{d}{dt}(\Delta s) = J_U X_U + J_m X_m \tag{16.74}$$

Comparison of eqns (16.73) and (16.74) enables the forces and fluxes to be defined, giving

$$
\begin{aligned}
J_U &= \Delta\dot{U} \quad X_U = \Delta\left(\frac{\partial S}{\partial U}\right)_m \\
J_m &= \Delta\dot{m} \quad X_m = \Delta\left(\frac{\partial S}{\partial m}\right)_U
\end{aligned}
\tag{16.75}
$$

From the First and Second Laws, for a system at constant pressure,

$$T\,ds = du + p\,dv - \mu\,dm \tag{16.76}$$

Hence

$$\Delta\left(\frac{\partial s}{\partial u}\right)_m = \Delta\left(\frac{\partial S}{\partial U}\right)_m = \Delta\left(\frac{1}{T}\right) = -\frac{\Delta T}{T^2} \tag{16.77}$$

$$\Delta\left(\frac{\partial s}{\partial m}\right)_U = -\Delta\left(\frac{\mu}{T}\right) = -\frac{T\Delta\mu - \mu\Delta T}{T^2} = -\frac{\Delta\mu}{T} + \mu\frac{\Delta T}{T^2} \tag{16.78}$$

Substituting for

$$\Delta\mu = v\,\Delta p - s\,\Delta T \tag{16.79}$$

and

$$h = \mu + Ts = u + pv \tag{16.80}$$

gives

$$\Delta\left(\frac{\partial s}{\partial m}\right)_U = -v\frac{\Delta p}{T} + s\frac{\Delta T}{T} + h\frac{\Delta T}{T^2} - Ts\frac{\Delta T}{T^2} = -v\frac{\Delta p}{T} + h\frac{\Delta T}{T^2} \tag{16.81}$$

Hence substituting these values in eqn (16.8),

$$
\begin{aligned}
J_U &= L_{11} X_U + L_{12} X_m \\
J_m &= L_{21} X_U + L_{22} X_m
\end{aligned}
$$

gives

$$J_U = -L_{11} \frac{\Delta T}{T^2} + L_{12} \left(-v \frac{\Delta p}{T} + h \frac{\Delta T}{T^2} \right)$$

$$= \left(\frac{hL_{12} - L_{11}}{T^2} \right) \Delta T - \frac{L_{12}}{T} v \, \Delta p \tag{16.82}$$

and

$$J_m = -\frac{L_{21}}{T^2} \frac{\Delta T}{T^2} + L_{22} \left(-v \frac{\Delta p}{T} + h \frac{\Delta T}{T^2} \right)$$

$$= \left(\frac{hL_{22} - L_{21}}{T^2} \right) \Delta T - \frac{L_{22}}{T} v \, \Delta p \tag{16.83}$$

If the flow of internal energy which occurs due to the diffusion is defined as $(J_u / J_m)_{\Delta T = 0} = U^*$ then

$$U^* = \left(\frac{J_U}{J_m} \right)_{\Delta T = 0} = \frac{L_{12} v \, \Delta p}{T} \frac{T}{L_{22} v \, \Delta p} \tag{16.84}$$

i.e.

$$L_{12} = L_{22} U^* = L_{21} \tag{16.85}$$

In the stationary state, $J_m = 0$, i.e.

$$\frac{\Delta p}{\Delta T} = \left(\frac{hL_{22} - L_{21}}{T^2} \right) \frac{T}{L_{22} v} = \frac{h - \dfrac{L_{21}}{L_{22}}}{vT} = \frac{h - U^*}{vT} \tag{16.86}$$

Now, the term $h - U^*$ may be defined as $-Q^*$, and the significance of this is that it drives the system to generate a pressure gradient due to the temperature gradient. Hence, if two systems are connected together with a concentration (or more correctly, a chemical potential) gradient then a pressure difference will be established between the systems until $h = U^*$.

The pressure difference is defined as

$$\frac{\Delta p}{\Delta T} = -\frac{Q^*}{vT} \tag{16.87}$$

The significance of this result will be returned to later.

16.8.3 THE UNCOUPLED EQUATIONS OF DIFFUSION

It is also possible to derive the result in eqn (16.87) in a more easily understandable manner, if the fluxes and forces defined above are simply accepted without proof.

Diffusion

For the uncoupled process of diffusion, it was shown in eqn (16.78) that the thermodynamic force for component i, in one-dimensional diffusion, is

$$X_i = -T \frac{\partial(\mu_i/T)}{\partial x} \tag{16.88}$$

where μ_i = chemical potential.

Evaluating the differential in eqn (16.88) gives, as shown in eqn (16.78)

$$X_i = -T \left[\frac{T \dfrac{\partial \mu}{\partial x} - \mu \dfrac{\partial T}{\partial x}}{T^2} \right]$$

$$= -\frac{\partial \mu}{\partial x} + \frac{\mu}{T} \frac{\partial T}{\partial x} \tag{16.89}$$

Equation (16.89) applies for a single component only, but it is possible to derive the equations for multicomponent mixtures.

If Onsager's rule for entropy flux is applied to this situation

$$T\theta = \Sigma \, JX$$

Thus

$$T\theta = -J_m \frac{\partial \mu}{\partial x} + \frac{J_m}{T} \mu \frac{\partial T}{\partial x} \tag{16.90}$$

Hence the entropy production per unit volume, θ, is given as

$$\theta = \underbrace{-\frac{J_m}{T} \frac{\partial \mu}{\partial x}}_{\substack{\text{Due to isothermal} \\ \text{diffusion}}} + \underbrace{\frac{J_m}{T^2} \mu \frac{\partial T}{\partial x}}_{\substack{\text{Due to temperature} \\ \text{gradient}}} \tag{16.91}$$

Coupled diffusion and heat pr~~ocesses~~

Such processes are also refe   *Coupled diffusion and heat*   nsider the
system shown in Fig 16.7. The   *properties*

heat

$$J_Q = L_{11} X_1 + L_{12} X_2 \tag{16.92}$$

diffusion (*mass transfer*)

$$J_m = L_{21} X_1 + L_{22} X_2 \tag{16.93}$$

The thermodynamic forces   fer process,
$X_1 = -1/T(dT/dx)$ (see secti $\partial(\mu/T)/\partial x$
(see section 16.8.2).

Thus, from eqn (16.92)

$$J_Q = -\frac{L_{11}}{T}\frac{dT}{dx} - L_{12}T\frac{d(\mu/T)}{dx}$$ (16.94)

and, from eqn (16.93)

$$J_m = -\frac{L_{21}}{T}\frac{dT}{dx} - L_{22}T\frac{d(\mu/T)}{dx}$$ (16.95)

Consider the steady state, i.e. when there is no flow between the two vessels and $J_m = 0$. The effect on J_Q can be evaluated from eqn (16.95)

$$-\frac{L_{21}}{T}\frac{dT}{dx} = L_{22}T\frac{d(\mu/T)}{dx}$$ (16.96)

or

$$d(\mu/T) = -\frac{L_{21}}{L_{22}}\frac{dT}{T^2}$$ (16.97)

Now $(\mu/T) = f(T, p) = [\mu/T](T, p)$, and hence

$$d(\mu/T) = \left(\frac{\partial(\mu/T)}{\partial T}\right)_p dT + \left(\frac{\partial(\mu/T)}{\partial p}\right)_T dp$$ (16.98)

The specific enthalpy, h, can be related to this parameter by

$$h = -T^2\left(\frac{\partial(\mu/T)}{\partial T}\right)$$ (16.99)

Rearranging eqn (16.99) gives

$$\left(\frac{\partial(\mu/T)}{\partial T}\right)_p = -\frac{h}{T^2}$$ (16.100)

The term

$$\left(\frac{\partial(\mu/T)}{\partial p}\right)_T = \frac{1}{T}\left(\frac{\partial\mu}{\partial p}\right)_T,$$

because $T = $ constant, and from the definition of chemical potential, μ, the term

$$d\mu = dg = Vdp + s\,dT$$ (16.101)

Thus

$$\left(\frac{\partial\mu}{\partial p}\right)_T = v$$ (16.102)

Substituting for

$$\left(\frac{\partial(\mu/T)}{\partial T}\right)_p$$

and

$$\left(\frac{\partial(\mu/T)}{\partial p}\right)_T$$

in eqn (16.98) gives

$$d(\mu/T) = -\frac{h}{T^2}\,dT + \frac{v}{T}\,dp \qquad\qquad (16.103)$$

Hence from eqns (16.97) and (16.103)

$$-\frac{h}{T^2}\,dT + \frac{v\,dp}{T} = -\frac{L_{21}}{L_{22}}\frac{dT}{T^2} \qquad\qquad (16.104)$$

giving

$$v\,dp = \left\{h - \frac{L_{21}}{L_{22}}\right\}\frac{dT}{T} \qquad\qquad (16.105)$$

If both vessels were at the same temperature then $dT/dx = 0$ and

$$J_Q|_T = -L_{12}T\,\frac{d(\mu/T)}{dx} \qquad\qquad (16.106)$$

$$J_m|_T = -L_{22}T\,\frac{d(\mu/T)}{dx} \qquad\qquad (16.107)$$

Thus, as shown in eqn (16.84)

$$\left(\frac{J_Q}{J_m}\right)_T = \frac{L_{12}}{L_{22}} \qquad\qquad (16.108)$$

The ratio $(J_Q/J_m)_T$ is the energy transported when there is *no heat flow through thermal conduction*. Also from Onsager's reciprocal relation $(J_Q/J_m)_T = L_{21}/L_{22}$. If this ratio is denoted by the symbol U^* then eqn (16.105) can be written (eqn (16.86))

$$\frac{dp}{dT} = \frac{(h - U^*)}{vT}$$

Now the difference $h - U^*$ was denoted $-Q^*$, which is called the *heat of transport*. Using this definition, eqn (16.86) may be rewritten (eqn (16.87))

$$\frac{dp}{dT} = \frac{Q^*}{vT}$$

The significance of Q^*

Q^* is the heat transported from region I to region II by diffusion of the fluid. It is a measure of the difference in energy associated with the gradient of chemical potential. In a simple case it is possible to evaluate Q^*. Weber showed by kinetic theory that when a gas passes from a vessel into a porous plate it has a decrease in the energy it carries. The energy carried by the gas molecule in molecular flow through the passages in the porous medium is smaller by $RT/2$ than it was when the motion was random.

Hence $Q^* = -RT/2$ for flow through a porous plug. The energy, Q^*, is liberated when the molecules enter the plug and the same amount is absorbed when the molecules emerge from the plug. When Q^* is dependent on temperature, as in the thermal effusion case, it would appear reasonable that if Q^* is liberated on entering the plug at temperature T then $Q^* + dQ^*$ would be absorbed on the molecule leaving the plug, when the general temperature is $T + dT$. This is erroneous and a qualitative explanation follows. Q^* is the amount by which the mean energy per unit mass of molecules that are in the process of transit through the plug exceeds the mean energy of the bulk of the fluid. Hence, if Q^* were not applicable to both sides of the plug then the principle of conservation of energy would be contravened. The difference of average values of the bulk fluid energy on either side of the plug has been taken into account by the different values of the enthalpy, h, on each side of the plug.

Now, if

$$Q^* = -RT/2 \tag{16.109}$$

and for a perfect gas

$$pv = RT \tag{16.110}$$

then

$$\frac{dp}{dT} = -\frac{Q^*}{vT} = \frac{RT}{2vT} = \frac{R}{2v} = \frac{p}{2T} \tag{16.111}$$

Integrating across the plug from side I to side II,

$$\ln \frac{p_{II}}{p_I} = \frac{1}{2} \ln \frac{T_{II}}{T_I} \tag{16.112}$$

or

$$\frac{p_{II}}{p_I} = \sqrt{\frac{T_{II}}{T_I}} \tag{16.113}$$

This is called the *Knudsen* equation for molecular flow. From eqn (16.87) over an increment of distance, Δx

$$\Delta p = -\frac{Q^*}{vT} \Delta T$$

This means that if Q^* is negative, as in the thermal effusion case, the pressure increment is of the same sign as the temperature increment and molecules *flow* from the cold side to the hot side.

16.8.4   THERMAL TRANSPIRATION

It will be shown in this section, without recourse to Onsager's relationship, that $L_{12} = L_{21}$. This derivation will be based on statistical mechanics, and this is the manner by which Onsager derived his relationship. The problem considered is similar to that discussed in section 16.8.3, when flow through a porous plug was examined. In this case the flow will be assumed to occur along a small bore pipe in which the cross-sectional area is small compared with the mean free path of the molecules; this is what happens when gas flows through the pores in a porous membrane.

It can be shown from statistical thermodynamics that the number of particles striking a unit area in unit time is

$$\frac{\dot{n}_c}{A} = \frac{p}{(2\pi kTm)^{1/2}} \tag{16.114}$$

where $\dot{n}_c$ = number of particles striking wall
  A = area of wall
  p = 'pressure' of gas
  k = Boltzmann's constant
  T = temperature
  m = mass of 'particle'

Consider a hole of area A, then $\dot{n}_c$ would be the number of particles passing through this hole. These particles would carry with them kinetic energy. The hole also makes a selection of the atoms which will pass through it and favours those of a higher energy because the frequency of higher velocity particles hitting the wall is higher. It can be shown that the energy passing through the hole is given by

$$\frac{\dot{E}}{A} = p\left(\frac{2kT}{\pi m}\right)^{1/2} \tag{16.115}$$

Equations (16.114) and (16.115) may be written in the form

$$\frac{\dot{n}_c}{A} = \left(\frac{p}{\tilde{N}kT}\right)\left(\frac{kT}{2m}\right)^{1/2}\left(\frac{\tilde{N}m^{1/2}}{\pi^{1/2}}\right) \tag{16.116}$$

$$\frac{\dot{E}_c}{A} = \left(\frac{p}{\tilde{N}kT}\right)\left(\frac{kT}{2m}\right)^{1/2}\left(\frac{3}{2}kT\right)\left(\frac{4\tilde{N}}{3\pi^{1/2}}\right) \tag{16.117}$$

where $\tilde{N}$ = Avogadro's number.

The groups of terms can be defined in the following way:

$$\left(\frac{p}{\tilde{N}kT}\right) \equiv \text{density}$$

$$\left(\frac{kT}{2m}\right)^{1/2} \equiv \text{velocity}$$

$$\frac{3}{2}kT \equiv \text{energy}$$

$\tilde{N}m^{1/2}/\pi^{1/2}$ and $4\tilde{N}/3\pi^{1/2}$ are weighting factors.

Thus

$$\frac{\dot{n}_c}{A} = \text{density} \times \text{mean velocity}$$

$$\frac{\dot{E}}{A} = \text{density} \times \text{mean velocity} \times \text{mean energy}$$

The weighting factors show that the particles passing through area A carry with them an unrepresentative sample of the energy of the system from which they come.

This example can be expanded into an apparatus that contains two systems separated by a porous wall. It will be assumed that the pores in the wall are much shorter than the mean free path of the molecules. The system is shown in Fig 16.8. The gases on either side of the partition are the same, but one side of the system is heated and the other side is cooled.

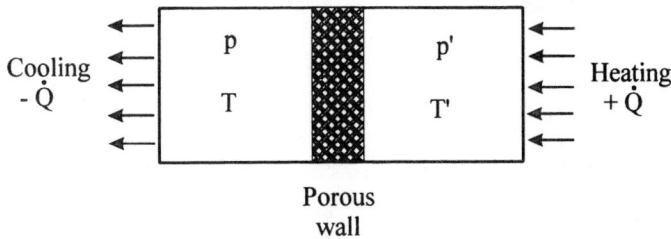

Porous
wall

Fig. 16.8 Arrangement for flow through a porous plug

From eqn (16.114) the net flux of molecules passing through the wall is given by

$$\frac{\dot{n}'_c - \dot{n}_c}{A} = (2\pi km)^{-1/2}\left[\frac{p'}{\sqrt{T'}} - \frac{p}{\sqrt{T}}\right] \tag{16.118}$$

and the net energy passing through the wall is given from eqn (16.115) as

$$\frac{\dot{E}' - \dot{E}}{A} = \left(\frac{2k}{\pi m}\right)^{1/2}[p'T'^{1/2} - pT^{1/2}] \tag{16.119}$$

It is possible to substitute for the pressure of a monatomic gas by the expression

$$p = e^{-\alpha}h^{-3}(2\pi m)^{3/2}\beta^{-5/2} \tag{16.120}$$

and the temperature by

$$T = 1/k\beta \tag{16.121}$$

Then eqns (16.118) and (16.119) become

$$\frac{\dot{E}}{A} = (4\pi mh^{-3})[e^{-\alpha}\beta^{-3} - e^{-\alpha'}\beta'^{-3}] \tag{16.122}$$

$$\frac{\dot{n}_c}{A} = (2\pi mh^{-3})[e^{-\alpha}\beta^{-2} - e^{\alpha'}\beta'^{-2}] \tag{16.123}$$

If the state denoted by the primed symbols is constant, then differentiating gives

$$d\left(\frac{\dot{E}}{A}\right) = (2\pi m h^{-3})\beta^{-3}e^{-\alpha}[-6\beta^{-1}\,d\beta - 2\,d\alpha] \tag{16.124}$$

$$d\left(\frac{\dot{n}_c}{A}\right) = (2\pi m h^{-3})\beta^{-3}e^{-\alpha}[-2\,d\beta - \beta\,d\alpha] \tag{16.125}$$

Equations (16.124) and (16.125) may be written in the form

$$d\left(\frac{\dot{E}}{A}\right) = L_{\dot{E},\beta}\,d\beta + L_{\dot{E},\alpha}\,d\alpha \tag{16.126}$$

and

$$d\left(\frac{\dot{n}_c}{A}\right) = L_{\dot{n}_c,\beta}\,d\beta + L_{\dot{n}_c,\alpha}\,d\alpha \tag{16.127}$$

where

$$L_{\dot{E},\beta} = -12\pi m\beta^{-4}h^{-3}e^{-\alpha}$$

$$L_{\dot{E},\alpha} = -4\pi m\beta^{-3}h^{-3}e^{-\alpha}$$

$$L_{\dot{n}_c\beta} = -4\pi m\beta^{-3}h^{-3}e^{-\alpha}$$

$$L_{\dot{n}_c\alpha} = -2\pi m\beta^{-2}h^{-3}e^{-\alpha}$$

Hence $L_{\dot{E},\alpha} = L_{\dot{n}_c\beta}$, which is equivalent to the result given by Onsager, that $L_{12} = L_{21}$ and has been proved from Statistical thermodynamics.

16.9 Concluding remarks

This chapter extended 'thermodynamics' into a truly dymanic arena. The processes involved are non-equilibrium ones which are in a steady, dynamic state. The concepts of entropy generation and coupled phenomena have been introduced, and Onsager's reciprocal relation has been used to enable the latter to be analysed.

Thermoelectric phenomena have been considered and the coupling between them has been described. The major thermoelectric effects have been defined in terms of the thermodynamics of the device. Coupled diffusion and heat transfer processes have been introduced and analysed using these techniques, and the conjugate forces and fluxes have been developed. Finally, statistical thermodynamics was used to demonstrate that Onsager's reciprocal relation can be proved from molecular considerations.

PROBLEMS

1 The emf of a copper–iron thermocouple caused by the Seebeck effect, with a cold junction at 0°C, is given by

$$\varepsilon = a_1 t + \frac{a_2}{2}t^2 + \frac{a_3}{3}t^3\ \mathrm{V}$$

where $a_1 = -13.403 \times 10^{-6}$ V/°C
$a_2 = +0.0275 \times 10^{-6}$ V/(°C)^2
$a_3 = +0.00026 \times 10^{-6}$ V/(°C)^3
$t =$ temperature (°C)

If the hot junction is at $t = 100$°C, calculate

(a)　the Seebeck emf;
(b)　the Peltier effect at the hot and cold junctions;
(c)　the net Thornson emf;
(d)　the difference between the entropy of transport of the copper and iron.

$$[1.116 \text{ mV}; 3.66 \text{ mV}; 3.00 \text{ mV}; -8.053 \times 10^{-6} \text{ V}]$$

2　The emf of a copper–iron thermocouple with its cold junction at 0°C is given by

$$\varepsilon = -13.403t + 0.0138t^2 + 0.0001t^3 \ \mu\text{V}$$

where $t =$ temperature (°C).

Show that the difference in the Thomson coefficient for the two wires is

$$7.535 + 0.1914t + 0.0006t^2 \ \mu\text{V}/°C$$

3　If a fluid, consisting of a single component, is contained in two containers at different temperatures, show that the difference in pressure between the two containers is given by

$$\frac{dp}{dT} = \frac{h - u^*}{vT}$$

where $h =$ specific enthalpy of the fluid at temperature T,
　　　$u^* =$ the energy transported when there is no heat flow through thermal conduction,
　　　$v =$ specific volume,
　　　$T =$ temperature.

4　A thermocouple is connected across a battery, and a current flows through it. The cold junction is connected to a reservoir at 0°C. When its hot junction is connected to a reservoir at 100°C the heat flux due to the Peltier effect is 2.68 mW/A, and when the hot junction is at 200°C the effect is 4.11 mW/A. If the emf of the thermocouple due to the Seebeck effect is given by $\varepsilon = at + bt^2$, calculate the values of the constants a and b. If the thermocouple is used to measure the temperature effect based on the Seebeck effect, i.e. there is no current flow, calculate the voltages at 100°C and 200°C.

$$[5.679 \times 10^{-3} \text{ mV/K}; 7.526 \times 10^{-6} \text{ mV/(K)}^2; 0.6432 \text{ mV}; 1.437 \text{ mV}]$$

5　A pure monatomic perfect gas with $c_p = 5\Re/2$ flows from one reservoir to another through a porous plug. The heat of transport of the gas through the plug is $-\Re T/2$. If the system is adiabatic, and the thermal conductivities of the gas and the plug are negligible, evaluate the temperature of the plug if the upstream temperature is 60°C.

$$[73°C]$$

6 A thermal conductor with constant thermal and electrical conductivities, k and λ respectively, connects two reservoirs at different temperatures and also carries an electrical current of density, J_I. Show that the temperature distribution for one-dimensional flows is given by

$$\frac{d^2T}{dx^2} - \frac{J_I\sigma}{k}\frac{dT}{dx} + \frac{J_I^2}{\lambda} = 0$$

where σ is the Thomson coefficient of the wire.

7 A thermal conductor of constant cross-sectional area connects two reservoirs which are both maintained at the same temperature, T_0. An electric current is passed through the conductor, and heats it due to Joulean heating and the Thomson effect. Show that if the thermal and electrical conductivities, k and λ, and the Thomson coefficient, σ, are constant, the temperature in the conductor is given by

$$T - T_0 = \frac{J_I k}{\lambda\sigma L}\left(\frac{x}{L}\right) - \frac{J_I k}{\lambda\sigma L(e^{J_I\sigma L/k} - 1)}(e^{J_I\sigma L(x/L)/k} - 1)$$

Show that the maximum temperature is achieved at a distance

$$\left(\frac{x}{L}\right) = \frac{k}{J_I\sigma L}\ln\left\{\frac{k(e^{J_I\sigma L/k} - 1)}{J_I\sigma L}\right\}$$

Evaluate where the maximum temperature will occur if $J_I\sigma L/k = 1$, and explain why it is not in the centre of the bar. Show that the maximum temperature achieved by Joulean heating alone is in the centre of the conductor.

$$[x/L = 0.541]$$

17

Fuel Cells

Engineering thermodynamics concentrates on the production of work through cyclic devices, e.g. the power to drive a vehicle as produced by a reciprocating engine; the production of electricity by means of a steam turbine. As shown previously, these devices are all based on converting part of the Gibbs function of the fuel to useful work, and if an engine is used some energy must be 'thrown away': all engines are limited by the Carnot, or Second Law, efficiency.

However, there are some devices which are capable of converting the Gibbs energy of the fuel directly to electricity (a form of work); these are called *fuel cells*. The advantage of a fuel cell is that it is not a heat engine, and it is not limited by the Carnot efficiency. The thermodynamics of fuel cells will be developed below.

The concept of the fuel cell arises directly from the operating principle of the electric cell, e.g. the Daniell cell, and as early as 1880 Wilhelm Ostwald wrote: 'I do not know whether all of us realise fully what an imperfect thing is the most essential source of power which we are using in our highly developed engineering – the steam engine.' He realised that chemical processes could approach efficiencies of 100% in galvanic cells, and these were not limited by the Carnot efficiency. Conversion efficiencies as high as 60–80% have been achieved for fuel cells, whereas the practical limit even for sophisticated rotating machinery is not much above 50%. A further benefit of the fuel cell is that its efficiency is not reduced by part load operation, as is the case for all heat engines. Hence, if a fuel cell operating on hydrocarbon fuels can be achieved it will improve 'thermal efficiency' significantly and reduce pollution by CO_2 and NO_x. The current situation is that successful commercial fuel cells are still some way from general use but small ones have been used in specialist applications (e.g. space craft) and large ones are being developed (e.g. 1 MW by Tokyo Gas, Japan). Figure 17.1 shows a proposal for a fuel cell to power a motor vehicle using methanol (CH_3OH) as its primary fuel. In this case the methanol is used to generate hydrogen for use in a hydrogen–oxygen fuel cell. Figure 17.1(a) shows the hydrogen generator which is based on the water gas reaction, and Fig 17.1(b) depicts the fuel cell itself. This arrangement has the advantage that the fuel can be carried in its liquid phase at atmospheric conditions, which is much more convenient than carrying hydrogen either in gas, liquid or hydride form.

The theory of fuel cells can be developed from the previously derived thermodynamic principles, and it shows how equilibrium reversible thermodynamics can be interwoven with irreversible thermodynamics. Before developing the theory of the fuel cell itself it is

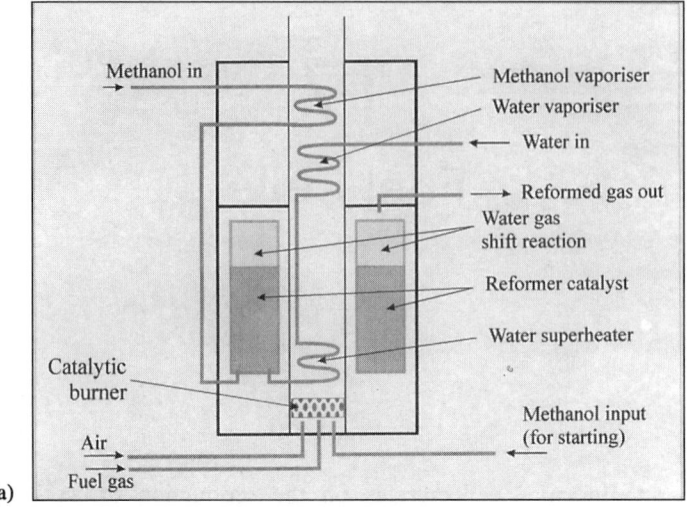

(a)

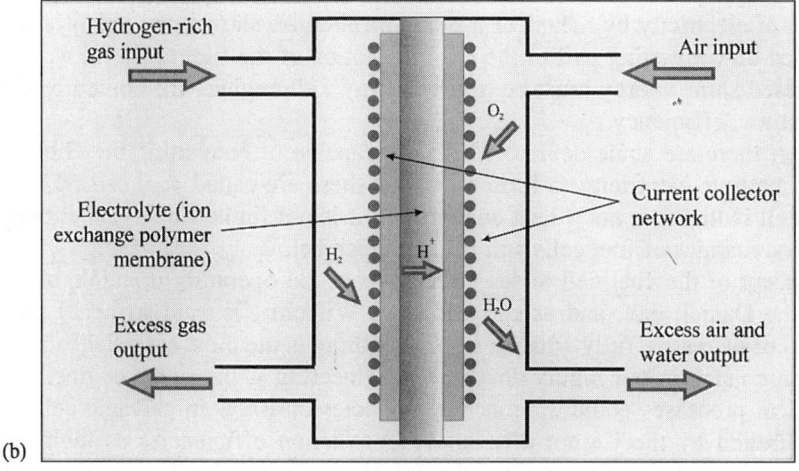

(b)

Fig. 17.1 Proposed hydrogen fuel cell for vehicle applications

necessary to consider simpler electrical cells. A good place to start is the Daniell cell which produces electricity through consuming one electrode into solution and depositing onto the other one.

17.1   Electric cells

A schematic of a Daniell cell is shown in Fig 17.2. This can be represented by the convention

$$Zn \,|\, ZnSO_4 \,|\, CuSO_4 \,|\, Cu \tag{17.1}$$

where the | represents an interface, or phase boundary. The convention adopted in representing cells in this way is that the electrode on the right-hand side of eqn (17.1) is positively charged relative to that on the left if the reaction takes place spontaneously.

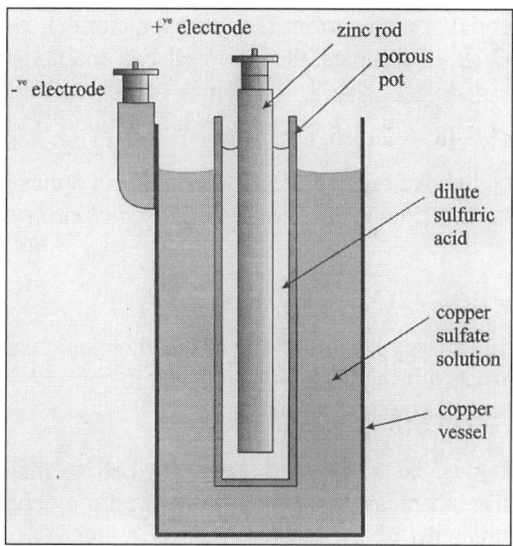

Fig. 17.2 Schematic diagram of a Daniell cell

To understand how the Daniell cell produces a potential difference and a current it is necessary to consider the basic reaction involved,

$$Zn + CuSO_4(aq) \rightarrow Cu + ZnSO_4(aq) \tag{17.2}$$

which indicates that the zinc reacts with the copper sulfate solution to produce copper and zinc sulfate solution. The notation *aq* indicates an aqueous solution of the salt. If the reaction in eqn (17.2) takes place in a constant volume container using 1 kmol Zn (45 kg Zn) then 214 852 kJ of heat must be transferred from the container to maintain the temperature of the system at 25°C. This reaction is similar to a combustion reaction, and must obey the First Law of Thermodynamics

$$\delta Q = dU + \delta W \tag{17.3}$$

where $\delta W = 0$ in this case. Hence

$$\delta Q_{oc} = -214\ 852\ \text{kJ/kmol Zn} = u_{Cu} + u_{ZnSO_4(aq)} - u_{Zn} - u_{CuSO_4(aq)} \tag{17.4}$$

The reaction described above is basically an *open circuit reaction* for a Daniell cell, and δQ_{oc} represents the open circuit energy released. If, as is more usual, the reaction takes place as work is being taken from the cell then supposing a current I flows for time, t, then, from Faraday's laws of electrolysis, the ratio of the amount of substance (Zn) dissolved to the valency of the element, n/z, is proportional to the electric charge passed, i.e.

$$n/z \propto Q, \qquad \Rightarrow n/z = FQ \tag{17.5}$$

When $n/z = 1$ kmol, $Q = 96\ 485$ kC (where kC denotes kcoulomb), and hence $F = 96\ 485$ kC/kmol. This is known as the Faraday constant, and is the product of Avogadro's number and the charge on a proton. Now, if the potential between the electrodes is E_{oc} then the work done is

$$\delta W = 2E_{oc}F \text{ V kC/kmol} = 2E_{oc}F \text{ kJ/kmol} \tag{17.6}$$

In a Daniell cell the potential at zero current (i.e. on open circuit), which is called the *emf*, $E_{oc} = 1.107$ V at 25°C. If it is assumed that the cell can maintain this potential at low currents, then

$$\delta W = 2 \times 1.107 \times 96\ 485 = 213\ 618\ \text{kJ/kmol}$$

If the reaction described above takes place isothermally in a closed system then it must obey the First Law, which this time is applied to the *closed circuit* system and gives

$$dU = \delta Q - \delta W$$
$$= \delta Q_R - \delta W = \delta Q_R - 213\ 618\ \text{kJ/kmol} \tag{17.7}$$

Now the change in internal energy is simply due to the chemical changes taking place, and for an isothermal reaction must be equal to δQ_{oc} defined in eqn (17.4), giving

$$\delta Q_R = -214\ 852 + 213\ 618 = -1234\ \text{kJ/kmol} \tag{17.8}$$

This indicates that heat must be transferred *from* the cell to maintain the temperature constant. This heat transfer is a measure of the change of entropy contained in the bonds of the product ($ZnSO_4$) compared with the reactant ($CuSO_4$), and

$$\Delta S = -1234/298 = -4.141\ \text{kJ/K} \tag{17.9}$$

In this approach the Daniell cell has been treated as a thermodynamic system – a black box. It is possible to develop this approach further to evaluate the electrical performance of the cell. Suppose that an amount of substance of Zn, dn, enters the solution at the negative pole, then it will carry with it a charge of $zF\ dn$, where z is the valency (or charge number of the cell reaction) of the Zn, and is 2 in this case. The Coulombic forces in the cell are such that an equal and opposite charge has to be absorbed by the copper electrode, and this is achieved by the copper ions absorbing electrons which have flowed around the external circuit. The maximum work that can be done is achieved if the cell is reversible, and the potential is equal to the open circuit potential; thus

$$\delta W = zFE\ dn \tag{17.10}$$

However, the total work that could be obtained from a cell if it changed volume would be

$$\delta W = zFE\ dn + p\ dV \tag{17.11}$$

thus, applying the First Law, and assuming the processes are reversible gives

$$dU = \delta Q - \delta W = T\ dS - p\ dV - zFE\ dn \tag{17.12}$$

and hence the electrical work output is

$$-zFE\ dn = dU + p\ dV - T\ ds = dG = G_2 - G_1 \tag{17.13}$$

For a cell which is spontaneously discharging, $G_2 < G_1$, and hence

$$\delta W = -dG \tag{17.14}$$

The equations derived above define the operation of the Daniell cell from a macroscopic viewpoint. It is instructive to examine the processes which occur at the three interfaces shown in eqn (17.1). Hence

$$\left.\begin{array}{ll} Zn \rightarrow Zn^{++} + 2e & \text{at zinc electrode} \\ Zn^{++} + CuSO_4 \rightarrow ZnSO_4 + Cu^{++} & \text{in solution} \\ Cu^{++} + 2e \rightarrow Cu & \text{at copper electrode} \end{array}\right\} \tag{17.15}$$

This means that the zinc is 'dissolved' by the sulfuric acid at the zinc electrode and a zinc anion enters the solution. Meanwhile, two electrons are left on the zinc electrode (because the valency of zinc is 2) and these are free to travel around the circuit, but cause the zinc electrode to be at a negative potential, i.e. it is the cathode. The zinc anion reacts with the copper sulfate in solution to form zinc sulfate and release a copper ion which migrates to the copper electrode, where it withdraws electrons from the electrode giving it a positive potential. Hence, the Daniell cell consists of electrons (negative charges) travelling around the outer circuit, from the cathode to the anode, while positive ions travel through the solution from cathode to anode. (Note: the convention for positive electric current is in the opposite direction to the electron flow; the current is said to flow from the anode to the cathode.) The net effect is to maintain the potential difference between the electrodes constant for any given current: this is a state of dynamic equilibrium. It can be seen that the electrochemical cell is a situation governed by thermodynamic equilibrium and steady state (irreversible) thermodynamics (see Chapter 16).

The reactions defined in eqn (17.15) resulted in electrons flowing from the zinc to the copper (this would be defined as a current flowing from the copper [anode] to the zinc [cathode]), and the potential on the anode would be higher than the cathode. If the cell were connected to a potential source (e.g. a battery charger), such that the potential difference of the source were slightly higher than the cell emf, then the current flow could be reversed and the reaction would become

$$
\left.
\begin{array}{ll}
Cu \rightarrow Cu^{++} + 2e & \text{at Cu electrode} \\[2mm]
Cu^{++} + ZnSO_4 \rightarrow Zn^{++} + CuSO_4 & \text{in solution} \\[2mm]
Zn^{++} + 2e \rightarrow Zn & \text{at Zn electrode}
\end{array}
\right\}
\tag{17.16}
$$

Hence, the Daniell cell is *reversible* if the current drawn from (or fed to) it is small. The Daniell cell can be used to 'generate' electricity, by consuming an electrode, or to store electricity.

Although the Daniell cell was one of the early examples of a device for generating electricity, it is relatively difficult to analyse thermodynamically because it has electrodes of different materials. A simpler device will be considered below to develop the equations defining the operation of such cells, but first it is necessary to introduce another property.

17.1.1 ELECTROCHEMICAL POTENTIAL

To be able to analyse the cell in more detail it is necessary to introduce another parameter, the *electrochemical potential*, $\bar{\mu}$, of ions. This is related to the chemical potential by

$$
\bar{\mu}_i = \mu_i + z_i \psi F
\tag{17.17}
$$

where

z_i = valency of (or charge on) the ion $(+/-)$;

ψ = inner electric potential of the phase;

F = Faraday constant (96 485 kJ/kmol).

It was previously stated that the chemical potential was a 'chemical pressure' to bring about a change of composition to achieve a state of chemical equilibrium. In a similar manner the electrochemical potential is an 'electrical pressure' that pushes the system towards electrical equilibrium. The electrochemical potential also contains a term for the change in capacity to do work through electrical processes, introduced to the system by the introduction of a unit of mass.

The value of ψ is the potential of the phase, and is obtained from electrostatic theory. It is the work done in bringing a unit positive charge from infinity to within the phase. Hence, $\bar{\mu}$ is the sum of the chemical potential, which has been based on the 'chemistry' of the material, and an electrical potential. It defines the energy that can be obtained from mechanical, chemical and electrical processes.

17.1.2   THERMODYNAMIC ORIGIN OF EMF

Consider a simple galvanic cell, such as that shown in Fig 17.3. This consists of two electrodes, A and B, made of different materials, an electrolyte, and two connecting wires of identical material. The cell can be represented as

$$Cu' \,|_{x_6} A \,|_{x_5} \text{electrolyte} \,|_{x_4} B \,|_{x_3} Cu'' \qquad (17.18)$$

The junctions of interest here are X_5 and X_4; junctions X_6 and X_3 are potentially thermocouples (because they are junctions of two dissimilar metals) but these effects will be neglected here.

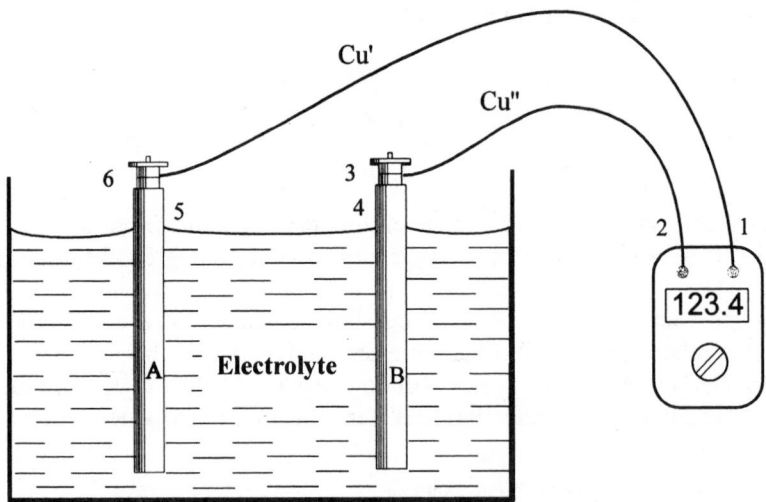

Fig. 17.3 Simple galvanic cell

Considering the electrochemical potential between points 1 and 2

$$\bar{\mu}_1 - \bar{\mu}_2 = \bar{\mu}_1 - \bar{\mu}_6 - (\bar{\mu}_5 - \bar{\mu}_6) - (\bar{\mu}_4 - \bar{\mu}_5) - (\bar{\mu}_3 - \bar{\mu}_4) - (\bar{\mu}_2 - \bar{\mu}_3)$$

$$= \bar{\mu}_{Cu'} - \bar{\mu}_A - (\bar{\mu}_{elec} - \bar{\mu}_A) - (\bar{\mu}_{elec} - \bar{\mu}_{elec}) - (\bar{\mu}_B - \bar{\mu}_{elec}) - (\bar{\mu}_{Cu''} - \bar{\mu}_B)$$

$$= \bar{\mu}_{Cu'} - \bar{\mu}_{Cu''}$$

$$= \mu_{Cu'} + z_{Cu'} \psi_{Cu'} F - (\mu_{Cu''} + z_{Cu''} \psi_{Cu''} F)$$

$$= z_{Cu} F (\psi_{Cu'} - \psi_{Cu''}) \qquad (17.19)$$

The difference in chemical potential of the two electrodes $(\mu_{Cu'} - \mu_{Cu''})$ is zero because both electrodes are of the same composition and are at the same temperature. Equation

(17.19) defines the maximum work that can be obtained from the cell by a reversible process, i.e. a small flow of current. The difference, $\psi_{Cu'} - \psi_{Cu''}$, is equivalent to the open circuit potential, or emf, E_{oc} and hence

$$E_{oc} = \frac{\bar{\mu}_{Cu'} - \bar{\mu}_{Cu''}}{z_{Cu}F} = V_{Cu'} - V_{Cu''} \tag{17.20}$$

17.2 Fuel cells

The above sections have considered how electrical work can be obtained by the transfer of material from one electrode to another. In the Daniell cell the Zn electrode is consumed, and ultimately the cell would cease to function. The situation in the Daniell cell can be reversed by recharging.

If it is desired to manufacture a cell which can operate without consuming the electrodes then it is necessary to supply the fuel along an electrode: such a device is called a fuel cell. An example of a fuel cell is one which takes hydrogen (H_2) and combines it with chlorine (Cl_2) to form hydrochloric acid (HCl).

A practical cell similar to that in Fig 17.3 can be constructed in the following way:

$$^-Cu' \,|\, Pt \,|\, H_2(g; p) \,|\, HCl(aq; c) \,|\, Cl_2(g; p) \,|\, Pt \,|\, Cu''\,^+ \tag{17.21}$$

where $(g; p)$ denotes a gas at pressure, p

and $(aq; c)$ denotes an aqueous solution with concentration, c

This is basically a cell using platinum electrodes (Pt) in an electrolyte of dilute hydrochloric acid (HCl). The platinum acts both as the electrodes and a catalyst, and when it is dipped into the dilute hydrochloric acid the system acts like an electric cell and hydrogen is evolved at the left-hand wire and chlorine at the right-hand one; electricity flows around the circuit and the energy is derived by breaking the bonds in the HCl molecules.

If the cell is to operate as a fuel cell then it will not break down the hydrochloric acid, as described above, but it will have hydrogen and chlorine delivered as fuels, which will be combined in the fuel cell to produce hydrochloric acid and electrical power. The operation is then defined by the reaction

$$H_2(g; p) + Cl_2(g; p) \longrightarrow 2HCl(aq; c) \tag{17.22}$$

where p is the pressure of the gases in the cell. The reactions at the interfaces are of interest because they produce the electrons and govern the potential. At the left-hand electrode, if the process is considered to be reversible

$$\frac{1}{2} H_2(g) + Pt + aq \longleftrightarrow Pt(e) + H^+(aq) \tag{17.23}$$

This reaction indicates that for hydrogen gas (H_2) to be evolved at the electrode a hydrogen ion (H^+) must be released into the solution and an electron must be left on the electrode. Now, it is not obvious from eqn (17.23) which direction the reaction will go; this is governed by considering the pair of processes occurring at the platinum electrodes. At the right-hand electrode

$$\frac{1}{2} Cl_2(g) + Pt + aq \longleftrightarrow Pt^+ + Cl^-(aq) \tag{17.24}$$

Equations (17.23) and (17.24) in combination show that for the reaction shown in eqn (17.22) the electron current must flow from the Pt-H$_2$ electrode to the Pt-Cl$_2$ electrode. Hence, the Pt-H$_2$ is a cathode and the Pt-Cl$_2$ electrode is an anode. It is apparent that when the cell is operating open circuit it will build up a potential difference between the electrodes. However, once a current starts to flow, that potential difference will be affected by the ability of the H$^+$ and Cl$^-$ ions to undertake reaction (17.22): this then introduces irreversible thermodynamics.

First, consider the device operating open circuit. It is in a state of equilibrium and $\Delta G)_{p,T} = 0$ for an infinitesimal process. If it is assumed that the equilibrium can be perturbed by an amount of substance $1/2 \, dn$ of hydrogen and chlorine being consumed at the electrodes, then these will generate dn H$^+$ and Cl$^-$ in the electrolyte and transfer $dn \, F$ of charge from Cu' to Cu". Thus

$$Cu''(e) \rightarrow Pt(e) \tag{17.25}$$

$$Pt(e) \rightarrow Cu'(e) \tag{17.26}$$

The reactions taking place in the cell can now be examined to give

$$
\left.
\begin{array}{ll}
(1) \quad \text{eqn (25)} & \Delta G_1 = \bar{\mu}_{Pt(e)} - \bar{\mu}_{Cu''(e)} \\[2ex]
(2) \quad \text{eqn (24)} & \Delta G_2 = \bar{\mu}_{Cl^-_{aq}} - \bar{\mu}_{Pt(e)} - \dfrac{1}{2}\mu_{Cl_2} - \mu_{Pt} \\[2ex]
(3) \quad \text{eqn (23)} & \Delta G_3 = \bar{\mu}_{H^+_{aq}} + \bar{\mu}_{Pt(e)} - \dfrac{1}{2}\mu_{H_2} - \mu_{Pt} \\[2ex]
(4) \quad \text{eqn (26)} & \Delta G_4 = \bar{\mu}_{Cu'(e)} - \bar{\mu}_{Pt(e)}
\end{array}
\right\} \tag{17.27}
$$

The change of molar Gibbs function across the cell is

$$\Delta G = \Delta G_1 + \Delta G_2 + \Delta G_3 + \Delta G_4 \tag{17.28}$$

which gives

$$\Delta G = \bar{\mu}_{H^+_{aq}} + \bar{\mu}_{Cl^-_{aq}} - \frac{1}{2}\mu_{H_2} - \frac{1}{2}\mu_{Cl_2} + \bar{\mu}_{Cu'(e)} - \bar{\mu}_{Cu''(e)} \tag{17.29}$$

This equation consists of a mixture of chemical potential (μ) and electrochemical potential ($\bar{\mu}$) terms. Expanding the latter gives

$$
\begin{aligned}
\bar{\mu}_{H^+_{aq}} &= \mu_{H^+_{aq}} + F\psi_{soln} \\
\bar{\mu}_{Cl^-_{aq}} &= \mu_{Cl^-_{aq}} + F\psi_{soln} \\
\bar{\mu}_{Cu'(e)} &= \mu_{Cu'} + F\psi_{Cu'} \\
\bar{\mu}_{Cu''(e)} &= \mu_{Cu''} + F\psi_{Cu''}
\end{aligned}
\tag{17.30}
$$

These terms may be substituted into eqn (17.29) to give

$$\Delta G = \mu_{H^+_{aq}} + \mu_{Cl^-_{aq}} - \frac{1}{2}\mu_{H_2} - \frac{1}{2}\mu_{Cl_2} + F(\psi_{Cu'} - \psi_{Cu''}) \tag{17.31}$$

Now, at equilibrium $\Delta G)_{p,T} = 0$ and $F(\psi_{Cu'} - \psi_{Cu'}) = FE$, thus (introducing the valency, z, to maintain generality)

$$zFE = -\left(\mu_{H^+_{aq}} + \mu_{Cl^-_{aq}} - \frac{1}{2}\mu_{H_2} - \frac{1}{2}\mu_{Cl_2}\right)$$

$$= -\left(\mu^0_{H^+_{aq}} + \mu^0_{Cl^-_{aq}} - \frac{1}{2}\mu^0_{H_2} - \frac{1}{2}\mu^0_{Cl_2}\right) - \Re T \ln\left(\frac{a_{H^+} a_{Cl^-}}{p_{rH_2}^{1/2} p_{rCL_2}^{1/2}}\right) \qquad (17.32)$$

where $p_r = p/p_0$ and $a =$ the activity coefficient of the particular phase. The activity coefficient, a, can be defined in such a way that the chemical potential *of a solution* is

$$\mu_i = \mu_i^* + \Re T \ln a_i x_i \qquad (17.33)$$

In eqn (17.33) μ_i^* is a function of temperature and pressure alone, and a_i takes account of the interaction between the components in the solution. It can be shown that at low concentrations the activity coefficients are approximately equal to unity. If the standard chemical potential terms (μ^0) at temperature T are denoted ΔG_T^0 then

$$E = \frac{-\Delta G_T^0}{zF} - \frac{\Re T}{zF} \ln\left(\frac{a_{H^+} a_{Cl^-}}{p_{rH_2}^{1/2} p_{rCl_2}^{1/2}}\right) \qquad (17.34)$$

where the term $-\Delta G_T^0/zF$ is called the standard emf of the cell, E^0.

Equations (17.32) and (17.34) show quite distinctly that the processes taking place in the fuel cell are governed by similar equations to those describing combustion. The equations are similar to those for dissociation, which also obeys the law of mass action. In the case of gaseous components the activity can be related to the partial pressures by $a_i = p_i/p_0$, where $p_i =$ partial pressure of component i.

17.2.1 EXAMPLE: A HYDROGEN–OXYGEN FUEL CELL

Consider a fuel cell in which the fuel is hydrogen and oxygen, and these gases are supplied down the electrodes. A suitable electrolyte for this cell is aqueous potassium hydroxide (KOH). The cell can be defined as

$$H_2 \mid KOH(aq) \mid O_2 \qquad (17.35)$$

The basic reactions taking place at the electrodes are

$$H_2 + 2OH^- \rightarrow 2H_2O + 2e \quad \text{at hydrogen electrode}$$

$$\frac{1}{2}O_2 + H_2O + 2e \rightarrow 2OH^- \qquad \text{at oxygen electrode} \qquad (17.36)$$

and these can be combined to give the overall reaction

$$H_2 + \frac{1}{2}O_2 \rightarrow H_2O \qquad (17.37)$$

It can be seen that the potassium hydroxide (KOH) does not take part in the reaction, and simply acts as a medium through which the charges can flow. It is apparent from eqn

(17.36) that the valency (z_i) of hydrogen is 2. From eqn (17.34) the cell open circuit potential, the standard emf, is given by

$$E^0 = -\frac{\Delta G^0_{298}}{z_{H_2}F} = -\frac{-(-30\ 434.3 - 0.5 \times 52\ 477.3 - [-239\ 081.7 - 46\ 370.1])}{2 \times 96\ 485}$$

$$= 1.185\ \text{V} \tag{17.38}$$

The standard emf in eqn (17.38) has been evaluated by assuming that the partial pressures of the gases are all 1 atmosphere. In an actual cell at equilibrium they will be controlled by the chemical equation, and the actual emf achievable will be given by an equation similar to eqn (17.34) which includes the partial pressures of the constituents,

$$E = \frac{-\Delta G^0_T}{z_i F} - \frac{\Re T}{z_i F} \ln\left(\frac{\prod\limits_{\text{products}} p_r^v}{\prod\limits_{\text{reactants}} p_r^v}\right) = E^0 - \frac{\Re T}{z_i F} \ln\left(\frac{\prod\limits_{\text{products}} p_r^v}{\prod\limits_{\text{reactants}} p_r^v}\right) \tag{17.39}$$

Equation (17.39) can be written in a shorter form as

$$E = \frac{-\Delta G^0_T}{z_i F} - \frac{\Re T}{z_i F} \ln\left(\prod p_r^v\right) = E^0 - \frac{\Re T}{z_i F} \ln\left(\prod p_r^v\right) \tag{17.39a}$$

where the stoichiometric coefficients, v, are defined as positive for products and negative for reactants. Hence the emf of the hydrogen–oxygen fuel cell is

$$E = E^0 - \frac{\Re T}{z_i F} \ln\left(\frac{p_{rH_2O}}{p_{rH_2} p_{rO_2}^{1/2}}\right) \tag{17.40}$$

Considering again the term for E^0, it is directly related to the equilibrium constant, K_{p_r}, by

$$E^0 = -\frac{\Delta G}{zF} = \frac{\Re T}{zF} \ln K_{p_r} \tag{17.41}$$

giving

$$E = \frac{\Re T}{z_i F} \ln K_{p_r} - \frac{\Re T}{z_i F} \ln\left(\frac{p_{rH_2O}}{p_{rH_2} p_{rO_2}^{1/2}}\right) = \frac{\Re T}{z_i F}\left\{\ln K_{p_r} - \ln\left(\frac{p_{rH_2O}}{p_{rH_2} p_{rO_2}^{1/2}}\right)\right\} \tag{17.42}$$

Examination of eqn (17.42) shows that when the constituents of the cell are in equilibrium, which is defined by

$$K_{p_r} = \prod_{\text{products}} p_r^v \Big/ \prod_{\text{reactants}} p_r^v$$

the potential output of the cell is zero. Hence, like a combustion process, the fuel cell only converts chemical bonds into another form of energy when it is transferring from one equilibrium state to another. A combustion process converts the bond energy of the reactants into thermal energy which can then be used to drive a powerplant; the fuel cell converts the bond energy of the fuel *directly* into electrical energy, and the amount of electricity produced is not constrained by the Carnot efficiency. The Second Law does limit the energy output but only defines the equilibrium condition. A fuel cell is basically a direct conversion device which is constrained mainly by the First Law.

An interesting difference between the fuel cell and the combustion process is that many fuel cells attempt to release the *higher* enthalpy of reaction, whereas combustion usually releases the *lower* enthalpy (or internal energy) of reaction. In the case of the H_2–O_2 fuel cell the product (H_2O) is in the liquid phase and not the vapour phase as is usual after combustion. However, the potential to do work is not changed by the phase of the products (the chemical potential of H_2O at any particular temperature is the same in the liquid and vapour phases). This means that while fuel cells are controlled largely by the First Law, they have thermal efficiencies significantly below 100% based on the higher enthalpy of reaction: they have to transfer away the difference between the higher and lower enthalpy of reaction in the form of heat if they are going to operate isothermally.

17.2.2 EFFECT OF TEMPERATURE ON FUEL CELL OPERATION

The effect of the temperature of operation on the emf of the cell can be evaluated by differentiating eqn (17.42) with respect to T, in this case keeping the pressure constant. This gives

$$\left.\frac{\partial E}{\partial T}\right|_p = \frac{\partial}{\partial T}\left[\frac{\Re T}{z_i F}\left\{\ln K_{p_r} - \ln\left(\prod p_r^\nu\right)\right\}\right] = \frac{\partial}{\partial T}\left[\frac{\Re T}{z_i F}\ln K_{p_r} - \frac{\Re T}{z_i F}\ln\left(\prod p_r^\nu\right)\right]$$

$$= \frac{\Re}{z_i F}\ln K_{p_r} + \frac{\Re T}{z_i F}\frac{\partial}{\partial T}(\ln K_{p_r}) - \frac{\Re}{z_i F}\ln\left(\prod p_r^\nu\right)$$

$$= \frac{\Re T}{z_i F}\frac{\partial}{\partial T}(\ln K_{p_r}) + \frac{E}{T} \tag{17.43}$$

Now, from the Van't Hoff equation (12.101),

$$\frac{d}{dT}(\ln K_{p_r}) = \frac{Q_p}{\Re T^2} \tag{17.44}$$

Substituting from eqn (17.44) into eqn (17.43) gives

$$\left(\frac{\partial E}{\partial T}\right)_p = \frac{\Re T}{z_i F}\frac{Q_p}{\Re T^2} + \frac{E}{T} = \frac{Q_p}{z_i F T} + \frac{E}{T} \tag{17.45}$$

Equation (17.45) shows that the change of emf is related to the heat of reaction, Q_p. Now, for exothermic reactions, Q_p is negative, and this means that the emf of a cell based on an exothermic reaction decreases with temperature. This is in line with the effect of temperature on the degree of dissociation in a combustion process, as discussed in Chapter 12.

17.2.3 EXAMPLES

Example 1

A Pb–Hg fuel cell operates according to the following equation:

$$Pb + Hg_2Cl_2 \longrightarrow PbCl_2 + 2Hg$$

and the heat of reaction, $Q_p = -95\ 200$ kJ/kmol Pb. The cell receives heat transfer of 8300 kJ/kmol Pb from the surroundings. Calculate the potential of the cell, and evaluate the electrical work produced per kilogram of reactants, assuming the following atomic weights: Pb – 207; Hg – 200; Cl – 35.

Solution

This problem can be solved by a macroscopic approach to the cell as a closed system. Applying the First Law gives

$$-zFE\ dn = dU - \delta Q_{trans} + T\ dS - p\ dV$$
$$= dG - \delta Q_{trans}$$

Now this value of energy transfer has to be used in eqn (17.34) in place of ΔG_T^0, giving

$$E = -\frac{(\Delta G_T^0 - \delta Q_{trans})}{zF} = -\frac{-95\ 200 - 8300}{2 \times 96\ 487} = 0.5363\ \text{V}$$

The quantity of work obtained per kg of reactants is given in the following way. Assuming the potential of the cell is not reduced by drawing a current, then the charge transferred through the cell per kmol Pb is $2F$ (because the valency of lead is 2). Hence the work done per kmol Pb is

$$\delta W = 2 \times 96\ 485 \times 0.5363 = 103\ 500\ \text{kJ/kmol Pb}$$

giving the work per unit mass of reactants as

$$\frac{\delta W}{m_{react}} = \frac{103\ 500}{207 + 2 \times (200 + 35)} = 152.9\ \text{kJ/kg}$$

Example 2

A hydrogen–oxygen fuel cell operates at a constant temperature of 227°C and the hydrogen and oxygen are fed to the cell at 40 bar, and the water is taken from it at the same pressure. Evaluate the emf at this condition, and the heat transfer from the cell.

Solution

This question requires the application of eqn (17.42), giving

$$E = \frac{\mathcal{R}T}{z_i F}\left\{\ln K_{p_r} - \ln\left(\frac{p_{rH_2O}}{p_{rH_2}p_{rO_2}^{1/2}}\right)\right\}$$

From tables at 500 K, $K_{Pr} = 7.92127 \times 10^{22}$ bar$^{-1/2}$. Hence the open circuit potential is

$$E = \frac{8.3143 \times 500}{2 \times 96\ 487}\left\{\ln(7.92127 \times 10^{22} \times 1^{1/2}) - \ln\left(\frac{40 \times 1^{1/2}}{40 \times 40^{1/2}}\right)\right\}$$

$$= 1.176\ \text{V}$$

The maximum work output from the fuel cell based on eqn (17.14) is

$$\dot{W} = -\dot{m}\Delta G_T^0 = \dot{m}\{-239\ 081.7 - 86\ 346 - (0.5 \times [-95\ 544] + [-58\ 472])\}$$
$$= \dot{m}(-219.18)\ \text{MJ/kmol H}_2$$

Also, at 500 K the lower enthalpy of reaction of the process is 244.02 MJ/kmol, which gives a higher enthalpy of reaction of

$$Q_p = -(244.02 \times 10^3 + 18 \times 1831) = -277.0 \times 10^3 = -277.0\ \text{MJ/kmol H}_2$$

This is equal to the change of enthalpy of the working fluid as it passes through the cell and hence applying the steady flow energy equation to the fuel cell gives

$$\dot{Q} - \dot{W} = \dot{m}(\Delta h)$$

giving

$$\dot{Q} = \dot{m}(\Delta h) + \dot{W} = -277.0 + 219.18 = -57.82\ \text{MJ/kmol H}_2$$

Thus, the efficiency of the fuel cell, based on the higher enthalpy of reaction, is

$$\eta = \frac{219.18}{277.0} = 0.7913$$

If it were evaluated from the lower enthalpy of reaction, which is usually used to calculate the efficiency of powerplant, then the value would be

$$\eta = \frac{219.18}{244} = 0.8983$$

Example 3

The operating pressure of the fuel cell in Example 2 is changed to 80 bar. Calculate the change in emf.

Solution

If the pressure is raised to 80 bar, this only affects the pressure term because $K_{p_r} = f(T)$, and hence

$$E_{80} - E_{40} = \frac{\Re T}{z_i F} \left\{ \left[\ln K_{p_r} - \ln\left(\frac{80 \times 1^{1/2}}{80 \times 80^{1/2}}\right) \right] - \left[\ln K_{p_r} - \ln\left(\frac{40 \times 1^{1/2}}{40 \times 40^{1/2}}\right) \right] \right\}$$

$$= \frac{\Re T}{z_i F} \ln(2^{1/2}) = 0.0075\ \text{V}$$

giving

$$E_{80} = 1.176 + 0.0075 = 1.183\ \text{V}$$

This shows that the potential of the hydrogen–oxygen fuel cell increases with pressure. This is similar to the effect of pressure on the hydrogen–oxygen combustion reaction, when dissociation decreased from 0.0673 to 0.0221 as the pressure was increased from 1 bar to 100 bar for the stoichiometric combustion of methane in Chapter 12. Hence, if the amount of substance in the reactants is not equal to the amount of substance in the products then the emf of the cell will be a function of the pressure of operation.

17.3   Efficiency of a fuel cell

The fuel cell is not 100% efficient, and it is possible to define its efficiency by the following equation:

$$\eta = \frac{\text{Maximum useful work output}}{\text{Heat of reaction}} = \frac{W_{\text{use}}}{Q_{\text{p}}} \tag{17.46}$$

Now the maximum useful work output obtainable is defined by the change of Gibbs function, i.e.

$$
\begin{aligned}
W_{\text{use}} &= \Delta G = G_{\text{products}} - G_{\text{reactants}} \\
&= \sum_{\text{products}} H_{\text{f}}^{0} - \sum_{\text{reactants}} H_{\text{f}}^{0} - T\left(\sum_{\text{products}} S_{\text{f}}^{0} - \sum_{\text{reactants}} S_{\text{f}}^{0} \right)
\end{aligned}
\tag{17.47}
$$

and

$$Q_{\text{p}} = \sum_{\text{products}} H_{\text{f}}^{0} - \sum_{\text{reactants}} H_{\text{f}}^{0} \tag{17.48}$$

Thus, the efficiency is

$$\eta = 1 + \frac{T\left(\displaystyle\sum_{\text{products}} S_{\text{f}}^{0} - \sum_{\text{reactants}} S_{\text{f}}^{0} \right)}{\displaystyle\sum_{\text{products}} H_{\text{f}}^{0} - \sum_{\text{reactants}} H_{\text{f}}^{0}} = 1 - \frac{T\,\Delta S}{Q_{\text{p}}} = 1 + \frac{T\,\Delta S}{Q_{\text{p}}'} \tag{17.49}$$

The last expression in eqn (17.49) has been written in terms of the calorific value of the fuel because this is usually positive, and hence the efficiency is defined by the sign of the change of entropy. If the entropy change is positive then the efficiency is greater than unity; if it is negative then it is less than unity. Consider now the hydrogen–oxygen reaction described above. If the fuel cell is maintained at the standard temperature of 25°C, and the processes are assumed to be isothermal, then

$$
\begin{aligned}
\Delta G &= (h_{\text{f}}^{0})_{H_2O} - T(s_{H_2O} - s_{H_2} - 0.5 s_{O_2}) \\
&= -241\,820 - 298(188.71 - 130.57 - 0.5 \times 205.04) \\
&= -228\,594.8 \text{ kJ/kmol}
\end{aligned}
\tag{17.50}
$$

and $Q_{\text{p}} = -241\,820$ kJ/kmol. Hence the efficiency is

$$\eta = \frac{228\,594.8}{241\,820} = 0.945 \tag{17.51}$$

This efficiency can be evaluated using eqn (17.49) to give

$$\eta = 1 + \frac{T\,\Delta S}{Q_{\text{p}}'} = 1 + \frac{298(188.71 - 130.57 - 0.5 \times 205.04)}{241820} = 1 - 0.0547 = 0.945 \tag{17.52}$$

This is exactly the same value as was obtained in Chapter 2 on exergy and availabilty. The rational efficiency of this, and any, fuel cell would be 100%. The efficiency of this cell is less than 100% because the change of entropy as the reactants change to products is negative. Hence, if the value of $T\Delta S < 0$ then the cell will be less than 100% efficient, and this is because 'heat' equivalent to $T\Delta S$ must be transferred with the surroundings, i.e. to the surroundings. If $T\Delta S > 0$ then the efficiency can be greater than 100%, and this comes about because energy has to be transferred to the fuel cell from the surroundings to make the process isothermal.

17.4   Thermodynamics of cells working in steady state

The previous theory did not relate to a cell in which current was being continually drawn from it. Under such circumstances, concentration gradients might be set up in the cell, and it will be governed by the theories of irreversible thermodynamics (Chapter 16). It can be shown for a fuel cell that a possible force term is the gradient of electrochemical potential, which is similar to the use of the chemical potential gradient when considering diffusion processes (note that when considering thermal and mass diffusion in combination the gradient of the ratio of chemical potential and temperature (μ/T) was used.) Hence, the force term is

$$\frac{d}{dx}\bar{\mu}_i = \frac{d}{dx}(\mu_i + F\psi) = \frac{d\mu_i}{dx} + F\frac{d\psi}{dx} \tag{17.53}$$

In this analysis the flux term is taken to be the movement of charge per unit area through the cell. Since the charge is carried on the ions travelling through the cell it can be directly related to the transfer of the ions from one electrode to the other, and this is equivalent to current density (e.g. amp/m²). Hence, the current in the cell is

$$J = \left(\sum_{\substack{\text{all} \\ \text{ions}}} J_i\right)F \tag{17.54}$$

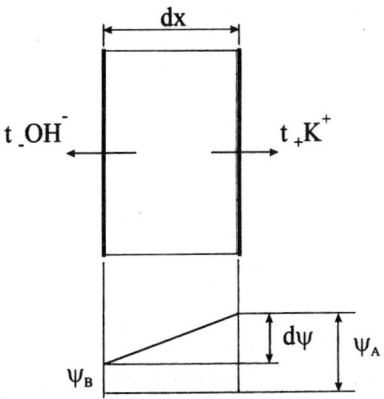

Fig. 17.4 Ions in cell

If the hydrogen–oxygen cell is taken as an example, it is apparent that the charge is carried across the cell by OH⁻ ions being absorbed on the hydrogen electrode, and being produced on the oxygen electrode. These OH^- ions are produced in the electrolyte (KOH) as the KOH molecule spontaneously dissociates into K^+ and OH^- ions (in weak aqueous solutions), so while the negative ions travel in one direction the positive ones go in the opposite direction. This can be considered schematically as in Fig 17.4, which represents a thin element of the electrolyte. Ions are flowing through the electrolyte, and t_+ represents the transport number of the cations, while t_- represents that of the anions. If the electrochemical potentials of the two ions on side A of the element are

$$\bar{\mu}_{K^+} = \mu_{K^+} + F\psi_A$$
$$\bar{\mu}_{OH^-} = \mu_{OH^-} - F\psi_A$$

(17.55)

then it is possible to evaluate the change in Gibbs function, dG, when an infinitesimal number of ions move through the element, giving

$$-dG = t_+ \, d\bar{\mu}_{K^+} - t_- \, d\bar{\mu}_{OH^-}$$

(17.56)

Now when the system is in the steady state, dG must be zero, and hence, substituting from eqn (17.55)

$$-t_+ \, d\mu_{K^+} - t_+ F \, d\psi + t_- \, d\mu_{OH^-} - t_- F \, d\psi = 0$$

(17.57)

Since $t_+ + t_- = 1$ then

$$F \, d\psi = -t_+ \, d\mu_{K^+} + t_- \, d\mu_{OH^-}$$

(17.58)

Applying Onsager's relationship, $J_i = \sum_{i=1}^{k} L_{ik} X_k$, gives

$$\left. \begin{aligned} J_{K^+} &= -L_{11} \, d\mu_+ - L_{12} \, d\mu_- - L_{13} \, d\psi \\ J_{OH^-} &= -L_{21} \, d\mu_+ - L_{22} \, d\mu_- - L_{23} \, d\psi \\ J &= -L_{31} \, d\mu_+ - L_{32} \, d\mu_- - L_{33} \, d\psi \end{aligned} \right\}$$

(17.59)

The term L_{33} is equivalent to the conductance of the electrolyte, k, and by the Onsager reciprocal relationship,

$$L_{32} = L_{23} = -\frac{t_- k}{F}$$

$$L_{31} = L_{13} = +\frac{t_+ k}{F}$$

(17.60)

Hence, the current, J, is given by

$$J = -\frac{t_+ k}{F} \, d\mu_+ + \frac{t_- k}{F} \, d\mu_- - k \, d\psi$$

(17.61)

If the circuit connecting the electrodes is open then no current flows and $J = 0$, giving the result obtained in eqn (17.58). In the case where a current flows, $J > 0$, then there will be a reduction in the potential available at the electrodes because some of the 'force' that can be generated in the cell has to be used to propel the ions through the concentration gradients that exist.

A more comprehensive analysis of the irreversible thermodynamics of the cell would take into account the temperature gradients that might be set up in operating the cell, and another group of equations for thermal conduction would have to be added to the equations in the matrix of eqn (17.59).

17.5 Concluding remarks

It has been shown that fuel cells offer the potential of converting the bond energies of a reaction *directly* to electrial power: this enables the restrictions of the Second Law to be avoided. The theory of fuel cells has been developed from basic thermodynamic principles, and a new property called electrochemical potential has been introduced. Finally, the application of irreversible thermodynamics to the fuel cell has demonstrated where some of its shortcomings will lie.

PROBLEMS

Assume the value of the Faraday constant, $F = 96\ 487$ kC/kmol.

1 An electric cell has the following chemical reaction:

$$Zn(s) + 2AgCl(s) = ZnCl_2 + 2Ag(s)$$

and produces an emf of 1.005 V at 25°C and 1.015 V at 0°C, at a pressure of 1 bar. Estimate the following parameters at 25°C and 1 bar:

(a) change in enthalpy during the reaction,
(b) amount of heat absorbed by the cell, per unit amount of zinc, to maintain the temperature constant during reversible operation.

$$[-216\ 937 \text{ kJ/kmol Zn}; -23\ 002 \text{ kJ/kmol Zn}]$$

2 An electric cell is based on the reaction $Pb + Hg_2Cl_2 \longrightarrow PbCl_2 + 2Hg$. If the enthalpy of reaction for this reaction, Q_p, at 25°C is −95 200 kJ/kg Pb, calculate the emf and rate of change of emf with temperature at constant pressure if the heat transfer to the cell is 8300 kJ/kg Pb.

$$[0.5363 \text{ V}; 1.44 \times 10^{-4} \text{ V/K}]$$

3 Calculate the emf of a hydrogen–oxygen fuel cell operating reversibly if the overall reduction in Gibbs energy is 238 MJ/kmol H_2. If the cell operates at 75% of the reversible emf due to internal irreversibilities, calculate the magnitude and direction of the heat transfer with the surroundings, assuming that the enthalpy of reaction at the same conditions is −286 MJ/kmol H_2.

$$[1.233 \text{ V}; -53750 \text{ kJ/kg } H_2]$$

4 An ideal, isothermal, reversible fuel cell with reactants of oxygen and hydrogen, and a product of water operates at a temperature of 400 K and a pressure of 1 bar. If the operating temperature increases to 410 K, what must be the new pressure if the open circuit voltage is maintained constant? The values of K_{p_r} are 1.8134×10^{29} at 400 K and 2.4621×10^{28} at 410 K.

$$[2.029 \text{ bar}]$$

5 A hydrogen–oxygen fuel cell is required to produce a constant voltage and operate over a pressure range of 0.125 to 10 bar. The datum voltage is 1.16 V at a temperature of 350 K and pressure of 10 bar. If all the streams are at the same pressure evaluate the range of temperature required to maintain the voltage at this level, and show how the operating temperature must vary with pressure.

Assume that the enthalpy of reaction of the cell, $Q_p = -286\ 000$ kJ/kmol H_2, and that this remains constant over the full temperature range. The valency, $z_i = 2$.

[317.4 K]

6 A hydrogen–oxygen fuel cell operates at a temperature of 450 K and the reactants and products are all at a pressure of 3 bar. Due to internal resistances the emf of the cell is only 70% of the ideal value. Calculate the 'fuel consumption' of the ideal cell and the actual one in g/kW h.

An alternative method of producing electrical power from the hydrogen is to burn it in an internal combustion engine connected to an electrical generator. If the engine has a thermal efficiency of 30% and the generator is 85% efficient, calculate the fuel consumption in this case and compare it with that of the fuel cell. Explain why one is higher than the other.

Table P17.6 Table of values of enthalpy and Gibbs energy at $T = 450$ K

Species	Enthalpy, h (kJ/kmol)	Gibbs energy, g (kJ/kmol)
Oxygen, O_2	13 125	−84 603
Hydrogen, H_2	12 791	−51 330
Water, H_2O	−223 672	−314 796

[32.25 g/kW h;   46 g/kW h;   116.2 g/kW h]

Bibliography

This bibliography lists some of the references that have been used in preparing the book, and on which some of the ideas were based. In general it is simply a bibliography and not a set of specific references, although some material has been drawn from the material in this list; where that is the case a specific reference is made in the text. There is not a bibliography for all chapters because the information is subsumed into the books relating to adjacent chapters.

Chapter 1   State of equilibrium

Atkins, P W, 1994: *The Second Law – energy, chaos, and form*. Scientific American Books

Benson, R S, 1977: *Advanced Engineering Thermodynamics*. Pergamon

Denbigh, K G, 1981: *The Principles of Chemical Equilibrium*. Cambridge University Press

Hatsopoulos, G N and Keenan, J H, 1972: *Principles of General Thermodynamics*. John Wiley

Keenan, J H, 1963: *Thermodynamics*, John Wiley, New York, 14th printing

Chapter 2   Availability and exergy

Ahearn, J E, 1980: *The Exergy Method of Energy Systems Analysis*. John Wiley

Chin, W W, and El-Masri, M A, 1987: Exergy analysis of combined cycles: Part 2 – analysis and optimisation of two-pressure steam bottoming cycles, *Transactions of the American Society of Mechanical Engineers*, **109**, 237–243

El-Masri, M A, 1987: Exergy analysis of combined cycles: Part I – air-cooled Brayton-cycle gas turbines, *Transactions of the American Society of Mechanical Engineers*, **109**, 228–236

Goodger, E M, 1979: *Combustion Calculations*. Macmillan

Haywood, R W, 1980: *Equilibrium Thermodynamics for Engineers and Scientists*, John Wiley

Heywood, J B, 1988: *Internal Combustion Engine Fundamentals*. McGraw-Hill

Horlock J H, and Haywood, R W, 1985: Thermodynamic availability and its application to combined heat and power plant, *Proceedings of Institution of Mechanical Engineers*, **199**, C1, 11–17

Keenan, J H, 1963: *Thermodynamics*. New York: John Wiley, 14th printing

Kotas, T J, 1995: *The Exergy Method of Thermal Plant Analysis*. Florida: Kreiger

Moran, M J, and Shapiro, H N, 1988: *Fundamentals of Engineering Thermodynamics*. Wiley International Edition

Patterson, D J, and Van Wylen, G J, 1964: A digital computer simulation for spark-ignited engine cycles, *Society of Automotive Engineers Progress in Technology Series, 7*, Society of Automotive Engineers

Rant, Z, 1956: Exergie, ein neues Wort fur 'technische Arbeitsfahigkeit' (Exergy, a new word for 'technical work capacity'), *Forsch Gebeite Ingenieurwes*, **22**, 36

Rogers, G F C, and Mayhew, Y R, 1988 *Thermodynamic and transport properties of fluids*. SI units; Third edition. Oxford, Blackwell

Tsatsaronis, G and Winfold M, 1985: Exergoeconomic analysis and evaluation of energy-conversion plants: Part I – A new general methodology; Part II – Analysis of a coal-fired steam power plant, *Energy*, **10**, 69–80, and 81–94

Chapter 3:   Pinch technology

Barclay, F J, 1995: *Combined Power and Process – an exergy approach*. London, Mechanical Engineering Publications

Linnhoff, B, and Senior, P R, 1983: Energy targets clarify scope for better heat integration. *Process Engineering*, Mar, 29–33

Linnhoff, B, and Turner, J A, 1981: Heat recovery networks: new insights yield big savings. *Chemical Engineering*, Nov, 56–70

Smith, R, 1995: *Chemical Process Design*. McGraw-Hill

Chapter 4:   Rational efficiency

Haywood, R W, 1975: *Analysis of Engineering Cycles*. Pergamon

Haywood, R W, 1980: *Equilibrium Thermodynamics for Engineers and Scientists*. John Wiley

Horlock, J H, 1987: *Co-generation – combined heat and power*. Pergamon

Chapter 5:   Endoreversible engines

Bejan, A, 1988: *Advanced Engineering Thermodynamics*. Wiley International Edition

El-Masri, M A, 1985: On thermodynamics of gas turbine cycles: Part I – Second law analysis of combined cycles. *Transactions of the American Society of Technical Engineers*, **107**, 880–889

Chapter 6:   Relationships between properties

This subject is covered in most of the general texts on engineering thermodynamics. The reader is referred to these to obtain additional information or to see different approaches.

Chapter 7:   Equations of state

This subject is covered in most of the general texts on engineering thermodynamics. The reader is referred to these to obtain additional information or to see different approaches.

Chapter 8:   Liquefaction of gases

Haywood, R W, 1972: *Thermodynamic Tables in SI Units.* Cambridge University Press
Haywood, R W, 1975: *Analysis of Engineering Cycles.* Pergamon
Haywood, R W, 1980: *Equilibrium Thermodynamics for Engineers and Scientists.* John
  Wiley

Chapter 9:   Thermodynamic properties of ideal gases

Benson, R S, 1977: *Advanced Engineering Thermodynamics.* Pergamon
JANAF, 1971: *Thermochemical Tables.* Washington, DC: National Bureau of Standards
  Publications, NSRDS-N35 37

Chapter 10:   Thermodynamics of combustion

Glassman, I, 1986: *Combustion.* New York: Academic Press
Goodger, E M, 1979: *Combustion Calculations.* Macmillan
Strahle, W C, 1993: *An Introduction to Combustion.* Gordon and Breach

Chapter 11:   Chemistry of combustion

Benson, R S, 1977: *Advanced Engineering Thermodynamics.* Pergamon
Benson, S W, and Buss, J H, 1958: Additivity rules for the estimation of molecular
  properties, and thermodynamic properties. *J. Chem Phys*, **29**, No 3
Goodger, E M, 1979: *Combustion Calculations.* Macmillan
Kuo, K K, 1986: *Principles of Combustion.* Wiley International Edition

Chapter 12:   Chemical equilibrium and dissociation

Benson, R S, Annand, W J D, and Baruah P C, 1975: A simulation model including intake
  and exhaust systems for a single cylinder four-stroke cycle spark ignition engine.
  International Journal of Mechanical Sciences, **17**, 97–124
Denbigh, K G, 1981: *The Principles of Chemical Equilibrium.* Cambridge: Cambridge
  University Press
Horlock, J H, and Winterbone, D E, 1986: *The Thermodynamics and Gas Dynamics of
  Internal Combustion Engines, vol II.* Oxford: Oxford University Press
Heywood, J B, 1988: *Internal Combustion Engine Fundamentals.* McGraw-Hill
Lavoie, G A, Heywood, J B, and Keck, J C, 1970: Experimental and theoretical study of
  nitric oxide formation in internal combustion engines, *Combustion Science and
  Technology*, **1**, 313–326
Vickland, C W, Strange, F M, Bell, R A, and Starkman, E S A, 1962: A consideration of
  high temperature thermodynamics of internal combustion engines. *Society of Automotive
  Engineers Transactions*, **70**, 785–795

Chapter 14:   Chemical kinetics

Annand, W J D, 1974: Effects of simplifying kinetic assumptions in calculating nitric
  oxide formation in spark ignition engines. *Proceedings of Institution of Mechanical
  Engineers*, **188**, (41/74), 431–436

Daneshyar, H, and Watfa, M, 1974: Predicting nitric oxide and carbon monoxide concentrations in spark ignition engines. *Proceedings of Institution of Mechanical Engineers*, **188**, (41/74), 437–443

Glassman, I, 1986: *Combustion*. New York: Academic Press

Heywood, J B, 1988: *Internal Combustion Engine Fundamentals*. McGraw-Hill

Heywood, J B, Faye, J A, and Linden, L H, 1971: Jet aircraft air pollutant production and dispersion. *AIAA Journal*, **9** (5), 841–850

Lavoie, G A, Heywood, J B, and Keck, J C, 1970: Experimental and theoretical study of nitric oxide formation in internal combustion engines. *Combustion Science & Technology*, **1**, 313–326

Chapter 15:   Combustion and flames

Abdel-Gayed, R G, Bradley, D, and Laws, M, 1987: Turbulent burning velocities: a general correlation in terms of straining rates. *Proceedings Royal Society*, London, A414, 389–413

Bradley, D, Lau, A K, and Lawes, M, 1992: Flame stretch rate as a determinant of turbulent burning velocity. *Philosophical Transactions Royal Society*, London, A338, 359–387

Damkohler, Z, 1940: *Elektrochem*, **46**, 601 (Translated as NACA Technical Memo 1112 – The effects of turbulence on flame velocities in gas mixtures)

Gaydon, A G, and Wolfhard, H G, 1978: *Flames: their structure, radiation and temperature*. London: Chapman & Hall

Glassman, I, 1986: *Combustion*. New York: Academic Press

Heikal, M R, Benson R S, and Annand W J D, 1979: A model for turbulent burning speed in spark-ignition engines. *Proceedings of Conference of Institution of Mechanical Engineers on Fuel Economy and Emissions of Lean-burn engines*, London

Heywood, J B, 1988: *Internal Combustion Engine Fundamentals*. McGraw-Hill

Kuehl D K, 1962: Laminar burning velocities in propane-air mixtures. *Eighth International Symposium on Combustion*

Kuo, K K, 1986: *Principles of combustion*. Wiley International Edition

Lancaster, D R, Kreiger, R B, Sorenson, S C, and Hull W, 1976: Effects of turbulence on spark-ignition engine combustion. *Society of Automotive Engineers*, 760160, Detroit

Lewis, B, and von Elbe, G, 1961: *Combustion, Flames and Explosions of Gases*, 2nd edn. New York: Academic Press

Mallard, E and le Chatelier, H L, 1883: *Ann Mines*, **4**, 379

Metgalchi, M, and Keck, J C, 1980: Laminar burning velocity of propane-air mixtures at high temperature and pressure. *Combustion & Flame*, **38**, 143–154

Metgalchi, M, and Keck, J C, 1982: Burning velocities of mixtures of air with methanol iso-octane, and indolene at high pressure and temperature. *Combustion & Flame*, **48**, 191–210

Strahle, W C, 1993: *An Introduction to Combustion*. Gordon and Breach

Whitehouse, N D, and Way, R J B, 1970: Rate of heat release in diesel engines and its correlation with fuel injection data. *Proceedings of Institution of Mechanical Engineers*, **184**, 3J, 17–27

Zel'dovitch, Y B, 1948: *J Phys Chem USSR*, **22** (1)

Zel'dovitch, Y B, and Frank-Kamenetsky, D A, 1938: *Compt Rend Acad Sci USSR*, **19**, 693

Zel'dovitch, Y B, and Semenov, N, 1940: *Journal of Experimental and Theoretical Physics USSR*, **10**, 1116 (Translated as NACA Tech Memo 1084)

Chapter 16: Irreversible thermodynamics

Bejan, A, 1988: *Advanced Engineering Thermodynamics*, Wiley International Edition

Benson, R S, 1977: *Advanced Engineering Thermodynamics*. Pergamon

de Groot, S R, 1951: *Thermodynamics of Irreversible Processes*. North Holland

Denbigh, K G, 1951: *Thermodynamics of the Steady State* Methuens Monographs on Chemical Subjects

Fick, A, 1856: *Die medizinische Physik.*

Hoase, R: *Thermodynamics of Irreversible Processes*. Addison-Wesley

Lee, J F, and Sears, F W, 1963: *Thermodynamics*. Addison-Wesley

Onsager, L, 1931a: Reciprocal relations in irreversible processes, part I. *Phys Rev*, **37**, 405–426

Onsager, L, 1931b: Reciprocal relations in irreversible processes, part II. *Phys Rev*, **38**, 2265–2279

Tribus, M, 1961: *Thermodynamics and Thermostatitics*. Van Nostrand

Yougrau, W, Merwe, A, and Raw, R, 1966: *Treatise on Irreversible and Statistical Thermophysics*. Macmillan

Chapter 17: Fuel cells

Berger, C (Ed), 1968: *Handbook of Fuel Cell Technology*. Prentice-Hall

Denbigh, K G, 1981: *The Principles of Chemical Equilibrium*. Cambridge University Press

Hart, A B, and Womack, G J, 1967: *Fuel Cells*. London: Chapman and Hall

Haywood, R W, 1980: *Equilibrium Thermodynamics for Engineers and Scientists*. John Wiley

Liebhafsky, H A, and Cairns, E J, 1968: *Fuel Cells and Fuel Batteries*. Wiley

Index

Index of tables of properties